Proceedings in Adaptation, Learning and Optimization

Volume 4

Series editors

Yew Soon Ong, Nanyang Technological University, Singapore
e-mail: asysong@ntu.edu.sg

Meng-Hiot Lim, Nanyang Technological University, Singapore
e-mail: emhlim@ntu.edu.sg

About this Series

The role of adaptation, learning and optimization are becoming increasingly essential and intertwined. The capability of a system to adapt either through modification of its physiological structure or via some revalidation process of internal mechanisms that directly dictate the response or behavior is crucial in many real world applications. Optimization lies at the heart of most machine learning approaches while learning and optimization are two primary means to effect adaptation in various forms. They usually involve computational processes incorporated within the system that trigger parametric updating and knowledge or model enhancement, giving rise to progressive improvement. This book series serves as a channel to consolidate work related to topics linked to adaptation, learning and optimization in systems and structures. Topics covered under this series include:

- complex adaptive systems including evolutionary computation, memetic computing, swarm intelligence, neural networks, fuzzy systems, tabu search, simulated annealing, etc.
- machine learning, data mining & mathematical programming
- hybridization of techniques that span across artificial intelligence and computational intelligence for synergistic alliance of strategies for problem-solving
- aspects of adaptation in robotics
- agent-based computing
- autonomic/pervasive computing
- dynamic optimization/learning in noisy and uncertain environment
- systemic alliance of stochastic and conventional search techniques
- all aspects of adaptations in man-machine systems.

This book series bridges the dichotomy of modern and conventional mathematical and heuristic/meta-heuristics approaches to bring about effective adaptation, learning and optimization. It propels the maxim that the old and the new can come together and be combined synergistically to scale new heights in problem-solving. To reach such a level, numerous research issues will emerge and researchers will find the book series a convenient medium to track the progresses made.

More information about this series at http://www.springer.com/series/13543

Jiuwen Cao · Kezhi Mao
Erik Cambria · Zhihong Man
Kar-Ann Toh

Editors

Proceedings of ELM-2014 Volume 2

Applications

 Springer

Editors
Jiuwen Cao
Institute of Information and Control
Hangzhou Dianzi University
Zhejiang
China

Kezhi Mao
School of Electrical and Electronic
    Engineering
Nanyang Technological University
Singapore
Singapore

Erik Cambria
School of Computer Engineering
Nanyang Technological University
Singapore
Singapore

Zhihong Man
Faculty of Engineering and Industrial
    Sciences
Swinburne University of Technology
Hawthorn Victoria
Australia

Kar-Ann Toh
School of Electrical and Electronic
    Engineering
Yonsei University
Seoul
Korea, Republic of (South Korea)

ISSN 2363-6084   ISSN 2363-6092   (electronic)
Proceedings in Adaptation, Learning and Optimization
ISBN 978-3-319-36685-2   ISBN 978-3-319-14066-7   (eBook)
DOI 10.1007/978-3-319-14066-7

Springer Cham Heidelberg New York Dordrecht London
© Springer International Publishing Switzerland 2015
Softcover reprint of the hardcover 1st edition 2015

Printed on acid-free paper

Springer International Publishing AG Switzerland is part of Springer Science+Business Media
(www.springer.com)

Contents

Applications with ELM

Using Extreme Learning Machine for Filamentous Bulking Prediction in Wastewater Treatment Plants

Yuchao Zhao[1], Zhengchao Xie[2], and Inchio Lou[2,*]

[1] School of Environment, Beijing Normal University, Beijing, China
zhaoy@bnu.edu.cn
[2] Faculty of Science and Technology, University of Macau, Macau SAR
{zxie,iclou}@umac.mo

Abstract. Sludge bulking is the most common solids settling problem in wastewater treatment plants, resulting in the wastewater treatment efficiency decreasing and the water quality in the effluent deteriorating. Previous studies showed that the mechanisms have not yet been completely understood to form the deterministic cause-effect relationship. In this study, Extreme Learning Machine (ELM) was identified using the data from Chongqing wastewater treatment plant (CQWWTP), including temperature, pH, biochemical oxygen demand (BOD), chemical oxygen demand (COD), suspended solids (SS), ammonia (NH_4^+), total nitrogen (TN), total phosphorus (TP), and mixed liquor suspended solids (MLSS). The models were subsequently used to predict the sludge volume index (SVI), the indicator of the bulking occurrence. Results showed that the model has the prediction power R^2 of 0.85, which providing a useful guide for practical sludge bulking control.

Keywords: Extreme learning machine, Sludge bulking, activated sludge, prediction model.

1 Introduction

Sludge bulking is the most common solid separation problem in activated sludge problem, which is caused by the excessive growth of filamentous bacteria extending outside the flocs, thus interfering with the settling of activated sludge. Bulking leads to high level of total suspended solids in effluent that exceeds the discharge permit limitation and subsequently loses activated sludge in the aeration basin, resulting in the deterioration of wastewater treatment process [1].

So far, there is no single or combined proposed mechanisms can explain completely the sludge bulking problem; for example, the uncertainty about the factors triggering the filaments growth is still unclear. The current efforts to study sludge bulking problem rely mostly on experimental observation of filamentous bacteria population in the system, while some experimental results could lead to contradictory conclusions. Thus it is difficult to formulate deterministic mathematical models for predicting the filaments population, though some existing models were developed [2-3].

---

* Corresponding author.

© Springer International Publishing Switzerland 2015
J. Cao et al. (eds.), *Proceedings of ELM-2014 Volume 2,*
Proceedings in Adaptation, Learning and Optimization 4, DOI: 10.1007/978-3-319-14066-7_1

Developing a model that could predict the potential for bulking in real time with reasonable accuracy is of great practical importance, as it can be used to improve the treatment plant efficiency and cost saving [4-5]. The complexity of the problem can be overcome by applying data-driven model for the whole system, rather than the breaking down of the system into small components described individually.

Computational artificial intelligence techniques have been developed as the efficient tools in recent years for prediction. Previous studies [6-7] have used the principle component regression (PCR), i.e., principal component analysis (PCA) followed by multiple linear regressions (MLR), to predict the sludge volume index (SVI), the fundamental index of sludge. However, the intrinsic problem of PCR is that the variables data set used as the input of the model have high complex non-linearity, expecting that PCR alone is inadequate for prediction and the prediction results were unsatisfactory. With the development of artificial intelligence models, artificial neural network (ANN) such as back propagation (BP) was applied to predict the sludge volume index. ANN is a well-suited method with self-adaptability, self-organization and error tolerance, which is better than PCR for non-linear simulation. ANN has been used for predicting the sludge volume index. However, this method has such limitations as requirement of a great amount of training data, difficulty in tuning the structure parameter that is mainly based on experience, and its "black box" nature that is difficult to understand and interpret the data [7-9].

Considering the drawbacks of the both methods, recently extreme learning machine (ELM) is thought as the best solution. ELM is a simple and efficient learning algorithm that was developed recently. In the name of ELM, extreme means that its learning speed is extremely fast while it has higher generalization than the gradient-descent based learning [10]. Furthermore, ELM can be used to solve issues like local minima, improper learning rate and over-fitting which are very possible in traditional ANN. Examples [10-12] also showed that ELM possesses a superior performance than other conventional algorithms on different benchmark problems from both regression and classification areas. Recent studies [13-16] furthered the application or combination of ELM on various areas including biology and social sciences. However, by far, as the best knowledge of authors, there is no existing application of using ELM on prediction or forecast the sludge bulking.

2   Materials and Methods

2.1   Study Area

CQWWTP (29.601615in latitude and 106.634133 E in longitude is designed to have a capacity of an average flow rate of 300,000m^3/d and about 750,000 person equivalents (in carbon, nitrogen, and phosphorus). CQWWTP uses conventional A/A/O (anaerobic/anoxic/aerobic) treatment processes that are susceptible to sludge bulking. It was reported that 36% of sludge experience bulking in the year of 2010, and the situation appeared to be worsening in recent years, particularly in the springs.

Fig. 1 showed the change of SVIs over time, which clearly indicates that the bulking mostly happened in the springs from Jan. to April, with the SVIs greater than 150 mL/g. On the other hand, bulking levels were low in the summers from July to September, with the SVI around 50mL/g.

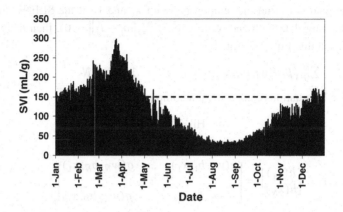

Fig. 1. Change of SVIs over time in 2010

2.2   Modeling Approaches

- Extreme Learning Machine

ELM originally was proposed as a learning scheme for single-hidden-layer feed-forward neural networks (SLFNs). Then, it was extended to the generalized SLFNs where the hidden layer needs not be neuron alike [17-18]. In the past, gradient descent based approaches were used for feed-forward neutral networks, and all parameters need to be tuned which usually take a long time. While for ELM which the basic idea is that, the model has only one hidden layer, and the parameters of this hidden layer, including the input weights and biases of the hidden nodes, need not to be tuned. On the contrary, these hidden nodes parameters are assigned randomly, which means that they may be independent of the training data [18]. After these input weights and hidden layer biases are assigned randomly, SLFNs can be treated as linear system and output weights which link hidden layer to the output layer can be calculated using generalized inverse operation [19]. References [11-12] proposed and proved the theory of ELM. In order to make it clear on ELM and its application in water treatment prediction, here the fundamental theory of ELM [20] will be briefly re-introduced first as follows:

Consider a training data set D of N arbitrary distinct samples $(x_i t_{i)}$, where $\{x_i\} \in R^m$ is the m×1 input vector and $t_i \in R^n$ is the n×1 target vector. Standard SLFNs with $\tilde{N}$ nodes and activation function g(x) are mathematically modeled a

$$\sum_{i=1}^{\tilde{N}} \beta_i\, g_i\left(x_j\right) = \sum_{i=1}^{\tilde{N}} \beta_i g(w_i, b_i, x_j) = o_j, \qquad 1 \le j \le N \tag{1}$$

where w_i is weight vector connecting the ith hidden node and the input nodes, β_i is the weight vector connecting the ith hidden node and the output nodes, and b_i is the threshold of the ith hidden node.

Since the goal is to find the relation between x_i and t_i, if the SLFNs can approximate the training data with zero error (i.e., $\sum_{j=1}^{\tilde{N}} \|o_j - t_j\| = 0$), then there exists β_i, w_i and b_i such that Eq. (2) is satisfied.

$$\sum_{i=1}^{\tilde{N}} \beta_i g(w_i, b_i, x_j) = t_j, \qquad 1 \le j \le N \tag{2}$$

The above N equations can be written compactly as

$$H\beta = T \tag{3}$$

where

$$H = \begin{bmatrix} h(x_1) \\ \vdots \\ h(x_N) \end{bmatrix} = \begin{bmatrix} g(w_1, b_1, x_1) & \cdots & g(w_{\tilde{N}}, b_{\tilde{N}}, x_1) \\ \vdots & \ddots & \vdots \\ g(w_1, b_1, x_N) & \cdots & g(w_{\tilde{N}}, b_{\tilde{N}}, x_N) \end{bmatrix}_{N \times \tilde{N}} \tag{4}$$

$$\text{and} \quad T = \begin{bmatrix} t_1^T \\ \vdots \\ t_N^T \end{bmatrix}_{N \times n} \tag{5}$$

H is called the hidden layer output matrix of SLFN. $h(x)=g(w_1,b_1,x), \ldots, g(w_{\tilde{N}}, b_{\tilde{N}}, x)$ is called the hidden layer feature mapping. The i th column of H is the ith hidden node output with respect to inputs $x_1, x_2, \ldots, x_N$ The ith row of H is the hidden layer feature mapping with respect to the ith input x_i.

According to the proofs in [10-11], if the activation function is infinitely differentiable, the input weight vectors w_i and hidden layer biases b_i can be randomly assigned. Moreover, these parameters are not necessarily tuned and the hidden layer output matrix H can actually remain unchanged once random values have been assigned in the beginning of learning.

Different from traditional learning algorithms, ELM tends to reach not only the smallest training error but also the smallest norm of output weights [21]:

$$Min = \|H\beta - T\| \ and \ \beta \tag{6}$$

Then, if the number $\tilde{N}$ of hidden neurons is equal to the number N of distinct training samples (i.e., $\tilde{N}$ =N), the matrix H is square and invertible, which means that the output weights β can be analytically calculated by simply inverting H, and thus the SLFNs can approximate these training samples with zero error. However, most of the times the number of hidden nodes is much less than the number of distinct training samples (i.e., $\tilde{N} \ll N$), and thus H is a non-square matrix and there may not exist β_i, w_i and b_i , and Eq. (3) cannot be satisfied. Fortunately, since w_i and b_i are fixed, Eq. (3) becomes a linear system, and the smallest norm least square method can be used instead of the standard optimization method to estimate the output weights.

$$\beta = H+T \tag{7}$$

where H^+ is the Moore–Penrose pseudo inverse of matrix H[, which can be calculated using the orthogonal projection method:

$$H^+ = (H^T H)^{-1} H^T \tag{8}$$

when $H^T H$ is singuar, or

$$H^+ = H^T (HH^T)^{-1} \tag{9}$$

when $H^T H$ is nonsingular
where the superscript T means matrix transposition.

Based on this learning algorithm, the training time can be extremely fast because only three calculation steps are required: 1. randomly assign hidden nodes parameters; 2. calculate the hidden layer output matrix H; 3. calculate the output weight β. Moreover, since the output weights are calculated analytically using inverse matrix, it ensures that the results are global and hence better prediction accuracy and generalization performance can be achieved. After training, the output function of ELM for an unseen vector X (take one output node case as an example) can be expressed as:

$$f(x) = h(x)\beta \tag{10}$$

- Performance Indicators
The performance of models was evaluated using the following indicators: square of correlation coefficient (R^2) that provides the variability measure for the data reproduced in the model; mean absolute error (MAE) and root mean square error (RMSE). The indicators were defined as below by Equation 11-15.

$$R^2 = 1 - \frac{F}{F_o} \tag{11}$$

$$F = \sum \left(Y_i - \hat{Y}_i \right)^2 \tag{12}$$

$$F_o = \sum \left(Y_i - \overline{Y}_i \right)^2 \tag{13}$$

$$MAE = \frac{1}{n} \sum_{i=1}^{n} (\hat{Y}_i - Y_i)^2 \tag{14}$$

$$RMSE = \sqrt{\frac{1}{n} \sum_{i=1}^{n} (\hat{Y}_i - Y_i)^2} \tag{15}$$

where n is the number of data; Y_i and $\overline{Y}_i$ are observation data and the mean of observation data, respectively, and $\hat{Y}_i$ is the modeling results.

3   Results and Discussion

3.1   Correlation Analysis

Correlation between SVIs and water parameters were analyzed to evaluate the influence of each parameter on the bulking level, which provides a measure of linear relationship between SVI and each parameter. The results (Table 1) showed that all the coefficients were greater than 0.15, indicating that all these parameters had high correlation with SVIs and thus included in the models as input variables. It is noted that high correlation coefficient (0.82) was found between SVI and temperature in the Prediction model, which was consistent with the observation that bulking of CQWWTP mostly occurs in the springs.

Table 1. Correlation coefficients between SVIs and water parameters in MSR

parameters	Temp	MLSS	BOD	COD	SS	pH	TP	TN	N-NH3
SVI	-0.82	-0.18	0.26	0.32	0.21	-0.23	0.18	0.28	0.19

3.2   Modeling Results

The performance of prediction was shown in Table 2. Using the Prediction model, the performance indexes for the training step were generally better than those for the testing step, with the R^2 of 0.851 for training and 0.841 for testing.

Table 2. Performance indexes of the Prediction and Forecast prediction models

	Prediction model	
Performance index	Accuracy performance (Training set)	Generalization performance (Testing set)
R^2	0.841	0.851
RMSE	26.705	26.7296
MAE	20.0821	20.9414

The observed data versus the modeling data were shown in Fig. 4, and the observed and modeling SVI change over time were listed in Fig. 2-3. These results confirmed that ELM can handle well the non-linear relationship between water parameters and SVI.

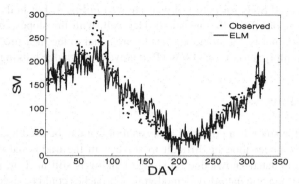

Fig. 2. Observed and predicted SVI for the training and validation data set of the prediction models

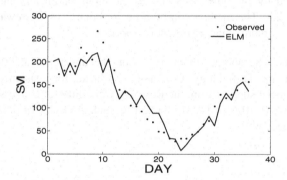

Fig. 3. Observed and predicted SVI for the testing data set of the prediction models

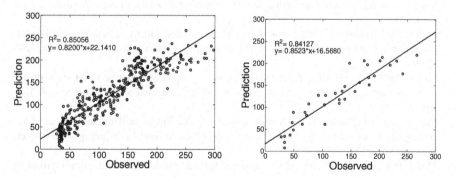

Fig. 4. ELM results for the training and validation (left) and testing (right) data set of the prediction model

In the models, no delay was observed for the prediction model in the training set data Fig. 2. The prediction of the testing set in Fig. 3 for the model exhibit over-estimates in the low SVI level region. In general, ELM was successful to predict the SVIs with a reasonable degree of accuracy for the prediction model. The modeling

SVIs versus observed SVIs for perdition was showed in Fig. 4. For both training and testing data, the model fitted the measured data well, with the slopes equal to 1 for both fitting curves, that is, the modeling results are equal to the measured data. It has provided enough information for CQWWTP to prevent the sludge bulking problems.

4   Conclusions

ELM applied in the study is a powerful analysis tool that can be used to solve a problem that is poorly understood or difficult to solve with the traditional deterministic relationship. The updated knowledge on sludge bulking is still unclear, and thus the unconventional systematic data-driven modeling approaches could be used to improve the prediction. The results showed the prediction power of SVI in our model is 0.85.

Though the ELM presented here is obtained from the CQWWTP, the technique can also be applied for the other WWTPs, as the input parameters and operational conditions are similar. The method can be used for control of wastewater treatment operation in order to improve the treatment performance.

Acknowledgments. The authors thank Han Lin, the undergraduate student in the Faculty of Science and Technology, University of Macau, for assistance in performing ELM. The financial support from the Research Committee at University of Macau under Grant no. MYRG106 (Y1-L3)-FST12-LIC and MYRG2014-000058-FST are gratefully acknowledged.

References

1. Jenkins, D., Richard, M.G., Digger, G.T.: Manual on the Caused and Control of Activated Sludge Bulking, Foaming and other Solids Separation Problems. Lewis Publishers, New York (2003)
2. Lou, I., De Los Reyes III, F.L.: Integrating decay, storage, kinetic selection, and filamentous backbone factors in a bacterial competition model. Water Environm. Res. 77, 287–296 (2005)
3. Lou, I., De Los Reyes III, F.L.: Substrate uptake tests and quantitative FISH show differences in kinetic growth of bulking and non-bulking activated sludge. Biotechnol. Bioeng. 92, 729–739 (2005)
4. Capodaglio, A.G., Jones, H.V., Novotny, V., Feng, X.: Sludge bulking analysis and forecasting: application of system identification and artificial neural computing technologies. Water Res. 25, 1217–1224 (1991)
5. Maier, H.R., Jain, A., Dandy, G.C., Sudheer, K.P.: Methods used for the development of neural networks for the prediction of water resource variables in river systems: current status and future directions. Environ.l Modell. Softw. 25, 891–909 (2010)
6. Camdevyren, H., Demyr, N., Kanik, A., Keskyn, S.: Use of principal component scores in multiple linear regression models for prediction of Chlorophyll-a in reservoirs. Ecol. Model. 181, 581–589 (2005)
7. Pallant, J., Chorus, I., Bartram, J.: Toxic cyanobacteria in water, SPSS Survival Manual (2007)

8. Hecht-Nielsen, R.: Kolmogorov's mapping neural network existence theorem. In: Proceedings of 1st IEEE International Jopint Conference of Neural Networks, New York (1987)
9. Lou, I., Zhao, Y.: Sludge bulking prediction using principle component regression and artificial neural network. Mathematical Problems in Engineering 2012, 17 pages (2012)
10. Huang, G.-B., Zhu, Q.-Y., Siew, C.-K.: Extreme learning machine: A new learning scheme of feed forward neural networks. In: IEEE International Conference on Neural Networks - Conference Proceedings, vol. 2, pp. 985–990 (2004)
11. Huang, G.-B., Zhu, Q.-Y., Siew, C.-K.: Extreme learning machine: Theory and applications. Neurocomputing 70, 489–501 (2006)
12. Huang, G.-B., Chen, L., Siew, C.-K.: Universal approximation using incremental constructive feedforward networks with random hidden nodes. IEEE Transactions on Neural Networks 17, 879–892 (2006)
13. Huang, G.-B.: An Insight into Extreme Learning Machines: Random Neurons, Random Features and Kernels. Cognitive Computation (in press, 2014)
14. Cao, J.W., Chen, T., Fan, J.: Fast Online Learning Algorithm for Landmark Recognition based on BoW Framework. In: Proceedings of the 9th IEEE Conference on Industrial Electronics and Applications, Hangzhou, China, June 9-12 (2014)
15. Cao, J.W., Xiong, L.: Protein Sequence Classification with Improved Extreme Learning Machine Algorithms. BioMed Research International 2014, 12 pages (2014)
16. Cao, J.W., Lin, Z., Huang, G.-B., Liu, N.: Voting based extreme learning machine. Information Sciences 185, 66–77 (2012)
17. Huang, G.-B., Chen, L.: Convex incremental extreme learning machine. Neurocomputing 70, 3056–3062 (2007)
18. Huang, G.-B., Chen, L.: Enhanced random search based incremental extreme learning machine. Neurocomputing 71, 3460–3468 (2008)
19. Huang, G.-B., Wang, D.H., Lan, Y.: Extreme learning machines: A survey. International Journal of Machine Learning and Cybernetics 2, 107–122 (2011)
20. Wong, K.I., Wong, P.K., Cheung, C.S., Vong, C.M.: Modeling and optimization of biodiesel engine performance using advanced machine learning methods. Energy 55, 519–528 (2013)
21. Huang, G.-B., Zhou, H., Ding, X., Zhang, R.: Extreme learning machine for regression and multiclass classification. IEEE Transactions on Systems, Man, and Cybernetics, Part B: Cybernetics 42, 513–529 (2012)

Extreme Learning Machine for Linear Dynamical Systems Classification: Application to Human Activity Recognition

Wen Wang[1,2], Lianzhi Yu[1], Huaping Liu[2], and Fuchun Sun[2]

[1] School of Optical-Electrical and Computer Engineering,
University of Shanghai for Science and Technology, Shanghai, China
[2] Department of Computer Science and Technology, Tsinghua University, State Key Laboratory of Intelligent Technology and Systems, TNLIST, Beijing, China

Abstract. This paper proposes a Extreme Learning Machine (ELM) recognition framework for human activities using essential dynamic characteristics of the activity. Raw activity time series are collected from inertial sensors embedded in smart phone.We model each activity sequence with a collection of linear dynamical system (LDS) models, each LDS model describing a small patch of the sequence. A codebook is formed using the K-medoids clustering algorithm and a Bag-of-Systems (BoS) is developed to represent the activity time series. Then use ELM to classify them. Great advantages of this method are that complicated statistical feature design procedure is avoided and the LDSs can well capture the dynamics of the activity. Our experiment validation on public dataset shows promising results.

Keywords: ELM, human activity recognition, linear dynamical systems, bag of systems.

1 Introduction

Recently, Extreme Learning Machine(ELM)[1],[2] has attracted more and more researchers' attention for its better performance than traditional parameters learning algorithm such as gradient descent algorithm in generalized single hidden layer feed-forward neural networks(SFLNs). In [1],[2] the authors have proved that ELM tends to have better scalability and achieve similar (for regression and binary class cases) or much better (for multi-class cases) generalization performance at much faster learning speed (up to thousands times) than traditional SVM. ELM has been used in several domains ranging from human action recognition[3],[4],[5], face recognition[6],[7], visual tracking[8] and so on.

On the other hand,in the last decade, human activity recognition has become an important emerging field of research within context-aware systems[9],[10]. Mobile phones or smart phones are rapidly becoming the central computer and communication device in people's lives. What's more, nowadays, smart phones are programmable and equipped with a growing set of cheap powerful sensors, such as accelerometer, digital compass, gyroscope, GPS, microphone, and camera[11].

© Springer International Publishing Switzerland 2015
J. Cao et al. (eds.), *Proceedings of ELM-2014 Volume 2*,
Proceedings in Adaptation, Learning and Optimization 4, DOI: 10.1007/978-3-319-14066-7_2

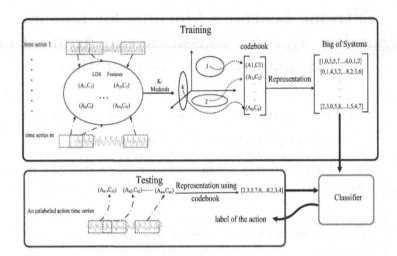

Fig. 1. Architecture of our framework. Top: Training progress. Bottom: Testing progress. Bottom right corner: ELM classifier.

For the sake of economy and portability, a lot of researches fix attention on the applications that can be built on smart phones.

Generally, two approaches have been proposed to extract features in time series data: statistical detectors and structure detectors[12]. Another important tool to deal with the time series is LDS. In[13], the LDS model has been successfully used for dynamic texture description. In[14], this model was developed for teaching control courses. Very recently, Ref.[15] proposed a Bag-of-Systems (BoS) model to describe complicated dynamic texture. His method was motivated by the popular Bag-of-Words (BoW) model, using un-ordered multiple local LDSs to represent a whole video sequence. Inspired by such works, we regard human activity time series as the output of an intrinsic dynamic system.

The main contribution of this paper is that BoS is developed to depict the characteristics of the time series collected by the smart phone sensors. To the best knowledge of the authors, this is the first time for such a method to be used for human activity recognition. Please note that in[16] the human activity recognition was addressed by k-nearest neighbor method. However, such a method depends on very complicated features, which should be designed by human. In this paper, such a tedious procedure is avoided and we just use the LDSs to represent the whole dynamics of the time series. In this regard, the proposed method is more principle and requires less feature design. The experiment results show that the proposed method can obtain comparable results with[16].

The rest of this paper is organized as follows. In Section 2, the overall architecture is illustrated. In Section 3 we categorize time series using proposed framework. Section 4 provides some experimental results. Finally, the conclusion is given in section 5.

2   Architecture

The framework for human activity recognition is inspired by the BoF approach to classify time series[17]and the bag of dynamical systems[15]in categorizing dynamic textures as shown in Fig.1. The main steps in our framework are as follows:

1. Extract LDS models from the training set.
2. Form codebook using K-medoids clustering algorithm.
3. Represent time series using the formed codebook.
4. Train ELM using the representation vectors and corresponding labels.
5. Given a new time series, infer which class it belongs to using the trained ELM.

3   Recognizing Human Activities Using a Bag of Linear Dynamical Systems

3.1   Brief Review about LDS

Assume that a time series $\{\xi_t\}_{t=1,\cdots\tau}, \xi_t \in \Re^m$ is a realization of a second-order stationary stochastic process[13]. In our paper, we assume that there exists symmetric positive-definite matrices $\mathbf{Q} \in \Re^{n \times n}$ and $\mathbf{R} \in \Re^{m \times m}$ such that

$$\begin{cases} \eta_{t+1} = \mathbf{F}\eta_t + v_t & v_t \sim \mathbf{N}(0, \mathbf{Q}) \\ \xi_t = \mathbf{G}\eta_t + \omega_t & \omega_t \sim \mathbf{N}(0, \mathbf{R}) \end{cases} \tag{1}$$

where $\eta_t \in \Re^n$ is the hidden state at time t, $\mathbf{F} \in \Re^{n \times n}$ models the dynamics of the hidden state, $\mathbf{G} \in \Re^{m \times n}$ maps the hidden state to the output of the system, v_t and ω_t are driven by Gaussian white noise.

It is well known that the choices of matrices of $\mathbf{F}, \mathbf{G}, \mathbf{Q}$ is not unique, but we can find a canonical model realization to represent each equivalence class[18]. Subspace methods calculate LDS parameters by first decomposing a matrix of observations to generate an estimate of the underlying state sequences. The most straightforward technique is singular value decomposition (SVD). SVD yields $\xi \approx \mathbf{U}\Sigma\mathbf{V}^T$ where $\mathbf{U} \in \Re^{m \times n}$ and $\mathbf{V} \in \Re^{\tau \times m}$ have orthonormal columns $\{u_i\}$ and $\{v_i\}$ and $\Sigma = diag\{\sigma_i, \cdots, \sigma_n\}$ contains the singular values. So we get the estimates of $\mathbf{G}$ and η where

$$\widehat{\mathbf{G}} = \mathbf{U} \qquad \widehat{\eta} = \Sigma\mathbf{V}^T \tag{2}$$

The least squares estimate of $\mathbf{F}$ is:

$$\widehat{\mathbf{F}} = \eta_{1:\tau}\eta^\dagger_{0:\tau-1} \tag{3}$$

where, † denotes the Moore-Penrose inverse. Stability is a desirable character-istic for linear dynamical systems, but in the above estimation procedure, the algorithm does not enforce stability. In[19], the author proposed a novel method for learning stable dynamical systems. Readers can read [19] for more details.So far, we have got the LDSs parameters$\{\mathbf{F}, \mathbf{G}\}$.

The LDS can be used to characterize the dynamics of the human activity. However, its representative capability is weak since it is a simple linear model, while practical time series always contain complicated dynamics. To tackle this problem, we extract multiple subsequences from the time series and use the BoS method to contract features for classification. In this section we first introduce the feature extraction, then the codebook design and the time series representa-tion. Finally we give the classification method.

3.2   Features Extraction

Feature extraction on windows with *50%* overlap has demonstrated success in previous works. The size of L is decided by the length of time series. In our experiment, there are *500* sample points, our results show that *L=100* works well (described in section 5). So for the training set, totally, N features are extracted. As described above, the features are a set of LDSs parameters.

3.3   Codebook Formation

In the traditional BoF framework, once the features and the corresponding de-scriptors are extracted, the descriptors are clustered into k groups using cluster-ing algorithm such as K-Means. The centers of the groups are selected to form the codebook. Unfortunately, in our paper, the descriptors are the LDSs param-eters, which lie in the non-Euclidean manifold. In order to solve this problem, we used a metric between two LDSs parameters, that is, Martin distance[20], [21]. In section 3.2, we extract N features $\mathbf{S} = \{\mathbf{F}_i, \mathbf{G}_i\}_{i=1}^{N}$ in the training set. Conse-quently, the distance matrix $\mathbf{D} \in \Re^{N \times N}$ can be calculated by Martins approach. A medoid of a set of data points is the data point that minimizes the sum of squared distances to all other data points. Therefore, the medoid is similar to the centroid of the data points, except that it is restricted to be one of the data points. Consequently, after running the K-medoids algorithm on the distance matrix D, we can directly obtain a codebook $\mathbf{W} = \{\mathbf{W}_1, \mathbf{W}_2, \cdots, \mathbf{W}_K\} \subset \mathbf{S}$

3.4   Activity Timer Series Representation

Each time series can be represented by the codebook. The simplest representa-tion is called Term Frequency (TF) [15] and is defined as:

$$r_j = \frac{c_j}{\sum_{j=1}^{K} c_j} \qquad j = 1, 2, \cdots, K \qquad (4)$$

where c_j is the times that codeword $\mathbf{W}_j$ *occurs* in a time series. Note that, we select the nearest codeword away from the feature descriptor as the *occurs* codeword. After the representation procedure, for each time series we can get a probability histogram.

3.5   Classification

In this section, a multi-class classifier should be designed for recognition. In our paper, the common used classifier ELM are adopted. Here we briefly introduce it.

Given N arbitrary distant features $\mathbf{r} = \{\mathbf{r}_i, \mathbf{t}_i\}_{i=1}^{N}$ where $\mathbf{r}_i \in \Re^p$ and $\mathbf{t}_i \in \Re^m$, standard SLFNs with n hidden nodes and activation function $g(x)$ can be modeled as:

$$\mathbf{H}\boldsymbol{\beta} = \mathbf{T} \tag{5}$$

where

$$\mathbf{H} = \begin{bmatrix} g(\mathbf{w}_1 \cdot \mathbf{r}_1 + b_1) & \cdots & g(\mathbf{w}_n \cdot \mathbf{r}_N + b_n) \\ \vdots & \cdots & \vdots \\ g(\mathbf{w}_1 \cdot \mathbf{r}_N + b_1) & \cdots & g(\mathbf{w}_n \cdot \mathbf{r}_N + b_n) \end{bmatrix}_{N \times n} \tag{6}$$

$$\boldsymbol{\beta} = \begin{bmatrix} \boldsymbol{\beta}_1^T \\ \vdots \\ \boldsymbol{\beta}_n^T \end{bmatrix}_{n \times m} \qquad \mathbf{T} = \begin{bmatrix} \mathbf{t}_i^T \\ \vdots \\ \mathbf{t}_n^T \end{bmatrix}_{N \times m} \tag{7}$$

$\mathbf{H}$ is called the hidden layer output matrix of the neural network where ith column of $\mathbf{H}$ is the output of ith hidden node with respect to inputs samples $\mathbf{r}_i, \cdots, \mathbf{r}_N$; $\mathbf{w}_i$ for $i = 1, \cdots, n$ is the weight vector connecting the ith hidden node and the input nodes; b_i for $i = 1, \cdots, n$ is the threshold of the ith node; $\boldsymbol{\beta}$ is the weight vector connecting the ith hidden node and the out nodes[1],[2].

Unlike traditional learning method for SLFNs, in ELM, $\mathbf{w}_i$ and b_i are not necessarily tuned and $\mathbf{H}$ can remain unchanged once random values have been assigned to these parameters in the beginning of learning[1]. The training of ELM requires minimization of the training error $||\mathbf{H}\boldsymbol{\beta} - \mathbf{T}||^2$ and the norm of the output weight $||\boldsymbol{\beta}||$. The smallest norm least-squares solution of (5) is :

$$\widehat{\boldsymbol{\beta}} = \mathbf{H}^\dagger \mathbf{T} \tag{8}$$

where

$$\mathbf{H}^\dagger = (\frac{\mathbf{I}}{C} + \mathbf{H}\mathbf{H}^T)^{-1}\mathbf{T} \tag{9}$$

is the Moor-Penrose generalized innverse of matrix $\mathbf{H}$ and C is a user-specified parameter that promotes the generalization performance.Then the output function of ELM can be written as

$$\mathbf{f}(\mathbf{r}) = \mathbf{h}(\mathbf{r})\mathbf{H}^T(\frac{\mathbf{I}}{C} + \mathbf{H}\mathbf{H}^T)^{-1}\mathbf{T} \tag{10}$$

If we don't know the feature mapping vector we can define a kernel matrix for ELM as follows:

$$\Omega_{ELM} = \mathbf{H}\mathbf{H}^T : \Omega_{ELM(i,j)} = \mathbf{h}(\mathbf{r}_i) \cdot \mathbf{h}(\mathbf{r}_j) = \boldsymbol{k}(\mathbf{r}_i, \mathbf{r}_j) \tag{11}$$

Then the output function of ELM can be written as

$$\mathbf{f}(\mathbf{r}) = \begin{bmatrix} \boldsymbol{k}(\mathbf{r}, \mathbf{r}_1) \\ \vdots \\ \boldsymbol{k}(\mathbf{r}, \mathbf{r}_N) \end{bmatrix} (\frac{\mathbf{I}}{C} + \Omega_{ELM})^{-1}\mathbf{T} \tag{12}$$

Human activity recognition is a multi-class classification problem. So the predicted label of the testing sample $\mathbf{r}$ is the index number of output node which has the highest output value.

$$label(\mathbf{r}) = arg \max_{i \in \{1, \cdots, m\}} f_i(\mathbf{r}) \tag{13}$$

4   Experiments and Results

4.1   DataSet

The activity recognition dataset is collected from 9 volunteers, and can be downed from CRCV in the university of Center Florida[1]. Nine activities *(Bike, Climbing, Descending, Gymbike, Jumping, Running, Standing, Treadmill, and Walking)* are performed and each activity is recorded for 5 times using the single *60Hz* IMU built in the phone. Three inertial sensors are used to collect data: accelerometer, gyroscope and magnetometer. In [16],the authors have proved that the data from magnetometer is not useful and the accelerometer data is most useful. So in our paper, we only adopt the accelerometer data. Each action sequence was trimmed to *8.33* seconds (*500* sample points).

4.2   Experiment Results

In our experiment, all of the accelerometer data in the database are used to construct the codebook.In this dataset, *L=100* works well. We also use the leave-one-subject-out validation test to evaluate the ability of classifiers to recognize

---

[1] http://crcv.ucf.edu/data/UCF-iPhone.php
[2] http://www.ntu.edu.sg/home/egbhuang/
[3] http://www.csie.ntu.edu.tw/~cjlin/libsvm/

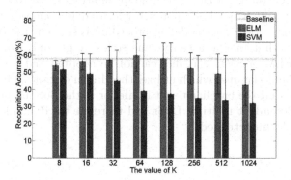

Fig. 2. Recognition accuracies using proposed BoS method and baseline method. Note that the error bar denotes the difference between highest value and mean value, and the green line denotes the baseline method(1-NN).

unacquainted actions. That is, for 9 subjects, in one trial, the features of one subject are used to test and other eight subjects' features are selected to train the classifier. Totally, nine trials of simulations are performed.

In this section, we compare the performance of Huang's KELM[2] and SVM[3] in our framework. The Gaussian RBF kernel $K(\mathbf{u}, \mathbf{v}) = exp(-\lambda||\mathbf{u} - \mathbf{v}||^2)$ are used in KELM and SVM.For KELM and SVM classifier, in order to get good generalization performance, parameter C (called regularization coefficientin int KELM and cost parameter in SVM) and kernel parameter λ should be chosen appropriately. We have tried 21 different C and 21 different λ. The 21 different values of C and λ are $\{2^{-10}, 2^{-9}, \cdots, 2^9, 2^{10}\}$. The average results of 9 trials are obtained for each combination of C and λ, and the performance is shown in Fig.2.

Fig.2 shows the performance when $K=$ *8, 16, 32, 64, 128, 256, 512, 1024,* respectively. K denotes the size of codebook.The experimental results indicate that an appropriate K leads to good performance. As K is more than *64*, the performance decreases. The reason is that a large K results too many codebook elements and two similar LDSs descriptors may be separated into different clusters, although they are both very close to the boundary. When $K=64$, our method performs the best for both classifier. The highest accuracy using the proposed method is about *72%*, which is a littler higher than the results *(68%)* in [16]. Although the performance improvement is not very significant, an important advantage lies in the fact that in the proposed method, no extra feature selection procedure is needed and the LDS naturally characterizes the intrinsic dynamics of the time series, on the contrary in [16], 13 ad-hoc features are designed by the human. In this sense, the proposed method gives a more principle solution to tackle this problem.

To show the advantage of BoS which adopt multiple LDSs to depict a whole time series, we design a baseline approach which adopts one single LDS model to depict the time series. The modeling procedure is same as the description in section 4, except that,for one activity time series we get only one LDS model. In this case, KELM or SVM classifier cannot be adopted and we use the nave k-nearest neighbour classifier which utilizes the Martin distance. Obviously, from Fig.2 we can see that the baseline result is inferior to the BoS method. The main reason is that BoS adopt multiple LDSs and therefore the dynamics of the time series can be well characterized.

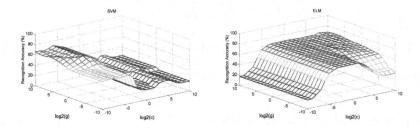

Fig. 3. Top: Performances(K=64) of SVM with Gaussian kernel is sensitive to the parameters (C, λ). Bottom: Performance (K=64) of ELM with Gaussian kernel is not very sensitive to the parameters (C, λ).

Overall, KELM performs better than SVM, though it does not get the highest results. Fig.2 shows that KELM get higher average recognition accuracy and lower standard error than SVM with different size of K. It can be seen from Fig.3 more intuitively that KELM can achieve good generalization performance while C or λ changes. On the contrary, the accuracy decreases quickly with the decrease of C, though SVM can get highest recognition accuracy.

In order to find which activities are harder to be recognized relatively, we analyzed the confusion matrix. Fig.4 shows the aggregate confusion matrix for the best overall recognition result *(71.36%)*. The confusion matrix gives information about the actual and predicted classifications done by classifier. We can see that different actions obtain different recognition accuracies. For example, *Standing* can be recognized completely right. *Bike* and *Climbing* obtains higher accuracy *(89% and 87%)* respectively.*Jumping* and *Running* are usually confused with each other. This result is reasonable, because the raw signals of *Jumping* and *Running* are indeed similar. In [16], *Climbing* and *Descending* are always confused even used the hierarchical classifier, but in our paper these two actions can be recognized accurately.

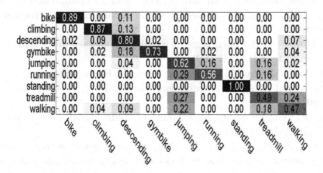

Fig. 4. Confusion matrix for best recognition result

5   Conclusion and Outlook

This paper proposes a novel framework for human activity recognition. The main difference between our method and previous work is that LDSs parameters are selected as features to represent activity time series. The average recognition results for 9 different human activities in the UCF-iPhone Data set using the proposed framework can be achieved to about 72%. A great advantage of this method is that the complicated feature design procedure is avoided and the LDSs can well capture the dynamics of the time series. To achieve better recognition results, in future studies, we will study the essential differences between the often confused activities such as *Walking, Treadmill* and *Jumping*. Then we will try to design special method to distinguish them based on the proposed framework.

Acknowledgement. This work was supported in part by the National Key Project for Basic Research of China under Grant 2013CB329403; in part by the National Natural Science Foundation of China under Grant 91120011 and Grant 61210013; in part by the Tsinghua Self-innovation Project under Grant 20111081111; and in part by the Tsinghua University Initiative Scientific Research Program under Grant 20131089295.

References

1. Huang, G., Zhu, Q., Siew, C.: Extreme learning machine: theory and applications. Neurocomputing 70, 489–501 (2006)
2. Huang, G., Zhou, H., Ding, X., Zhang, R.: Extreme Learning Machine for Regression and Multiclass Classification. IEEE Trans. on Systems, Man, and Cybernetics, Part B: Cybernetics 42, 513–529 (2012)
3. Iosifidis, A., Tefas, A., Pitas, I.: Dynamic action recognition based on denemes and extreme learning machine. Pattern Recognition Letters 34, 1890–1898 (2013)
4. Deng, W.Y., Zheng, Q.H., Wang, Z.M.: Cross-person activity recognition using reduced kernel extreme learning machine 53, 1–7 (2014)

5. Minhas, R., Baradaran, A., Seifzadeh, S., Wu, Q.M.J.: Human action recognition using extreme learning machine based on visual vocabularies. Neurocomputing 73, 1906–1917 (2010)
6. Zhong, W., Huang, G.B.: Face recognition based on extreme learning machine. Neurocomputing 74, 2541–2551 (2011)
7. Mohammed, A.A., Minhas, R., Wu, Q.M.J., et al.: Human face recognition based on multidimensional PCA and extreme learning machine. Pattern Recognition 44, 2588–2597 (2011)
8. Liu, H., Sun, F., Yu, Y.: Multitask extreme learning machine for visual tracking. Cognitive Computation (January 2014)
9. Cheng, H., Liu, Z., Zhao, Y., Ye, G., Sun, X.: Real world activity summary for senior home monitoring. Multimedia Tools and Applications, pp. 1–4 (July 2011)
10. Cheng, H., Liu, Z., Hou, L., Yang, J.: Sparsity induced similarity measure and its applications. IEEE Trans. on Circuits and Systems for Video Technology PP, 1 (2012)
11. Lane, N.D., Miluzzo, E., Hong, L., Peebles, D., Choudhury, T., Campbell, A.T.: A survey of mobile phone sensing. IEEE Trans. on Communications Magazine 48, 140–150 (2010)
12. Wang, J., Chen, R.H., Sun, X.P., She, M., Kong, L.X.: Generative models for automatic recognition of human daily activities from a single triaxial accelerometer. In: Proc: of Int. Conf. on Neural NetWorks (IJCNN), pp. 1–6 (June 2012)
13. Doretto, G., Chiuso, A., Wu, Y., Soatto, S.: Dynamic textures. International Journal of Computer Vision 51(2), 91–109 (2003)
14. Liu, H., Xiao, W., Zhao, H., Sun, F.: Learning and understanding system stability using illustrative dynamic texture examples. IEEE Trans. on Education 57(1), 4–11 (2014)
15. Vidal, R., Chaudhry, R., Vidal, R.: Categorizing dynamic textures using a bag of dynamical systems. IEEE Trans. on Pattern Analysis and Machine Intelligence 35(2), 342–353 (2013)
16. McCall, C., Reddy, K., Shah, M.: Macro-class selection for hierarchical K-NN classification of inertial sensor data. In: Proc. of 2nd Int. Conf. Pervasive and Embedded Computing and Communication Systems (PECCS), pp. 106–114 (February 2012)
17. Baydogan, M.G., Runger, G., Tuv, E.: A bag-of-features framework to classify time series. IEEE Trans. on Pattern Analysis and Machine Intelligence 35(11), 2796–2802 (2013)
18. Saisan, P., Doretto, G., Wu, Y., Soatto, S.: Dynamic texture recognition. In: Proc. of Computer Vision and Pattern Recognition (CVPR), vol. 2, pp. 58–63 (2001)
19. Siddiqi, S.M., Boots, B., Gordon, G.J.: A constraint generation approach to learning stable linear dynamical systems. In: Proc. of Neural Information Processing Systems, pp. 1329–1336 (December 2007)
20. Cock, K.D., Moor, B.D.: Subspace angles between ARMA models. Systems and Control Letters 46(4), 265–270 (2002)
21. Martin, R.J.: A metric for ARMA processes. IEEE Trans. on Signal Processing 48(4), 1164–1170 (2000)

Lens Distortion Correction Using ELM

Chang Liu and Fuchun Sun

State Key Laboratory of Intelligent Technology and Systems,
Tsinghua National Laboratory for Information Science and Technology,
Dept. of Computer Science and Technology, Tsinghua University, P.R. China, 100084
`cliu13@mails.tsinghua.edu.cn`,
`fcsun@mail.tsinghua.edu.cn`

Abstract. Lens distortion is one of the major issues in camera calibration since it causes the perspective projection of the camera model to no longer hold. Thus to eliminate lens distortion becomes an essential part of camera calibration. This paper proposes a novel method of correcting lens distortion by implementing extreme learning machine, a new learning algorithm for single-hidden layer feedforward networks. A camera calibration model which contains linear phase for calibration, and non-linear phase for lens distortion correction is introduced. The performance is evaluated in comparison with traditional learning methods, and the results show that the proposed model produces much better performance than that of the others.

Keywords: lens distortion correction, extreme learning machine, direct linear transformation, non-linear model, camera calibration, feedforward neural network.

1 Introduction

Camera calibration is the essential premise of stereo computer vision and three-dimensional reconstruction. The purpose of camera calibration is to reconstruct any world point from image points based on calibrated camera model. Plenty of camera calibration methods have been proposed, and they can be divided mainly into two categories: traditional calibration methods and self-calibration methods.

For a traditional method, a calibration object, whose structure of geometry is known in advance, is always needed. It constructs the constraints of the camera model parameters through the correspondence between the spatial point and the image points (two images for binocular vision), and then solves the optimization problem to obtain the model parameters. For example, the Direct Linear Transformation method of Abdel-Aziz and Karara [1] tries to obtain the camera model parameters by solving a set of linear equations; the two-step calibration method based on radial constrains given by Tsai [2] calculates the camera model parameters via RAC constrains; Zhang's new flexible camera calibration method [3] uses a calibration plane with axis Z in world coordinate equals zero to find the optimization solution.

© Springer International Publishing Switzerland 2015
J. Cao et al. (eds.), *Proceedings of ELM-2014 Volume 2,*
Proceedings in Adaptation, Learning and Optimization 4, DOI: 10.1007/978-3-319-14066-7_3

For a self-calibration method the calibration object is not required, since it directly performs calibration with the relationship of the corresponding points of multi images. For example, Faugeras proves the existence of quadratic nonlinear constraints in each two images, and uses LM algorithm to obtain the parameters [4]. Triggs implanted the consistency of the European transform of the absolute conic to perform the calibration [5].

Other methods include: vanishing points for orthogonal directions [6], self-calibration for varifocal cameras [7] and so on.

The procedure of camera calibration usually contains two phases: the linear phase which tries to find the projection relationship between the world points and the image points by determining the parameters of the camera model, and the non-linear phase which tries to solve the lens distortion of camera.

Extreme learning machine (ELM) is a new learning method for single-hidden layer neural network where the hidden layer neurons can be randomly generated independent of training data and application environment [8]. ELM has proved fine ability in terms of non-linear regression and classification. For example, Rong proposed a recognition scheme for identifying the aircrafts of different types using ELM to train the multiple single-hidden layer feedforward networks [9]. Sun introduced a sale forecasting model which solves a regression problem with the aid of ELM [10]. Xu Y raised a real time transient stability assessment model using ELM to gain better computation speed and accuracy [11] and so on.

This paper presents a new model for camera calibration and lens distortion correction based on ELM. To the best of our knowledge, the application of ELM has not been used in the literature before. The simulation results show that the proposed model performs well in three-dimensional reconstruction and outperforms other traditional regression methods, Back-Propagation Network and Support Vector Machine.

2   Extreme Learning Machine

A typical SLFN usually contains three layers: an input layer with n nodes, a single hidden layer with $\tilde{N}$ nodes and an output layer with m nodes.

For N arbitrary distinct samples $(\boldsymbol{x}_i, \boldsymbol{t}_i)$, where $\boldsymbol{x}_i = (x_{i_1}, x_{i_2}, \cdots, x_{i_n})^T$ and $\boldsymbol{t}_i = (t_{i_1}, t_{i_2}, \cdots, t_{i_m})^T$, the output function of the network is expressed as:

$$f_{\tilde{N}}(\boldsymbol{x}) = \sum_{i=1}^{\tilde{N}} \beta_i G(\boldsymbol{w}_i, b_i, \boldsymbol{x}_j) = \sum_{i=1}^{\tilde{N}} \boldsymbol{\beta}_i g(\boldsymbol{w}_i \cdot \boldsymbol{x}_j + b_i) = \boldsymbol{o}_j, j = 1, 2, \cdots, N \qquad (1)$$

where $\boldsymbol{\beta}_i = (\beta_{i_1}, \beta_{i_2}, \cdots, \beta_{i_m})^T$ is the weight vector from the hidden layer nodes to the output layer nodes for the ith sample; $\boldsymbol{w}_i = (w_{i_1}, w_{i_2}, \cdots, w_{i_n})^T$ denotes the weight vector from the input layer nodes to the hidden layer nodes and b_i is the bias of the ith hidden layer node. The activation function can be linear function or sigmoid function or other types of functions and here it is defined as linear function.

The traditional method of solving the SLFN is to minimize the cost function of error with gradient decent method. With back-propagation method, it updates the weight vector from learning. Two major problems of this method are, firstly, its convergence speed can be slow and secondly, it may stop at the local minima rather the optimal solution.

For a standard SLFN with $\tilde{N}$ nodes, it can approximate N samples with zero error, which implies $\sum_{j=1}^{\tilde{N}} \|o_j - t_j\|$. Given (1), it can also be expressed as:

$$\sum_{i=1}^{\tilde{N}} \beta_i g(w_i \cdot x_j + b_i) = t_j, j = 1,2,\cdots,N \tag{2}$$

Written in matrix form indicates:

$$H\beta = T \tag{3}$$

where

$$H(w_1, \cdots, w_{\tilde{N}}, b_1, \cdots, b_{\tilde{N}}, x_1, \cdots, x_{\tilde{N}})$$
$$= \begin{bmatrix} g(w_1 \cdot x_1 + b_1) & \cdots & g(w_{\tilde{N}} \cdot x_1 + b_1) \\ \vdots & \ddots & \vdots \\ g(w_1 \cdot x_{\tilde{N}} + b_1) & \cdots & g(w_{\tilde{N}} \cdot x_N + b_1) \end{bmatrix}_{N \times \tilde{N}}$$

$$\beta = [\beta_1^T, \beta_2^T, \cdots, \beta_{\tilde{N}}^T]_{\tilde{N} \times m}^T \text{ and } T = [t_1^T, t_2^T, \cdots, t_{\tilde{N}}^T]_{N \times m}^T. \tag{4}$$

H denotes the hidden layer output matrix; the ith column of H is the ith hidden node output with respect to inputs $x_1, x_2, \cdots, x_{\tilde{N}}$.

It has been proved by Huang G.B[8] in that if the activation function g is infinitely differentiable, then the input weight vector w and the hidden layer bias b can be randomly assigned. Thus solving the SLFN problem reduces to solve the output weight vector β, which can be accomplished by solving the Moore-Penrose of H from (3), noted as:

$$\beta = H^+T \tag{5}$$

where H^+ is the Moore-Penrose of H.

Thus, for a standard SLFN with infinitely differentiable activation function, the Extreme Learning Machine (ELM) algorithms is described as follows:

Algorithm ELM.

Input: training samples $X = \{(x_i, t_i) | x_i \in R^n, t_i \in R^m, i = 1,2,\cdots,N\}$
Output: output weight β
Algorithm steps:
      Step1: Random assign the input weight w and the hidden layer bias b
      Step2: Calculate the hidden layer output matrix H
      Step3: Calculate the output weight β

The Moore-Penrose of the output matrix H can always be obtained by singular value decomposition (SVD).

3   Lens Distortion in Camera Calibration

3.1   Linear Camera Model

Camera calibration means to calculate the intrinsic parameters and extrinsic parameters of the camera, in order to reconstruct any point in the camera image coordinate to the world 3D coordinate.

The theoretic camera model is the linear model, or pin-hole model, in which the image point and the world point comply with a linear transform. The pinhole camera model is based on the principle of collinearity, where each point in the object space is projected by a straight line through the projection center in to the image plane [12], as is shown in Figure 1.

For a point P in the world 3D coordinate $P(X_W, Y_W, Z_W)$, it can be represented with a linear transform:

$$s \begin{bmatrix} u \\ v \\ 1 \end{bmatrix} = \begin{bmatrix} f_x & 0 & u_0 & 0 \\ 0 & f_y & v_0 & 0 \\ 0 & 0 & 1 & 0 \end{bmatrix} \begin{bmatrix} \mathbf{R} & \mathbf{T} \\ \mathbf{0}^T & 1 \end{bmatrix} \begin{bmatrix} X_W \\ Y_W \\ Z_W \\ 1 \end{bmatrix} \tag{6}$$

where s is the scale factor, [u v 1]' is the homogeneous coordinates of point P', the corresponding point of P in the image plane; f_x and f_y denotes the scale factors along the coordinates u and v in the image plane; (u_0, v_0) is the principle point P_o of the image plane; $\mathbf{R}$ and $\mathbf{T}$ are the rotation matrix and the translation matrix respectively, which represents the relationship between the image coordinate and the world coordinate.

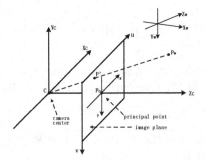

Fig. 1. Coordinates transform from image plane to 3D

(6) can also be expressed as:

$$s \begin{bmatrix} u \\ v \\ 1 \end{bmatrix} = \mathbf{M_1 M_2 X_W} \tag{7}$$

where $\mathbf{M_1} = \begin{bmatrix} f_x & 0 & u_0 & 0 \\ 0 & f_y & v_0 & 0 \\ 0 & 0 & 1 & 0 \end{bmatrix}$ is called the intrinsic matrix and $\mathbf{M_2} = \begin{bmatrix} \mathbf{R} & \mathbf{T} \\ \mathbf{0}^T & 1 \end{bmatrix}$ is called the extrinsic matrix. $\mathbf{X_W}$ denotes the homogeneous coordinates vector of 3D point $P(X_W, Y_W, Z_W)$.

3.2   Direct Linear Transformation and 3D Reconstruction with Binocular Vision

Direct Linear Transformation

Direct Linear Transformation (DLT) was proposed by Abdel-Aziz and Karara to calibrate the parameters of camera[1]. DLT establishes the geometric linear model of camera, which can be solved directly with linear equation system.

DLT is defined as (8):

$$s \begin{bmatrix} u \\ v \\ 1 \end{bmatrix} = \mathbf{K}(\mathbf{R}\ \mathbf{t}) \begin{bmatrix} X_W \\ Y_W \\ Z_W \\ 1 \end{bmatrix} = \mathbf{P}_{3x4} \begin{bmatrix} X_W \\ Y_W \\ Z_W \\ 1 \end{bmatrix} \tag{8}$$

Expand (8) and eliminate s presents:

$$p_{11}X_w + p_{12}Y_w + p_{13}Z_w + p_{14} - p_{31}uX_w - p_{32}uY_w - p_{33}uZ_w - p_{34}u = 0$$
$$p_{21}X_w + p_{22}Y_w + p_{23}Z_w + p_{24} - p_{31}vX_w - p_{32}vY_w - p_{33}vZ_w - p_{34}v = 0$$
$$\tag{9}$$

Suppose N 3D points and their corresponding image points are already known, then the linear equation system with 2*N equations:

$$\mathbf{AL = 0} \tag{10}$$

where $\mathbf{A}$ is a 2N*12 matrix, $\mathbf{L}$ is a vector consisting of elements from the perspective projection matrix $\mathbf{P}_{3x4}$. Thus, the camera calibration problem is reduced to solve a least square problem:

$$\boldsymbol{L_1} = (\boldsymbol{C}^T\boldsymbol{C})^{-1}\boldsymbol{C}^T\boldsymbol{B}$$

$$s.t.\ \ p_{34} = 1 \tag{11}$$

where $\boldsymbol{L_1}$ is a vector constructed by the first 11 elements of $\mathbf{L}$, and $\boldsymbol{C}$ is a matrix constructed by the first 11 columns of $\mathbf{A}$ and $\boldsymbol{B}$ is the 12th column of $\mathbf{A}$.

3D Reconstruction with Binocular Vision

Once the projection relationship is determined, with two cameras the world coordinate of a point can be reconstructed from two set of image coordinate gained from the images token by the two cameras.

3.3   Lens Distortion in Camera Calibration

The above section explains in detail the pin-hole model of camera. However, the pin-hole model is the ideal model for camera calibration since it assumes the projection from the world point to the image plane strictly follows the linear projection principle. In reality, yet, the linear projection principle does not usually hold because of the distortion of the camera.

The distortion of the camera is a phenomenon where straight lines can no longer remain straight after projection, and the further the line from the camera center, the more obvious the phenomenon is.

The most common camera distortions are the Barrel distortion, the Pincushion distortion and the Mustache distortion, and the mathematical model which describes the non-linear camera distortion is given by Atkison [14] and Weng [15]:

$$\bar{x} = x + \delta_x(x, y)$$
$$\bar{y} = y + \delta_y(x, y) \tag{12}$$

where $(\bar{x}, \bar{y})$ represents the ideal image coordinate of the world point P of the pin-hole model; (x, y) represents the actual image coordinate; δ_x and δ_y represents the non-linear distortion value, and can be given by:

$$\delta_x(x, y) = x(1 + k_1 r^2 + k_2 r^4 \cdots) + (p_1(2x^2 + r^2) + 2p_2 xy) + s_1 r^2$$
$$\delta_y(x, y) = y(1 + k_1 r^2 + k_2 r^4 \cdots) + (p_2(2x^2 + r^2) + 2p_1 xy) + s_2 r^2$$
$$\tag{13}$$

where $r^2 = x^2 + y^2$, and the first item is called the radial distortion, the second item decentering distortion and the third item thin prism distortion.

Although with (13) the non-linear projection problem can be alleviated to some extent, yet some issues still remain. For example, it is usually not clear to give the definite expression of the non-linear distortion given (13), since the types of distortion in which a camera can possess is beyond knowledge in advance. Moreover, even the types of distortion are confirmed, the degree of the radial distortion still needs further experiments. Tsai[3] points out that, to introduce too many non-linear arguments may cause the solution to be unstable rather than improve the accuracy. Thus the number of non-linear arguments can be a trade-off problem.

In the following section, a fast and explicit method of eliminating the non-linear distortion of camera using extreme learning machine, regardless of the types and arguments of distortion, is proposed, which proves to be simpler and with better performance through experiments.

4   Lens Distortion Correction Model using ELM

4.1   Model Structure

Recall in section 3.1, the purpose of camera calibration is to determine the intrinsic and the extrinsic parameters of the camera, and to amend non-linear lens distortion as well, in order to obtain correct projection relationship between the image points and the world 3D points; since only after the camera is correctly calibrated can it be used for 3D reconstruction. However, as discussed in section 3.3, the traditional ways to eliminate lens distortion bring about some other issues like the choice of the degree of the radial distortions and so on.

The above process may also be illustrated as the process to project the image points directly to the world 3D points, regardless of the intermediate process. The model contains a linear part and a non-linear part, which considers the calibration process as a black box. With this idea, the lens distortion correction model using ELM is proposed, and the algorithm of the model is detailed in section 4.2.

4.2   Algorithm Procedure

In order to obtain the world 3D coordinate, the algorithm is divided into two phases: the training phase and the testing phase. The training phase contains two stages, corresponding to the linear part and the non-linear part respectively. As discussed in chapter 3 and 4, the linear part is modeled based on the direct linear transformation, and the non-linear part is modeled based on the extreme learning machine. The algorithm procedure is detailed in Figure 2.

The training phase is composed with two stages: the linear stage and the non-linear stage, as is shown in Figure 2.

The linear stage takes the combination of image points coordinates from two cameras and the corresponding world points coordinates as the input training sample X_1. Note that the world coordinate (X_w, Y_w, Z_w) is pre-obtained and is regarded as the criteria for the expected output world coordinate. With DLT method, the perspective projection matrix P_1 and P_1 for the two cameras is calculated and then the training sample for 3D reconstruction can be constructed, whose world coordinate is marked as $(X^{(2)}, Y^{(2)}, Z^{(2)})$.

The non-linear stage utilizes the sample X_2 from linear stage and combines the corresponding world coordinate $(X_w^{(2)}, Y_w^{(2)}, Z_w^{(2)})$ as the training sample for the extreme learning machine, and at last the expected ELM output weight β.

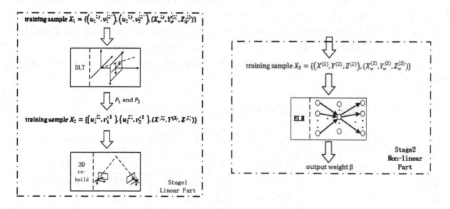

Fig. 2. Model Algorithm Procedure

5   Performance Evaluation

5.1   Dataset Description

In our experiment, the dataset is constructed manually with two cameras during the calibration and reconstruction procedure. Sixteen photos were taken by each camera, with ten gauge points on each picture. The 3D world coordinates were obtained by external calibration tool, and the world coordinates were determined with regard to a certain world point. Although the exact reference point cannot be obtained, the calibration and the reconstruction work can still be done without the knowledge of the point, as long as the relative relationship between the two cameras is known.

After eliminating the outlier points, a total dataset of 120 points is effective. Half of the dataset is used as training sample, which are selected randomly, leaving the rest as the testing sample.

5.2   Comparison Methods

The performance of the proposed ELM Lens Distortion Correction model is compared mainly with the popular algorithms for SLFN, namely the Back Propagation Network (BP network) and the Support Vector Machine (SVM). In the BP network, LM method is chosen and the activation function for hidden layer uses sigmoid function. In SVM, the simulation is carried out with LIBSVM [16], with RBF kernel function. In ELM, the All the three methods are carried out on a personal computer, which possesses a 16GB RAM, a 3.6GHz CPU. The simulation environment is MATLAB 2013b.

5.3   Performance Evaluation

MSE & Dev for Training & Testing

The number of hidden layer nodes is fixed at 40, since the size of the dataset is not large. Note the support vectors number for SVR is determined by LIBSVM tool. The number of the support vectors is 44, which also implies the rationality of the 40 nodes for ELM and BP. Table 3 shows the root mean square error and the standard deviations for the training phase and the testing phase of each algorithm. The performance is evaluated with the mean RMSE and DEV for X, Y and Z axis, and each algorithms is evaluated for 100 times and is averaged over the evaluated times.

From Table 1 it can be drawn that ELM generally performs better in terms of RMSE and DEV.

Table 1. Performance Comparison for ELM, BP and SVR

a. Performance for X-axis

Algorithms	Number of Nodes/SVs	Training		Testing	
		RMSE	Dev	RMSE	Dev
ELM	40	0.0128	0.0041	0.1252	0.1034
BP(LM)	40	0.0956	0.2359	0.2741	0.4515
SVR	44	0.1066	0.2810	0.2046	0.4253

Training & Testing Reconstruction Comparison

Figure 3 gives the reconstruction result figure for the above three algorithms. Note the red cross '+' stands for real outputs and the blue circle 'o' stands for the algorithms outputs, and only the testing samples of X-axis is presented in the figure. The results for Y and Z axis are similar with X axis, and are not shown for simplicity. From Figure 3, it can be drawn ELM performs better in reconstruction work than the other two algorithms.

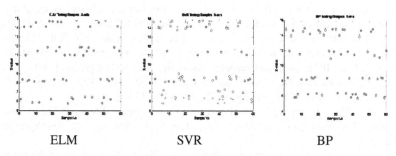

ELM   SVR   BP

Fig. 3. X-axis testing reconstruction

Training & Testing Accuracy of ELM

Figure 4 presents the testing accuracy of ELM model with regard to the change of the hidden layer nodes. It can be concluded that the performance gets better when the number of hidden layer nodes increases.

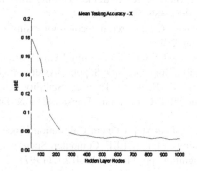

Fig. 4. ELM Testing Accuracy for X-axis

6   Conclusion

This paper proposes a novel method for camera calibration and lens distortion correction which employs the extreme learning machine. In the linear phase, the model uses DLT method to find the perspective projection relationship between the image coordinate and the world coordinate; in the non-linear phase, the model corrects the distortion by implementing ELM learning mechanism. The proposed model has been tested in real calibration work and the results show the validity of the model. Comparisons are made between the model and traditional regression models such as BP network and SVR. Yet future work on testing the robustness and fitness for big dataset is under investigation.

References

1. Abdel-Aziz, Y.I., Karara, H.M.: Direct Linear Transformation from Comparator Coordinates into Objectspace Coordinates in Close-range Photogrammetry. In: ASP Symp. on Close Range Photogrammetry. American Society of Photogrammetry, Falls Church (1971)
2. Tsai, R.Y.: An Efficient and Accurate Camera Calibration Technique for 3D Machine Vision. In: Proc. IEEE Conf. on Computer Vision and Pattern Recognition (1986)
3. Zhang, Z.: A Flexible New Technique for Camera Calibration. IEEE Transactions on Pattern Analysis and Machine Intelligence 22(11), 1330–1334 (2000)
4. Faugeras, O.D., Luong, Q.T., Maybank, S.J.: Camera Self-calibration: Theory and Experiments. In: Sandini, G. (ed.) ECCV 1992. LNCS, vol. 588, pp. 321–334. Springer, Heidelberg (1992)
5. Triggs, B.: Autocalibration and the Absolute Quadric. In: Proceedings of the IEEE Computer Society Conference on Computer Vision and Pattern Recognition, pp. 321–334 (1997)
6. Caprile, B., Torre, V.: Using Vanishing Points for Camera Calibration. International Journal of Computer Vision 4(2), 127–139 (1990)
7. Pollefeys, M., Koch, R., Van, G.L.: Self-calibration and Metric Reconstruction in spite of Varying and Unknown Intrinsic Camera Parameters. International Journal of Computer Vision 32(1), 7–25 (1999)
8. Huang, G.B., Zhu, Q.Y., Siew, C.K.: Extreme Learning Machine: Theory and Applications. Neurocomputing 70(1), 489–501 (2006)
9. Rong, H.J., Jia, Y.X., Zhao, G.S.: Aircraft recognition using modular extreme learning machine. Neurocomputing (2013)
10. Sun, Z.L., Choi, T.M., Au, K.F.: Sales Forecasting Using Extreme Learning Machine with Applications in Fashion Retailing. Decision Support Systems 46(1), 411–419 (2008)
11. Xu, Y., Dong, Z.Y., Meng, K.: Real-time Transient Stability Assessment Model Using Extreme Learning Machine. IET Generation, Transmission & Distribution 5(3), 314–322 (2011)
12. Heikkila, J., Silvén, O.: A Four-step Camera Calibration Procedure with Implicit Image Correction. In: Proceedings of the IEEE Computer Society Conference on Computer Vision and Pattern Recognition, pp. 1106–1112. IEEE (1997)
13. Source from, http://en.wikipedia.org/wiki/Distortion_(optics)
14. Atkinson, K.B.: Developments in Close Range Photogrammetry. Elsevier Science & Technology (1980)
15. Weng, J., Cohen, P., Herniou, M.: Camera Calibration with Distortion Models and Accuracy Evaluation. IEEE Transactions on Pattern Analysis and Machine Intelligence 14(10), 965–980 (1992)
16. Chang, C.C., Lin, C.J.: LIBSVM: a library for support vector machines. ACM Transactions on Intelligent Systems and Technology (TIST) 2(3), 27 (2011)

Pedestrian Detection in Thermal Infrared Image Using Extreme Learning Machine

Chunwei Yang[1,2], Huaping Liu[2,*], Shouyi Liao[1], and Shicheng Wang[1]

[1] High-Tech Institute of Xi'an, Xi'an, Shaanxi, 710025, P. R. China
[2] Department of Computer Science and Technology,
Tsinghua University, Beijing, China
hpliu@tsinghua.edu.cn

Abstract. Pedestrian detection in thermal infrared image is a challenging and hot topic. In this paper, a novel and robust pedestrian detection method based on Binarized Normed Gradients (BING) and Extreme Learning Machine (ELM) is proposed. The candidates are firstly generated using BING, and then ELM is used to classify a candidate to be pedestrian or not. Experiment results indicate that our method can obtain better performance than traditional Support Vector Machine (SVM).

Keywords: Pedestrian Detection, Thermal Infrared Image, Binarized Normed Gradients, Extreme Learning Machine.

1 Introduction

Pedestrian detection in thermal infrared image has attracted more and more attention in recent years, for its wide applications such as smart video surveillance, human-computer interfaces, drive assistance system, content based image/video indexing, and military applications. Compared with visible spectrum images, thermal infrared image offer superior performance to nighttime surveillance, and it is also applicable to daytime monitoring. However, infrared image has its own limitations. First, except for pedestrians, non-human objects (e.g. animals, cars, light poles) also produce additional bright areas, which makes it impossible to detect pedestrians only based on the brightness. Second, thermal sensors have low SNR and great noise due to the limitations in camera technology [1][2][3][4].

A considerable amount of previous work has addressed the problem of vision based pedestrian detection using visual or IR image. Ref. [1] presents a two-stage template-based method to detect people in thermal image, which initially performs a fast screening procedure using a generalized template to locate potential person locations and then employ AdaBoosted ensemble classifier to test the hypothesized person locations. Ref. [4] uses GMM background model to separate the foreground candidates from background and introduces a shape describer to construct the feature vector for pedestrian candidates, and a Support Vector Machine (SVM) classifier is trained to get the pedestrian targets. Ref. [5] presents a

---

* Corresponding author.

© Springer International Publishing Switzerland 2015
J. Cao et al. (eds.), *Proceedings of ELM-2014 Volume 2,*
Proceedings in Adaptation, Learning and Optimization 4, DOI: 10.1007/978-3-319-14066-7_4

pedestrian detection method based on shape and appearance, which introduces a layered representation and develops a generalized expectation-maximization algorithm to separate infrared image into background and foreground, then the detection is solved as multi scales template match problem based on the shape and appearance cues. Ref. [2] presents a double-density dual-tree complex wavelet transform (DD-DT CWT) and wavelet entropy based pedestrian detection method and uses SVM classifier to classify the true pedestrian regions. Ref. [3] proposes real-time pedestrian detection algorithm based on a robust representation of IR pedestrians by binary pattern features and a keypoint based sliding window SVM classifier.

Recently, a new fast neural learning algorithm called Extreme learning machine (ELM) [6][7] has been proposed for single-hidden-layer feedforward neural network (SLFN) which has overcome some challenging issues, such as slow learning speed, trivial human intervening and poor computational scalability and has been widely applied in many applications. Ref. [8] proposes a combination feature uniting PCA-HOG and seven discrete invariant moments and uses the ELM as a classifier to detect vehicle. Ref. [9] develops a driving simulation platform in a virtual scene and uses the ELM to detect the virtual roads and vehicles. Ref. [10] proposes a pedestrian detection method using multimodal histogram of oriented gradient (HOG) for pedestrian feature extraction and the ELM for classification. Ref. [11] proposes a recognition scheme for identifying the aircrafts of different types based on multiple modular ELM. Ref. [12] presents a traffic sign recognition based on locally normalized HOG descriptors and ELM. All of the above applications show the superiority of ELM compared with other classification algorithms.

In this paper, we develop a robust pedestrian detection method using BING as a candidate selector and ELM as a classifier for application in thermal infrared image. The remainder of this paper is organized in the following way. In section 2, we overview the proposed method for pedestrian detection. In section 3, we present the candidate generation method which is called BING. In section 4, we show the design of the ELM classifier. In section 5, we give the experimental results and conclusions are given in section 6.

2   Overview

Fig. 1 shows our proposed method which is divided into two stages: training and testing. The pipeline is as follows:

Step 1: BING is used for objectness estimation, which can generate candidates.

Step 2: The candidates are divided into two categories: positive and negative samples by the overlap areas of candidates and groundtruth, which more than 0.5 is positive samples and others are negative samples.

Step 3: The HOG and appearance features are extracted and the ELM is used to generate a classifier.

Step 4: The classifier is used to distinguish the testing images and finally we can obtain the location of the pedestrians.

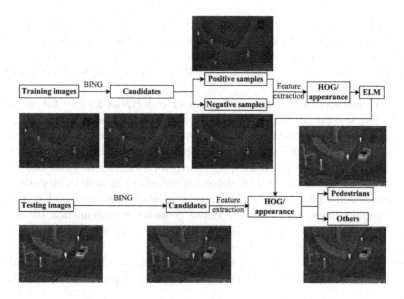

Fig. 1. Proposed pedestrian detection scheme

3   Candidate Generation

Inspired by the ability of human visual system which effectively perceives objects before identifying them [13][14], Ref. [15] introduces a simple 64D NG feature for efficiently capturing the objectness of an image window.

To find generic objects within an image, a predefined quantized window sizes (scales and aspect ratios) are scanned over. Each window is scored with a linear model $\mathbf{w} \in R^{64}$,

$$s_l = \langle \mathbf{w}, \mathbf{g}_l \rangle \ . \tag{1}$$

$$l = (i, x, y); \ . \tag{2}$$

where s_l, $\mathbf{g}_l$, i and (x, y) are filter score, NG feature, location, size and position of a window respectively. Using non-maximal suppression (NMS), a small set of proposals from each size i are selected. Some sizes (e.g. 18×300) are less likely than others to contain an object instance (e.g. 30×30). Thus the objectness score is defined as

$$\emptyset_l = v_i \cdot s_l + t_i \ . \tag{3}$$

where $v_i, t_i \in R$ are separately learnt coefficient and a bias terms for each quantised size i.

Objects are stand-alone things with well-defined closed boundaries and centers [16][17][18]. When resizing windows corresponding to real world objects to a small fixed size (e.g. 8×8), the norm (i.e. magnitude) of the corresponding image gradients becomes a good discriminative feature, because of the little variation that closed boundaries could present in such abstracted view. To utilize this observation for efficiently predicting the existence of object instances, the input image is firstly resized to different quantized sizes and calculate the normed gradient for each resized image. The values in an 8 × 8 region of these resized normed gradients maps are defined as a 64D normed gradients (NG) feature of its corresponding window.

The NG feature, as a dense and compact objectness feature for an image window, has several advantages. Firstly, no matter how an object changes its position, scale and aspect ratio, its corresponding NG feature will remain roughly unchanged because of the normalized support region of this feature. In other words, NG features are insensitive to change of translation, scale and aspect ratio, which will be very helpful for detecting objects of arbitrary categories. And these insensitivity properties are what a good objectness proposal generation method should have. Secondly, the dense compact representation of the NG feature makes it very efficient to be calculated and verified, thus having great potential to be involved in realtime applications.

The cost of introducing such advantages to NG feature is the loss of discriminative ability. Lucky, the resulted false-positives will be processed by subsequent ELM.

4   Design of ELM Classifier

4.1   Feature Extraction

In this paper, we use two features, i.e. HOG and appearance[12][19]. Before extracting features, the candidates obtained from BING should be resized to a unified scale. As to extraction of HOG, the resized candidates are divided into overlapping blocks and then each block is divided into non-overlapping cells. For each pixel of every cell, we compute the gradients using Gaussian smoothing followed by 1-D mask [-1, 0, 1]. A histogram of the gradient orientations of each cell is formed, and then weighted by the gradient magnitude. Finally, the histograms of the cells are concatenated to constitute a descriptor of this block. As to extraction of appearance, the grayvalue of each pixel of the resized candidate is obtained, and then is normalized by 2 norm. Finally, we reshape normalized pixels to a one-row matrix and appearance of the candidate is obtained.

4.2   Brief Introduction of ELM

Given N arbitrary distinct samples $(\mathbf{x}_k, \mathbf{t}_k) \in \mathbf{R}^n \times \mathbf{R}^m$, if a SLFN with $\tilde{N}$ hidden nodes can approximate these N samples with zero error, which implies that there exists $\boldsymbol{\beta}_i$, $\mathbf{c}_i$ and a_i such that

$$\sum_{i=1}^{\tilde{N}} \boldsymbol{\beta}_i G(\mathbf{x}_k; \mathbf{c}_i, a_i) = \mathbf{t}_k, \quad k = 1, \ldots, N . \tag{4}$$

where $\mathbf{c}_i$ is the weight vector connecting the i-th hidden node and the input nodes, $\boldsymbol{\beta}_i$ is the weight vector connecting the i-th hidden node and the output nodes, and a_i is the threshold of the i-th hidden node.

Eq. (4) can be written compactly as

$$\mathbf{H}\boldsymbol{\beta} = \mathbf{T} . \tag{5}$$

where

$$\mathbf{H}(c_1,\ldots,c_{\tilde{N}},a_1,\ldots,a_{\tilde{N}},\mathbf{x}_1,\ldots,\mathbf{x}_N) = \begin{bmatrix} G(\mathbf{x}_1;c_1,a_1) & \cdots & G(\mathbf{x}_1;c_{\tilde{N}},a_{\tilde{N}}) \\ \vdots & \ddots & \cdots \\ G(\mathbf{x}_N;c_1,a_1) & \cdots & G(\mathbf{x}_N;c_{\tilde{N}},a_{\tilde{N}}) \end{bmatrix}_{N \times \tilde{N}} . \tag{6}$$

$$\boldsymbol{\beta} = \begin{bmatrix} \boldsymbol{\beta}_1^T \\ \vdots \\ \boldsymbol{\beta}_{\tilde{N}}^T \end{bmatrix}_{\tilde{N} \times m} \quad and \quad \mathbf{T} = \begin{bmatrix} t_1^T \\ \vdots \\ t_{\tilde{N}}^T \end{bmatrix}_{N \times m}^T \tag{7}$$

$\mathbf{H}$ is called the hidden layer output matrix of the network [7][20,21]; the t-th column of $\mathbf{H}$ is the t-th hidden node's output vector with respect to inputs $\mathbf{x}_1, \mathbf{x}_2, \ldots, \mathbf{x}_N$ and the k-th row of $\mathbf{H}$ is the output vector of the hidden layer with respect to input $\mathbf{x}_k$.

For N arbitrary distinct samples $(\mathbf{x}_k, \mathbf{t}_k)$, in order to obtain arbitrarily small non-zero training error, one may randomly generate $\tilde{N}(\leq N)$ hidden nodes (with random parameters (c_i, a_i)). Eq. (5) then becomes a linear system and the output weight $\boldsymbol{\beta}$ are estimated as

$$\hat{\boldsymbol{\beta}} = \mathbf{H}^\dagger \mathbf{T} . \tag{8}$$

where $\mathbf{H}^\dagger$ is the Moor-Penrose generalized inverse [22] of the hidden layer output matrix $\mathbf{H}$. Calculation of the output weights is done in a single step here. Thus this avoids any lengthy training procedure where the network parameters are adjusted iteratively with appropriately chosen control parameters (learning rate, learning epochs, etc).

After training, we can get the input weights c_i, output weights $\boldsymbol{\beta}$ and the threshold a_i. As to the testing data, we can compute the corresponding hidden layer output matrix $\mathbf{H}$, and the target vector $\mathbf{Y}$:

$$\mathbf{Y} = \mathbf{H}\boldsymbol{\beta} . \tag{9}$$

Finally, the label of the test samples can be obtained by a maximum operation.

The ELM using HOG / appearance classification method is summarized as follows:

Step 1: Extract the feature (HOG / appearance) of the training images and testing images.

Step 2: For the training data, initiate the input weights c_i and the threshold a_i; compute the output matrix of the hidden layer $\mathbf{H}$ according to Eq. (6) and then compute the output weights $\boldsymbol{\beta}$ in terms of Eq. (8) .

Step 3: Given c_i and a_i, compute $\mathbf{H}$ of the testing data according to Eq. (6), and then compute $\boldsymbol{\beta}$ in terms of Eq. (8).

Step 4: Compute $\mathbf{Y}$ according to Eq. (9), and then determine the labels of the test samples via the maximum operation.

Fig. 2 shows the above mentioned process.

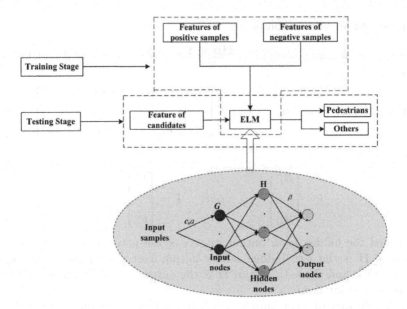

Fig. 2. The ELM based pedestrian detection method

5   Experiments

5.1   Dataset

In this section, we report our experimental results for OTCBVS benchmark -
OSU pedestrian database [1] which is acquired by Raytheon 300D thermal sensor.
There are 10 test sequences with a total of 284 frames and 984 people in OSU
thermal database. Each sequence contains 18-73 frames that are taken within
one minute but not temporally uniformly sampled. This database reasonably
covers a variety of environmental conditions such as rainy, cloudy and sunny
days. Example images are shown in Fig. 3.

Fig. 3. Representative frames of the thermal infrared images

5.2   Detection Results

We select 50% of the frames for training the system, and there are 492 people in both training and testing set. After objectness estimation using BING, for each image, 390 proposal windows are generated. In other words, there are 55380 samples in both training and testing set. Then we extract the features of HOG and appearance of the training and testing set.

We list the three original thermal infrared images and the results processed by BING and HOG based ELM respectively in Fig. 4. From Fig. 4, we can see that the candidates including pedestrians can be obtained by BING, and through ELM, we can obtain the locations of the pedestrians.

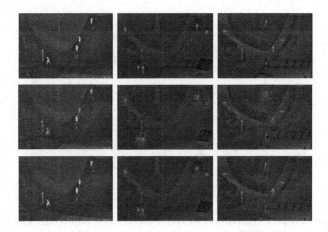

Fig. 4. Three frames of the thermal infrared images and corresponding results. The first row is the original images and the other two rows are candidates and ELM using HOG results respectively.

We compare our proposed HOG + ELM, appearance + ELM with the traditional HOG + SVM, appearance + SVM methods. The TP, FP, Sensitivity and Positive Predictive Value (PPV) of results are listed in Table 1. Here, Sensitivity and PPV is defined as follows:

$$Sensitivity = \frac{the\ number\ of\ pedestrians\ correctly\ detected}{the\ number\ of\ real\ pedestrians}\ . \tag{10}$$

$$PPV = \frac{the\ number\ of\ pedestrian\ correctly\ detected}{the\ number\ of\ candidates}\ . \tag{11}$$

where a high Sensitivity value corresponds to a high detection rate of people and a high PPV corresponds to a low number of false positives. From Table 1, we can see that HOG + ELM has higher TP, lower FP, higher Sensitivity and PPV than other methods. As to the features, HOG based methods obtain better performance than appearance based methods.

As to each sequence, we reports the Sensitivity and PPV of the four methods in detail which are shown in Fig. 5 and Fig. 6.

Table 1. The performance comparison of different methods

	HOG+ELM	HOG+SVM	appearance+ELM	appearance+SVM
TP	406	304	341	295
FP	40	54	40	54
Sensitivity	0.8258	0.6179	0.6904	0.5969
PPV	0.9187	0.8565	0.9059	0.8621

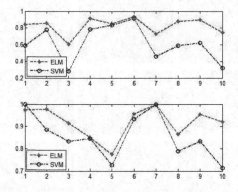

Fig. 5. The performance comparison of HOG based ELM and SVM. The top is Sensitivity and the bottom is PPV

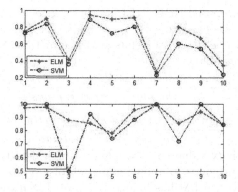

Fig. 6. The performance comparison of appearance based ELM and SVM. The top is Sensitivity and the bottom is PPV

6 Conclusions

We presented a pedestrian detection method in thermal infrared image based on BING and ELM. The candidates are selected by BING and the features of HOG and appearance are extracted, finally we use ELM to locate the pedestrians in thermal infrared image. The experimental results show that the proposed method outperforms the traditional SVM method. However, there also exists some problems. First, our method cannot distinguish pedestrians that are close together; second, some candidates, e.g. trees, cars, and street lamps will be thought to be pedestrian sometimes. We will keep optimizing our algorithm to solve the problems in our future work.

Acknowledgements. This work was supported in part by the National Key Project for Basic Research of China under Grant 2013CB329403; in part by the National Natural Science Foundation of China under Grant 91120011 and Grant 61210013; in part by the Tsinghua Self-innovation Project under Grant 20111081111; and in part by the Tsinghua University Initiative Scientific Research Program under Grant 20131089295.

References

1. Davis, J.W., Keck, M.A.: A Two-Stage Template Approach to Person Detection in Thermal Imagery. In: Workshop on Application of Computer Vision, pp. 364–369 (2005)
2. Li, J., Gong, W., Li, W., Liu, X.: Robust pedestrian detection in thermal infrared imagery using the wavelet transform. Infrared Physics & Technology 53, 267–273 (2010)
3. Sun, H., Wang, C., Wang, B., Naser, E.S.: Pyramic binary pattern features for real-time pedestrian detection from infrared videos. Neurocomputing 74, 797–804 (2011)
4. Wang, J., Chen, D., Chen, H., Yang, J.: On pedestrian detection and tracking in infrared videos. Pattern Recognition Letters 33, 775–785 (2012)
5. Dai, C., Zheng, Y., Li, X.: Pedestrian detection and tracking in infrared imagery using shape and appearance. Computer Vision and Image Understanding 106, 288–299 (2007)
6. Huang, G.B., Zhu, Q.Y., Siew, C.K.: Extreme learning machine: a new learning scheme of feedforward neural networks. In: Preceedings of IEEE International Joint Conference on Neural Networks, pp. 985–990 (2004)
7. Huang, G.B., Zhu, Q.Y., Siew, C.K.: Extreme learning machine: theory and applications. Neurocomputing 70, 489–501 (2006)
8. Li, X., Qu, S., Lei, C.: Vehicle detection from complex scenes based on combination features and ELM. In: IEEE International Conference on Signal Processing, Communication and Computing (ICSPCC), pp. 1–5 (2013)
9. Zhu, W., Miao, J., Hu, J., Qing, L.: Vehicle detection in driving simulation using extreme learming machine. Neurocomputing 128, 160–165 (2014)

10. Yang, K., Du, E.Y., Delp, E.J., Jiang, P., Jiang, F., Chen, Y., Sherony, R., Takahashi, H.: An extreme learning machine-based pedestrian detection method. In: 2013 IEEE Intelligent Vehicles Symposium, Gold Coast, Australia, pp. 1404–1409 (2013)
11. Rong, H.J., Jia, Y.X., Zhao, G.S.: Aircraft recognition using modular extreme learning machine. Neurocomputing 128, 166–174 (2014)
12. Sun, Z.L., Wang, H., Lau, W.S., Seet, G., Wang, D.: Application of BW-ELM model on traffic sign recognition. Neurocomputing 128, 153–159 (2014)
13. Koch, C., Ullman, S.: Shifts in selective wisual attention: towards the underlying neural circuitry. Human Neurbiology 4, 219–227 (1985)
14. Wolfe, J.M., Horowitz, T.S.: What attributes guide the deployment of visual attention and how do they do it? Nature Reviews Neuroscience 5, 495–501 (2004)
15. Cheng, M.M., Zhang, Z., Lin, W.Y., Torr, P.: BING: Binarized normed gradients for objectness estimation at 300fps. In: IEEE Computer Vision and Pattern Recognitin, CVPR (2014)
16. Alexe, B., Deselaers, T., Ferrari, V.: Measuring the objectness of image windows. IEEE Transactions on Pattern Analysis and Machine Intelligence (PAMI) 34, 2189–2202 (2012)
17. Forsyth, D.A., Malik, J., Fleck, M.M., Greenspan, H., Leung, T., Belongie, S., Carson, C., Bregler, C.: Finding pictures of objects in large collections of images. In: Ponce, J., Hebert, M., Zisserman, A. (eds.) ECCV-WS 1996. LNCS, vol. 1144, pp. 335–360. Springer, Heidelberg (1996)
18. Heitz, G., Koller, D.: Learning spatial context: Using stuff to find things. In: Forsyth, D., Torr, P., Zisserman, A. (eds.) ECCV 2008, Part I. LNCS, vol. 5302, pp. 30–43. Springer, Heidelberg (2008)
19. Dalal, N., Triggs, B.: Histogram of oriented gradients for human detection. In: Preceedings of the IEEE Computer Society Conference on Computer Vision and Pattern Recognitin (CVPR), pp. 886–893 (2005)
20. Huang, G.B., Babri, H.A.: Upper bounds on the number of hidden neurons in feedforward networks with arbitrary bounded nonlinear activation functions. IEEE Trans. Neural Networks 9, 224–229 (1998)
21. Huang, G.B., Zhu, Q.Y., Siew, C.K.: Real-time learning capability of neural networks. IEEE Trans. Neural Networks 17, 863–878 (2006)
22. Rao, C.R., Mitra, S.K.: Generalized inverse of matrices and its applications. Wiley, New York (1971)

Dynamic Texture Video Classification
Using Extreme Learning Machine

Liuyang Wang[1,2], Huaping Liu[1,2,*], and Fuchun Sun[1,2]

[1] Department of Computer Science and Technology,
Tsinghua University, Beijing, China
[2] State Key Laboratory of Intelligent Technology and Systems, Beijing, China
`hpliu@mail.tsinghua.edu.cn`

Abstract. Recognition of complex dynamic texture is a challenging problem and captures the attention of the computer vision community for several decades. Essentially the dynamic texture recognition is a multi-class classification problem that has become a real challenge for computer vision and machine learning techniques. Existing classifier such as extreme learning machine cannot effectively deal with this problem, due to the reason that the dynamic textures belong to non-Euclidean manifold. In this paper, we propose a new approach to tackle the dynamic texture recognition problem. First, we utilize the affinity propagation clustering technology to design a codebook, and then construct a soft coding feature to represent the whole dynamic texture sequence. This new coding strategy preserves spatial and temporal characteristics of dynamic texture. Finally, by evaluating the proposed approach on the DynTex dataset, we show the effectiveness of the proposed strategy.

Keywords: Extreme learning machine, affinity propagation, dynamic texture.

1 Introduction

Extreme Learning Machine (ELM), which was firstly proposed by Huang [1], has become an effective learning algorithm for various classification tasks. It works on a simple structure named Single-hidden Layer Feed-forward Neural networks (SLFNs) and randomly applies computational hidden nodes. This mechanism is different from the conventional learning of SLFNs. ELM yields better performance than other conventional learning algorithms in application with higher noise. It also has an extremely fast learning speed compared with traditional gradient-based algorithms. Furthermore, ELM technique successfully overcomes the difficulty of the curse of dimensionality [2,3]. Currently, the application scope of ELMs covers face recognition [4,5], action recognition [6], video concept detection [7], and so on. Ref.[8] gives a comprehensive survey about ELM and Ref.[9] provides a deep insight into ELM. For time-series, Ref.[10] developed time-series processing of large scale remote sensing data with ELM. Ref.[11] studied the

---

* Corresponding author.

© Springer International Publishing Switzerland 2015
J. Cao et al. (eds.), *Proceedings of ELM-2014 Volume 2,*
Proceedings in Adaptation, Learning and Optimization 4, DOI: 10.1007/978-3-319-14066-7_5

cross-person activity recognition using reduced kernel ELM. However, for general visual dynamic textures, which belong to the non-Euclidean manifold, how to realize effective classification using ELM is still an open problem.

Dynamic textures (DT) are video sequences of non-rigid dynamical objects that constantly change their shape and appearance over time. Some examples of dynamic textures are video sequences of fire, smoke, crowds, and traffic. These classes of video sequences are ubiquitous in our natural environment.

However, the classification of DT admits great challenging since DT include both spatial and temporal elements. To model DT, Ref.[12] developed a linear dynamic system (LDS) method. LDS can be used to model complex visual phenomena with a relatively low dimensional representation. However, the signal would rapidly decay. The work in [13] extended this work by introducing feedback control and modeled the system as a closed loop LDS. The feedback loop corrected the problem of signal decay. In [14], LDS was regarded as educational bridge to merge the gap between computer science and control engineering.

A difficult problem to use LDS for classification lies in the fact that LDS does not lie in an Euclidean space, and therefore many classification cannot be utilized. In [12], the Martin distance between LDS is adopted to compare the similarity between different DTs. Martin distance, is effective to evaluate the distance between LDS, but cannot be effectively used for video with multiple dynamic textures. Therefore, many works which utilize Martin distance are limited to investigate simple video with single DT.

Very recently, Ref.[15] proposed to categorize DT by using a novel Bag-Of-dynamic-Systems (BoS). It models each video sequence with a collection of LDSs, each one describing a small spatial-temporal patch extracted from the video. This BoS representation is analogous to the Bag-of-Words (BoW) representation for object recognition. This choice provides an effective strategy to deal with video sequences which are taken under different viewpoints or scales.

Because BoS model is similar to BoW, it naturally inherits the disadvantages of BoW. It is well known that BoW model assign only one codebook element to a descriptor, and therefore the quantization error is large. This usually degrades the classification performance. In [16] and [17], the sparse coding and local coding methods are proposed to address such problem. In such frameworks, more than one codebook element will be assigned to a descriptor and form coding vector. Such strategy can obviously improve the classification performance since the quantization error is attenuated. Unfortunately, neither sparse coding nor local coding can be used for BoS. The intrinsic reason is that both methods depends on the linear subspace assumption and used the linear reconstruction error to design the object function.

Another problem lies in the design of codebook. Conventional k-means or k-mediod clustering method requires the user to prescribe the size of the codebook. This is an non-trivial task.

In this paper, we present a new representation of DT for ELM classifier design. We use the LDS to model the spatial-temporal patches which is sampled from the original video sequence, and adopt the affinity propagation (AP) to design the codebook. AP algorithm is originally proposed in [18] to deal with exemplar extraction problem. The reason to adopt AP lies in two facts: (1) It can automatically determine the size of the codebook; (2) It can extract the existing DT exemplar to construct the codebook. After that, we develop a soft assignment coding method to design the DT representation. The experimental results show the advantage of the proposed method.

The rest of this paper is organized as follows: In Section 2 we give an introduction about ELM. Section 3 presents the details of the proposed coding method. Section 4 shows the experimental results. In Section 5 we give some conclusions.

2   ELM Classification

In this section, we provide a brief introduction of the ELM algorithm which will be used in dynamic texture classification. The more details can be found in [2,6].

In the case of multiple classes, the training data are denoted as $\{(\mathbf{u}_i, l_i)\}_{i=1}^N$, where $\mathbf{u}_i$ is the feature vector, l_i is the label and N is the number of the training samples. For each vector $\mathbf{u}_i$, l_i should be transformed to the vector $\mathbf{t}_i = [t_{i1}, ..., t_{iC}]^T$, where $t_{ik} = 1$ for the vector belonging to class k, i.e., when $l_i = k$, and $t_{ik} = -1$ otherwise.

The input weights of ELM are randomly chosen, while the output weights should be analytically calculated. Assume that the network's hidden layer consists of Q neurons and that $\mathbf{b} \in \mathbb{R}^Q$ is a vector containing the hidden layer neurons bias values. Function $G(\cdot)$ used for output calculation is denoted as $G(\mathbf{w}_j, b_j; \mathbf{u}_i)$, where $\mathbf{w}_j$ denotes the input weight. The hidden layer neurons outputs $\mathbf{G}$ can be represented as:

$$\mathbf{G} = \begin{bmatrix} G(\mathbf{w}_1, b_1; \mathbf{u}_1) & \cdots & G(\mathbf{w}_1, b_1; \mathbf{u}_N) \\ \vdots & \ddots & \vdots \\ G(\mathbf{w}_Q, b_Q; \mathbf{u}_1) & \cdots & G(\mathbf{w}_Q, b_Q; \mathbf{u}_N) \end{bmatrix} \in \mathbb{R}^{Q \times N}$$

The network's output vector corresponding to the training vector set $\{\mathbf{u}_i\}_{i=1}^N$ can be written in a matrix form as

$$\mathbf{O} = \mathbf{W}_o^T \mathbf{G} \qquad (1)$$

where $\mathbf{W}_o \in \mathbb{R}^{C \times Q}$ is the output weight.

By assuming that the network's predicted outputs $\mathbf{O}$ are equal to the network's desired outputs $\mathbf{T} = [\mathbf{t}_1, ..., \mathbf{t}_N]$, $\mathbf{W}_o$ can be analytically calculated by

$$\mathbf{W}_o = (\mathbf{G}\mathbf{G}^T)^{-1} \mathbf{G}\mathbf{T}^T \qquad (2)$$

Taking account of training errors, Ref.[2] have recently proposed an optimization based on regularized ELM algorithm formulated as follows:

$$\text{Minimize}: L_P = \frac{1}{2}\|\mathbf{W}_o\|^2 + C\frac{1}{2}\sum_{i=1}^N \|\xi_i\|^2 \qquad (3)$$

$$\text{Subject to}: \mathbf{g}_i^T \mathbf{W}_o = \mathbf{o}_i^T - \xi_i^T, i = 1, ..., N$$

where $\xi_i \in \mathbb{R}^C$ is a training error vector corresponding to training sample $\mathbf{u}_i$ and C is a parameter denoting the importance of the training error in the optimization problem. By adopting the above described optimization scheme, $\mathbf{W}_o$ can be calculated by:

$$\mathbf{W}_o = \mathbf{G} \left(\frac{1}{C}\mathbf{I} + \mathbf{G}^T\mathbf{G} \right)^{-1} \mathbf{T}^T \tag{4}$$

Finally, the test data can be introduced to the ELM network and be classified to the class corresponding to the highest networks output.

Further, we use the Kernel trick to deal with complex dynamic texture videos. According to [2], we define a kernel matrix for ELM as:

$$\Omega = \mathbf{G}^T\mathbf{G}, \tag{5}$$

where the $\{i, j\}$ element in Ω is $\mathbf{g}_i^T\mathbf{g}_j = \mathcal{K}(\mathbf{u}_i, \mathbf{u}_j)$. Then, the output function of ELM classifier can be written as

$$\mathbf{o}^T = \mathbf{g}^T\mathbf{W}_o = \mathbf{g}^T\mathbf{G} \left(\frac{1}{C}\mathbf{I} + \mathbf{G}^T\mathbf{G} \right)^{-1} \mathbf{T}^T = \begin{bmatrix} \mathcal{K}(\mathbf{u}, \mathbf{u}_1) \\ \vdots \\ \mathcal{K}(\mathbf{u}, \mathbf{u}_N) \end{bmatrix}^T \left(\frac{1}{C}\mathbf{I} + \Omega \right)^{-1} \mathbf{T}^T \tag{6}$$

In this work, we adopt the Gaussian kernel which is represented as

$$K(\mathbf{u}, \mathbf{v}) = exp(-\gamma\|\mathbf{u} - \mathbf{v}\|^2) \tag{7}$$

where γ is the prescribed parameter.

The whole procedure to utilize ELM is illustrated in Fig.1. In the following sections we will give more details.

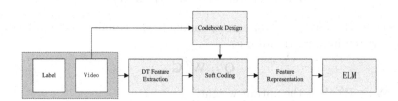

Fig. 1. The overview of the proposed method.

3   Video Representation

3.1   Dynamic Texture Modeling

This section summarizes the key concepts in dynamic texture. According to [12], an LDS model can be used to fit a small spatial-temporal patch. Assume the short spatial-temporal patches includes τ frames with resolution $m \times n$. The dynamic texture should obeys the standard state-space equation:

$$\begin{cases} \mathbf{x}(k+1) = \mathbf{A}\mathbf{x}(k) + \mathbf{u}(k), & \mathbf{u}(k) \sim N(0, \mathbf{Q}), \quad \mathbf{x}(0) = \mathbf{x}_0 \\ \mathbf{y}(k) = \mathbf{C}\mathbf{x}(k) + \mathbf{w}(k), & \mathbf{w}(k) \sim N(0, \mathbf{R}) \end{cases} \tag{8}$$

where $\mathbf{x}(k) \in \mathbb{R}^{n_x}$ is the hidden state variable vector, and $\mathbf{y}(k) \in \mathbb{R}^{n_y} (n_y = m \times n)$ is the data corresponding to the sequence of images. $\mathbf{A} \in \mathbb{R}^{n_x \times n_x}$, and $\mathbf{C} \in \mathbb{R}^{n_y \times n_x}$ are the parameter matrices. $\mathbf{u}(k) \in \mathbb{R}^{n_x}$ and $\mathbf{w}(k) \in \mathbb{R}^{n_y}$ are the zero-mean normally distributed random variables, which are used to compensate the modeling error.

Since $\mathbf{w}(k)$ and $\mathbf{u}(k)$ are modeling errors, they are expected to be negligible. Therefore the problem can be formulated as: Given the observed image sequence $\mathbf{y}(1), \mathbf{y}(2), ..., \mathbf{y}(\tau)(\tau > n_x)$, estimate the values of $\mathbf{A}$, $\mathbf{C}$ and $\mathbf{x}(k)$.

Denote $\mathbf{Y}_{1:\tau} = [\mathbf{y}(1), \mathbf{y}(2), ..., \mathbf{y}(\tau)] \in \mathbb{R}^{n_y \times \tau}$ and $\mathbf{X}_{1:\tau} = [\mathbf{x}(1), \mathbf{x}(2), ..., \mathbf{x}(\tau)]$ $\in \mathbb{R}^{n_x \times \tau}$. It can be reformulated as

$$\mathbf{Y}_{1:\tau} \approx \mathbf{C}\mathbf{X}_{1:\tau} \tag{9}$$

SVD can be performed on the matrix $\mathbf{Y}_{1:\tau}$ and the parameters are estimated as:

$$\mathbf{Y}_{1:\tau} \approx \mathbf{U}\Sigma\mathbf{V}^T$$
$$\mathbf{C} = \mathbf{U} \quad and \quad \mathbf{X}_{1:\tau} = \Sigma\mathbf{V}^T \tag{10}$$

Because $\mathbf{x}(k+1) \approx \mathbf{A}\mathbf{x}(k)$ for $k = 1, 2, ..., \tau - 1$, the estimation of $\mathbf{A}$ can be uniquely determined by solving the following least squares problem:

$$\mathbf{A} = \underset{\mathbf{A}}{\text{argmin}} \sum_{k=1}^{\tau-1} ||\mathbf{x}(k+1) - \mathbf{A}\mathbf{x}(k)||_2 \tag{11}$$

Finally, we can use the tuple $\mathbf{p} = (\mathbf{A}, \mathbf{C})$ to describe A DT spatial-temporal patch.

3.2 Codebook Design

To effectively code the obtained DT spatial-temporal patches, a reasonable codebook is needed. As the whole video is complicated , we sample non-overlapped spatial-temporal patches from videos and obtain the DT models of all the patches [19]. The codebook is generated from all the patch models in training data.

In this work, we adopt Affinity Propagation (AP) clustering algorithm [18] to find prototypes $\mathbf{V} = \{\mathbf{v}_1, ..., \mathbf{v}_K\}$ as the codebook. Such a method can automatically determine the codebook size K. Let $\mathbf{P} = \{\mathbf{p}_1, ..., \mathbf{p}_M\}$ be the set of all the DT patches. AP takes as input a collection of similarities between DT patches. The details of the algorithm are described in Algorithm.1.

A DT spatial-temporal patch $\mathbf{p}_i$ can be described as $\mathbf{p}_i = (\mathbf{A}_i, \mathbf{C}_i)$ by modeling dynamic textures. The similarities of $\mathbf{p}_i$ and $\mathbf{p}_j$ should be defined for AP

Algorithm 1. AP Clustering Algorithm

Input:
   $s(i, j)(1 \leq i, j \leq M)$: similarities and preference;
Output:
   $idx(i)(1 \leq i \leq M)$: indices of exemplars for each $\mathbf{p}_i$;
1: initial $a(i, j) = 0$;
2: **repeat**
3:    compute responsibilities $r(i, j)$ $(1 \leq i, j \leq M)$;
4:    $r(i, j) = s(i, j) - \max\limits_{j' \ s.t. \ j' \neq j} \left\{ a(i, j') + s(i, j') \right\}$
5:    compute availabilities $a(i, j)$ $(1 \leq i, j \leq M)$;
6:    $a(i, j) = \min \left\{ 0, r(j, j) + \sum\limits_{i' \ s.t. \ i' \notin \{i,j\}} \left\{ \max\{0, r(i', j)\} \right\} \right\}$ $\quad (i \neq j)$
7:    $a(j, j) = \sum\limits_{i' \ s.t. \ i' \neq j} \left\{ \max\{0, r(i', j)\} \right\}$
8:    Identifying exemplars;
9:    $idx(i) = \arg\max\limits_{1 \leqslant j \leqslant M} a(i, j) + r(i, j)$
10: **until** $idx(i)$ do not change.

---

clustering algorithm. One family of distances between two models is based on principal angles between specific subspaces derived from the models, namely the observability subspaces [15]. The observability subspaces is the range of the extended observability denoted by $O_\infty(\mathbf{p}_i) = \left[\mathbf{C}_i{}^T, (\mathbf{C}_i \mathbf{A}_i)^T, (\mathbf{C}_i \mathbf{A}_i{}^2)^T, ... \right]^T \in \mathbb{R}^{\infty \times n_x}$. Let θ_a be the a-th principal angles between the spaces. The Martin distance between $\mathbf{p}_i$ and $\mathbf{p}_j$ is defined as

$$d_M(\mathbf{p}_i, \mathbf{p}_j) = -\ln \prod_{a=1}^{n_x} \cos^2 \theta_a \qquad (12)$$

After we obtain the distance matrix $\mathbf{D} \in R^{M \times M}$ and its element $\mathbf{D}(i, j)$ is the Martin distance between $\mathbf{p}_i$ and $\mathbf{p}_j$, we set $s(i, j) = -\mathbf{D}(i, j)^2$ and $s(i, i) = \min\limits_{j=1,...,M} \left(-\mathbf{D}(i, j)^2 \right)$ for AP clustering algorithm. AP clustering provides two advantages: (1) the size of the codebook is automatically determined; (2) the result of the clustering is deterministic and do not depend on the initialization.

3.3   Soft Coding Feature Design

A popular method for coding is vector quantization which searches the nearest neighbor to represent the descriptor. Such a representation is usually called BoW. BoW quantizes a video into discrete "visual words", and then computes a histogram representation. One disadvantage of BoW is that it introduces significant quantization errors since only one element of the codebook is selected to represent the descriptor. To reduce the quantization error, we develop a soft coding approach to solve this problem.

Given a spatial-temporal patch which is denoted as $\mathbf{p}_i$, the corresponding feature vector is constructed as $\{\mu_{ik}\}_{k=1}^{K}$. In this representation, μ_{ik} is the membership value of sample $\mathbf{p}_i$ to the cluster identified by the center $\mathbf{v}_k^*$. The memberships can be obtained by

$$\mu_{ik} = \frac{e^{(-\beta \cdot d_M(\mathbf{p}_i, \mathbf{v}_k^*))}}{\sum\limits_{k'=1}^{K} e^{\left(-\beta \cdot d_M(\mathbf{p}_i, \mathbf{v}_{k'}^*)\right)}} \tag{13}$$

Please note that $\{\mu_{ik}\}_{k=1}^{K}$ is normalized to satisfy $\sum\limits_{k=1}^{K} \mu_{ik} = 1$.

The coding vector for the spatial-temporal patch $\mathbf{p}_i$ is then obtained as $\mathbf{c}_i = [\mu_{i1}, \mu_{i2}, ..., \mu_{iK}]^T \in \mathbb{R}^K$. For a single video sequence, if we extract N local DT descriptors, then we can get the codes $\mathbf{C} = [\mathbf{c}_1, ..., \mathbf{c}_N] \in \mathbb{R}^{K \times N}$. Then we need an operator to pool all codes in a video into a single vector $\mathbf{u_i} \in \mathbb{R}^K$. This pooling operation is defined as

$$\mathbf{u_i} = \mathbb{P}(\mathbf{C_i}) \tag{14}$$

where the pooling function $\mathbb{P}$ is defined for each column of $\mathbf{C}$. Each column of $\mathbf{u_i}$ corresponds to the responses of all the local descriptors in the specific video. Therefore, different pooling functions construct different image statistics. In this study, we select the average operator. Such strategy results in

$$\mathbf{u}_i(k) = \frac{\sum\limits_{j=1}^{N} |\mathbf{C}(k, j)|}{\sum\limits_{k'=1}^{K} \sum\limits_{j=1}^{N} |\mathbf{C}(k', j)|} \tag{15}$$

where $\mathbf{u}_i(k)$ is the k-th element of $\mathbf{u}_i(k)$ and $\mathbf{C}(k, j)$ is the matrix element in the k-th row and j-th column of $\mathbf{C}$.

4   Experimental Results

In this section, we present experimental results that validate the proposed algorithm. As indicated by [20], some existing DT data sets have a number of drawbacks such as the resolution is quite low; there is only a single occurrence per class and not enough classes are available for practical classification purposes. To tackle this problem, Ref.[20] developed the DynTex data set, which aims to serve as a reference database for dynamic texture research by providing a large and diverse database of high-quality dynamic textures. Dyntex provides 3 data sets for classification. In our experiment, we adopt the Gamma dataset for classification evaluation. It includes 10 classes such as sea, calm water, grass, etc. We randomly choose half of the videos in each class as the training data. The remaining videos are used for test purpose.

Table 1. Accuracy Comparisons

Method	Accuracy	Method	Accuracy
BoW-ELM	75.00%	*1-NN*	52.34%
BoW-SVM	75.78%	*3-NN*	49.22%
soft coding-ELM	88.28%	*5-NN*	40.62%
soft coding-SVM	82.81%		

In this work, the size of the codebook is never prescribed by the designer. After we run the AP procedure, we get a codebook for Gamma dataset. The obtained codebook size is 184.

As to the coding methods, we compare two methods: soft coding and hard coding. The hard coding method assigns only one codebook element to a local DT descriptor. It is equivalent to the usual BoW method. As to the classifiers, we compare three classifiers: ELM, SVM, and k-Nearest Neighbors(NN). Both ELM and SVM can be combined with two coding methods. For k-NN, we use a single DT model to model the video and use the Martin distance to measure the difference between videos. Such a method serves the role of baseline. In our implementation, we set $k = 1, 3$, and 5.

Table.1 lists the performance of all of the above methods. k-NN always obtains the worse results. This is not surprising because only a single DT is used to describe the video and therefore the modeling error will be significant. If we use BoW feature, both ELM and SVM obtain better results than k-NN, while ELM is a little worse than SVM. However, if we use soft coding method, the results of ELM will be dramatically increased and is superior to SVM. In Fig.2, we list the confusion matrice of ELM. From this figure we see the performance of ELM is rather good.

In the above comparison, the parameters of ELM and SVM are carefully tuned to get the best results. To show the influence of the parameters γ and C, we change the values of γ from 2^{-18} to 2^{18}, and the values of C from 2^{-18} to 2^{18}. The obtained performance results are listed in Figs.2. From this figure we see that the performance change of ELM is very smooth, while SVM is rather susceptible to the parameters.

Finally, to show the advantages of the obtained codebook, we also used the K-Medoids algorithm to design a codebook with the same size. Therefore we actually compare the following four methods: (1) AP-ELM(using AP and ELM); (2) AP-SVM(using AP and SVM); (3) KM-ELM(using K-Medoids and ELM); (4) KM-SVM(using K-Medoids and SVM).

We study the influence of the parameter β which plays important roles in the soft coding stage. In Fig.2 we list the accuracy versus the values of β. In addition to β, the other parameters in the algorithms are carefully-tuned to get the best results. From this figure we see that AP-ELM performs better than other methods and this shows that the AP codebook indeed helps ELM to get better results.

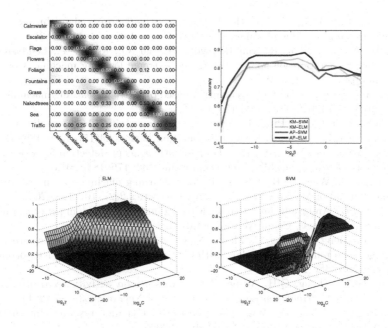

Fig. 2. 1st row: Confusion Matrix and Classification performance v.s. the parameter β; 2nd row: Classification performance v.s. the parameter γ and C

5 Conclusions

In this paper, the ELM classifier is developed to tackle the dynamic texture classification problem. Since the dynamic texture lies in the non-Euclidean space, we design a soft coding BoS representation for it. Such a representation can be used for ELM classifiers and obtains satisfactory performance on public datasets.

Acknowledgements. This work was supported in part by the National Key Project for Basic Research of China under Grant 2013CB329403; in part by the National Natural Science Foundation of China under Grant 91120011 and Grant 61210013; in part by the Tsinghua Self-innovation Project under Grant 20111081111; and in part by the Tsinghua University Initiative Scientific Research Program under Grant 20131089295.

References

1. Huang, G.B., Zhu, Q.Y., Siew, C.K.: Extreme learning machine: theory and applications. Neurocomputing 70(1), 489–501 (2006)
2. Huang, G.B., Zhou, H., Ding, X., Zhang, R.: Extreme learning machine for regression and multiclass classification. IEEE Transactions on Systems, Man, and Cybernetics, Part B: Cybernetics 42(2), 513–529 (2012)

3. Savitha, R., Suresh, S., Kim, H.: A meta-cognitive learning algorithm for an extreme learning machine classifier. Cognitive Computation, 1–11 (2013)
4. He, B., Xu, D., Nian, R., van Heeswijk, M., Yu, Q., Miche, Y., Lendasse, A.: Fast face recognition via sparse coding and extreme learning machine. Cognitive Computation, 1–14 (2013)
5. Zong, W., Huang, G.B.: Face recognition based on extreme learning machine. Neurocomputing 74(16), 2541–2551 (2011)
6. Iosifidis, A., Tefas, A., Pitas, I.: Dynamic action recognition based on dynemes and extreme learning machine. Pattern Recognition Letters 34(15), 1890–1898 (2013)
7. Lu, B., Wang, G., Yuan, Y., Han, D.: Semantic concept detection for video based on extreme learning machine. Neurocomputing 102, 176–183 (2013)
8. Huang, G.B., Wang, D.H., Lan, Y.: Extreme learning machines: a survey. International Journal of Machine Learning and Cybernetics 2(2), 107–122 (2011)
9. Huang, G.B.: An insight into extreme learning machines: random neurons, random features and kernels. Cognitive Computation, 1–15 (2014)
10. Chen, J., Zheng, G., Fang, C., Zhang, N., Chen, H., Wu, Z.: Time-series processing of large scale remote sensing data with extreme learning machine. Neurocomputing 128, 199–206 (2014)
11. Deng, W.Y., Zheng, Q.H., Wang, Z.M.: Cross-person activity recognition using reduced kernel extreme learning machine. Neural Networks 53, 1–7 (2014)
12. Doretto, G., Chiuso, A., Wu, Y.N., Soatto, S.: Dynamic textures. International Journal of Computer Vision 51(2), 91–109 (2003)
13. Yuan, L., Wen, F., Liu, C., Shum, H.-Y.: Synthesizing dynamic texture with closed-loop linear dynamic system. In: Pajdla, T., Matas, J(G.) (eds.) ECCV 2004. LNCS, vol. 3022, pp. 603–616. Springer, Heidelberg (2004)
14. Liu, H., Xiao, W., Zhao, H., Sun, F.: Learning and understanding system stability using illustrative dynamic texture examples. IEEE Transactions on Education 57(1), 4–11 (2014)
15. Ravichandran, A., Chaudhry, R., Vidal, R.: Categorizing dynamic textures using a bag of dynamical systems. IEEE Transactions on Pattern Analysis and Machine Intelligence 35(2), 342–353 (2013)
16. Yang, J., Yu, K., Gong, Y., Huang, T.: Linear spatial pyramid matching using sparse coding for image classification. In: IEEE Conference on Computer Vision and Pattern Recognition, pp. 1794–1801. IEEE (2009)
17. Wang, J., Yang, J., Yu, K., Lv, F., Huang, T., Gong, Y.: Locality-constrained linear coding for image classification. In: 2010 IEEE Conference on Computer Vision and Pattern Recognition (CVPR), pp. 3360–3367. IEEE (2010)
18. Frey, B.J., Dueck, D.: Clustering by passing messages between data points. Science 315(5814), 972–976 (2007)
19. Ravichandran, A., Chaudhry, R., Vidal, R.: Dynamic texture toolbox (2011), http://www.vision.jhu.edu
20. Péteri, R., Fazekas, S., Huiskes, M.J.: Dyntex: A comprehensive database of dynamic textures. Pattern Recognition Letters 31(12), 1627–1632 (2010)

Uncertain XML Documents Classification Using Extreme Learning Machine

Xiangguo Zhao*, Xin Bi, Guoren Wang, Chao Wang, and Hongbo Yang

College of Information Science and Engineering, Northeastern University, Liaoning,
Shenyang, China 110819
zhaoxiangguo@mail.neu.edu.cn

Abstract. Driven by the emerging network data exchange and storage, XML documents classification has become increasingly important. Most existing representation model and conventional learning algorithm are defined on certain XML documents. However, in many real-world applications, XML datasets contain inherent uncertainty, which brings greater challenges to classification problem. In this paper, we propose a novel solution to classify uncertain XML documents, including uncertain XML documents representation and two uncertain learning algorithms based on Extreme Learning Machine. Experimental results show that our approaches exhibit prominent performance for uncertain XML documents classification problem.

Keywords: ELM, Classification, XML, Uncertain Data.

1 Introduction

Classification over XML documents is a challenging and important task in XML data mining and management. In general XML documents classification problems, XML documents have to be transformed into a specific representation model, and then taken as the input to classifiers. Classifiers can be trained using various learning algorithms, among which Extreme Learning Machine (ELM) [3,4] shows good generalization performance and extreme learning speed in a variety of applications.

In real-world applications, due to the occurrence of inaccurate measurement, noises and incompleteness in data generation and collection, semantic and structural information of XML documents are usually uncertain.

Probabilistic XML data model was first studied in [9] to represent uncertainty in XML documents. Based on the expressiveness analysis of probabilistic XML models in [5], a full complexity analysis of queries and updates on probabilistic tree model *p-document* was given in [12,6]. The expressive power of p-documents was further studied in [1]. Based on those probabilistic XML models, in order to prune useless intermediate results, top-k ranking was introduced into uncertain XML query processing in [2,10,7,8,11].

---

* Corresponding author.

© Springer International Publishing Switzerland 2015   51
J. Cao et al. (eds.), *Proceedings of ELM-2014 Volume 2,*
Proceedings in Adaptation, Learning and Optimization 4, DOI: 10.1007/978-3-319-14066-7_6

However, to our best knowledge, this paper is the *first* one addressing the problem of uncertain XML documents classification. To address the above challenges, our contributions can be summarized as follows.

- *First* paper to discuss the uncertain XML documents classification problem.
- An enumerated instance appearance probability based uncertain ELM is proposed to demonstrate the problem definition.
- A sampling method based uncertain ELM is proposed to further improve the uncertain learning procedure.
- Extensive experiments are conducted to verify the effectiveness and efficiency of our approaches.

The rest of this paper is organized as follows. Section 2 introduces the problem definition. Section 3 gives a brief introduction of ELM. Section 4 presents the appearance probability based uncertain learning algorithm. Sampling based uncertain learning algorithm is proposed in section 5. Experimental results are analyzed in section 6, followed by the conclusion in section 7.

2   Problem Definition

2.1   Uncertain XML

An uncertain XML document is modeled as a probabilistic XML tree, i.e., p-document, which is a probability distribution over a space of ordinary documents. In p-document, the children of the distributional nodes are attached with appearance probabilities. According to the distributional node type, an uncertain XML document presented in p-document can derive a number of possible worlds. The appearance probability of a possible world is calculated according to the probability distribution of the p-document.

2.2   XML Representation Model

In this paper, we further revise our DSVM[14] to represent uncertain XML documents. Uncertain DSVM is descripted as

$$\mathbf{d}_{u_i^j} = \langle \mathbf{d}_{u_i^j}^1, \mathbf{d}_{u_i^j}^2, \dots, \mathbf{d}_{u_i^j}^n, p_{u_i^j} \rangle \tag{1}$$

where $\mathbf{d}_{u_i^j}$ is an $n+1$ dimensional feature vector of the jth possible world of the ith uncertain XML document. $p_{u_i^j}$ is the appearance probability of the possible world, $\mathbf{d}_{u_i^j}^k$ is the kth represent feature which is described as

$$\mathbf{d}_{u_i^j}^k = \sum_{j=1}^m (\mathrm{TF}(w_k, doc.e_j) \cdot \varepsilon_j) \cdot \mathrm{IDF}_{ex}(w_k, c) \cdot \rho_{CD} \tag{2}$$

where m is the number of elements in document doc, $doc.e_j$ is the jth element e_j of doc, ε_j, which is the unit vector of $doc.e_j$ in SLVM, is now the dot product of s-dimensional unit vector and s dimensional weight vector. $\mathrm{IDF}_{ex}(w_i, c)$ is the revised IDF. The factor ρ_{CD} is the distribution modifying factor. The detailed calculation of $\mathbf{d}_{u_i^j}^k$ can be found in [14].

2.3   Problem Definition

An uncertain object is often represented either *discretely* using a set of discrete values associated with probabilities, or *continuously* using a probability density function (pdf). In this paper, we follow the discrete model. In this problem, an uncertain object is an uncertain XML document, while an instance is a possible world derived from the uncertain XML document. Thus the formal definition of uncertain XML documents classification is given as Definition 1.

Definition 1 (Uncertain XML classification). *Given a training dataset* $\mathbb{D}$ *of labeled uncertain XML documents* $\langle U, c \rangle \in \mathbb{X} \times \mathbb{C}$, *where* $U \in \mathbb{X}$ *is an uncertain XML document and* $\mathbb{X}$ *is the document space;* $c \in \mathbb{C}$ *is the class label and* $\mathbb{C} = \{c_1, c_2, \ldots, c_m\}$ *is a fixed set of classes. The problem of* uncertain XML classification *is to learn a classifier or a classification function* $\phi : \mathbb{X} \to \mathbb{C}$ *using a specific learning algorithm, so that each instance* u_i^j *of uncertain object* $U_i \in \mathbb{D}$ *belongs to a class* $c_i \in \mathbb{C}$ *with a probability. The corresponding uncertain XML document* U_i *can be assigned to a class* c *if the probability of* U_i *for class* c *is the largest.*

3   Extreme Learning Machine

Given N arbitrary samples $(\mathbf{x}_i, \mathbf{t}_i) \in \mathbf{R}^{n \times m}$, ELM is modeled as

$$\sum_{i=1}^{L} \beta_i \, \mathrm{G}(\mathbf{w}_i \cdot \mathbf{x} + b_i) \tag{3}$$

where L is the number of hidden layer nodes, β_i is the output weight from the ith hidden node to the output node, $\mathrm{G}(\mathbf{x})$ is the activation function, $\mathbf{w}_i = [w_{i1}, w_{i2}, \ldots, w_{in}]^{\mathrm{T}}$ is the input weight vector, b_i is the bias of ith hidden node. $\mathbf{H}$ is the random feature mapping.

Since ELM is to minimize the training error and the norm of the output weights [3,4], the output weight β is calculated as

$$\beta = \mathbf{H}^{\dagger} \mathbf{T} \tag{4}$$

where $\mathbf{H}^{\dagger}$ is the Moore-Penrose Inverse of $\mathbf{H}$.

4   Appearance Probability based Uncertain Extreme Learning Machine

Previous work on classification using ELM is mainly over certain data. To our best knowledge, there is only *one* paper which studied classification problem based on ELM over uncertain data in [13]. However, the proposed pruning method is efficient for binary classification problems, but not for multi-class ones, since the pruning theorem cannot provide the lower bound of the probability of belonging to a specific class.

4.1   Classification Probabilities

Definition 2 (Instance classification probability). *Given an instance u_i^j, which belongs to class c, of an uncertain XML document, the* Instance classification probability *of u_i^j being classified to class c is the the appearance probability of u_i^j, that is $p_{u_i^j}^c = p_{u_i^j}$.*

Definition 3 (Uncertain object classification probability). *Given an uncertain object U_i, i.e. an uncertain XML document, which belongs to class c, we assume that all the instances in $S_{U_i}^c$ belong to class c. The* uncertain object classification probability *of U_i being classified to class c is the sum of instance classification probabilities of the instances in $S_{U_i}^c$, that is $p_{U_i}^c = \sum\limits_{u_i^j \in S_{U_i}^c} p_{u_i^j}^c$.*

Instance classification probability indicates that the probability of an instance being classified to a specific class is its appearance probability. While *uncertain object classification probability* defines that the probability of an uncertain object being classified to a specific class is the sum of the instance classification probability to the specific class of its instances.

4.2   PU-ELM

The PU-ELM is presented as Algorithm 1. During the training phase of PU-ELM (Lines 1-4), two new datasets D_{train}^E and D_{test}^E are first initiated (Line 1). For each uncertain object U_i, i.e., an uncertain XML document, all the instances of U_i are enumerated and saved in D_{train}^E (Lines 2,3). After all the XML documents are iterated, ELM algorithm is invoked to train a classifier with dataset D_{train}^E (Line 4).

During the testing phase (Lines 7-12), the output class labels of dataset D_{test}^E are calculated based on ELM theory (Line 7). For each instance u_i^j in D_{test}^E, we fetch the output class label c_{output} (Line 10) and add the appearance probability of u_i^j, i.e., $p_{u_i^j}$, to the uncertain object classification probability $p_{U_i}^{c_{out}}$ for class c_{out} (Line 11). The uncertain XML document U_i is classified to class c if $p_{U_i}^c$ is the largest in the set of p_{U_i} (Line 12).

5   Sampling Based Uncertain Extreme Learning Machine

In this section, Sampling based Uncertain Extreme Learning Machine (SU-ELM) is proposed. Different from PU-ELM and uncertain ELM algorithms proposed in [13], the computation theories of SU-ELM are based on estimator probabilities, instead of appearance probabilities. The probability of an uncertain XML document being classified to a class is estimated by the estimator.

Algorithm 1: Appearance Probability based Uncertain Extreme Learning Machine

---

1 Initiate empty datasets D_{train}^E and D_{test}^E;

2 foreach $U_i \in D_{train}$ **do**

3   Generate all the instances of U_i into D_{train}^E;

4 Train a classifier by calculating $\beta = \mathbf{H}^\dagger \mathbf{T}$ with D_{train}^E;

5 foreach $U_i \in D_{test}$ **do**

6   Generate all the instances $S_{U_i} = \{u_i^1, \ldots, u_i^m\}$ of U_i into D_{test}^E;

7 Calculate testing output $\mathbf{O} = \mathbf{H}\beta$ with D_{test}^E;

8 foreach $U_i \in D_{test}^E$ **do**

9   **foreach** $u_i^j \in S_{U_i}$ **do**

10   Get class label c_{out} of u_i^j in $\mathbf{O}$;

11   Calculate the uncertain object probability $p_{U_i}^{c_{out}} = p_{U_i}^{c_{out}} + p_{u_i^j}$;

12 U_i is classified into class $c = \arg \max_{c_{out} \in \mathbb{C}} p_{U_i}^{c_{out}}$;

5.1 Sampling

Different from sampling the general uncertain data, an uncertain XML document is presented by p-document, in which the probabilities are attached to distributional nodes. This procedure named *uncertain XML document sampling* (UXSampling) is presented as Algorithm 2.

While traversing an uncertain XML document presented as p-document, if the currently traversed node travNode is a distributed node (Line 3), we fetch all the children nodes childrenNodes of travNode (Line 4). If travNode is an IND node, we sample each node in childrenNodes (Line 7). If a node childNote is sampled existing, we continue to sample the children of childNote recursively (Lines 8-11). If travNode is a MUX node, we pick one or no node out of childrenNodes (Line 13). If no node is sampled, we do nothing; if a node childNode is sampled as existing, we sample the children of childNote recursively (Lines 14-17).

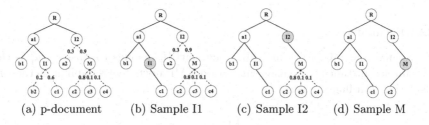

(a) p-document   (b) Sample I1   (c) Sample I2   (d) Sample M

Fig. 1. An example of an uncertain XML document

Algorithm 2: Uncertain XML Document Sampling

```
1  SamplePXML(travNode) begin
2     if !travNode.isLeaf() then
3        if travNode.type == "distributionalNodes" then
4           childrenNodes = travNode.getChildren();
5        if travNode.type == "IND" then
6           foreach childNode in childrenNodes do
7              Sample childNode;
8              if childNode.sampleStatus()==exist then
9                 nextChildrenNodes = childNode.getChildren();
10                foreach nextChildNode in nextChildrenNodes do
11                   samplePXML(nextChildNode);

12        if travNode.type == "MUX" then
13           childNode = sample one or no node out of childrenNodes;
14           if childNode.sampleStatus()==exist then
15              nextChildrenNodes = childNode.getChildren();
16              foreach nextChildNode in nextChildrenNodes do
17                 samplePXML(nextChildNode);
```

Example 1. Figure 1 shows a sampling procedure of an uncertain XML document presented as Figure 1(a). While traversing this p-document, the first visited distributional node is I1, which is the an IND node. Thus we sample its children b2 and c1 independently. We assume that b2 is sampled as nonexisting with a probability of $1 - 0.2 = 0.8$ and c1 as existing with a probability of 0.6. The sampling result of I1 is shown in Figure 1(b). In Figure 1(c), node I2 is visited and sampled as nonexisting with a probability of 1-0.3=0.7, and M as existing with a probability of 0.9. Then we recursively sample the children of node M, which is a distributional node of type MUX. Therefore, we pick a child node c2 out of all the three children with a probability of 0.8. The sampled instance is presented as Figure 1(d).

5.2   Estimator

We assume that an uncertain XML document U_i belongs to class c_s, u_i^j is the jth sampled instance of U_i using uncertain XML document sampling method in Algorithm 2, a flag $y_{u_i^j}^{c_t}$ is set as

$$y_{u_i^j}^{c_t} = \begin{cases} 1, \text{ if } c_s = c_t \\ 0, \text{ if } c_s \neq c_t \end{cases} \tag{5}$$

where c_t is the output class label of the classifier. During the training phase, if the class label of the uncertain object U_i is the same as the output class label of u_i^j, that is, $c_s = c_t$, the value of $y_{u_i^j}^{c_t}$ is 1, otherwise is 0. While during the testing phase, due to the unknown class label of U_i, for m classes, there are m flags $y_{u_i^j}^{c_1}, \ldots, y_{u_i^j}^{c_m}$. For each flag, let's say $y_{u_i^j}^{c_t}$, assuming u_i^j is classified to class c_s, if $c_s = c_t$, the value of $y_{u_i^j}^{c_t}$ is 1, otherwise is 0.

Then the *uncertain object classification probability*, i.e. the estimator $\widehat{Pr_B}$, of U_i for class c_s is defined as

$$Pr_B^{c_s} \approx \widehat{Pr_B^{c_s}} = \frac{\sum_{j=1}^{n} y_{u_i^j}^{c_s}}{n} \tag{6}$$

The estimator $Pr_B^{c_s}$ is an *unbiased* estimator, i.e., $E(\widehat{Pr_B^{c_s}}) = Pr_B^{c_s}$, thus the variance can be written as

$$Var(\widehat{Pr_B^{c_s}}) = \frac{1}{n}Pr_B^{c_s}(1 - Pr_B^{c_s}) \approx \frac{1}{n}\widehat{Pr_B^{c_s}}(1 - \widehat{Pr_B^{c_s}}) \tag{7}$$

5.3   SU-ELM

SU-ELM is shown as Algorithm 3. After two new datasets D_{train}^S and D_{test}^S are first initiated (Line 1), for each uncertain XML document $U_i \in D_{train}$, we generate a number of instances of U_i using sampling method presented as Algorithm 1 into D_{train}^S, which contain all the instances sampled from each uncertain object in D_{train} (Lines 2-4) and will be used as the input to the learning algorithm (Line 5).

Algorithm 3: Sampling based Uncertain Extreme Learning Machine

---

1 Initiate empty datasets D_{train}^S and D_{test}^S;
2 **foreach** $U_i \in D_{train}$ **do**
3   Sample U_i to generate instances;
4   Add all the instances into D_{train}^S;
5 Train a classifier by calculating $\beta = \mathbf{H}^\dagger \mathbf{T}$ with D_{train}^S;
6 **foreach** $U_i \in D_{test}$ **do**
7   Sample U_i to generate instances set $S_{U_i} = \{u_i^1, \ldots, u_i^m\}$;
8   Add all the instances into D_{test}^S;
9 Calculate testing output $\mathbf{O} = \mathbf{H}\beta$ with D_{test}^S;
10 **foreach** $U_i \in D_{test}^S$ **do**
11   **foreach** $u_i^j \in S_{U_i}$ **do**
12   Get class label c_{out} of u_i^j in $\mathbf{O}$;
13   Calculate the estimator probability $Pr_B^{c_{out}} = Pr_B^{c_{out}} + \frac{1}{n}$;
14   U_i is classified into class $c = \arg \max_{c_{out} \in \mathbb{C}} Pr_B^{c_{out}}$;

In the testing phase (Lines 6-14), for each uncertain object $U_i \in D_{test}$, we generate the instances of U_i by the procedure (Lines 6-8) as the same as the sampling method during the training phase. When the sampled testing dataset D_{test}^S is obtained, we get the testing result according to ELM theory (Line 9). Then we fetch the class label of each $u_i^j \in D_{test}^S$ (Line 12). Based on the estimator analysis in Section 5.2, we count the estimator flag $y_{u_i^j}^{c_s}$ of the uncertain object U_i for each class (Line 13). If there are m classes, for each U_i, there are m estimator probabilities, which are $Pr_B^{c_1}, \ldots, Pr_B^{c_m}$. The largest Pr_B, let's say $Pr_B^{c_k}$, indicates that the uncertain object U_i has the highest probability of being classified into class c_k. Therefore, according to the sampling theory, we assign the class label c_k to the uncertain object U_i (Line 14).

6   Experiments

6.1   Experiments Setup

All the experiments are conducted on a PC with 3.4 GHz Intel Core i7-3667U CPU and 8 GB RAM. The programs of uncertain XML documents preprocessing are implemented in C++. The classifier related algorithms, i.e., PU-ELM and SU-ELM, are implemented Matlab R2009a.

Three datasets of certain XML documents are used as original datasets, which are *Wikipedia XML Corpus* provided by INEX (10 classes and 100 XML pages in each class), *IBM DeveloperWorks* (6 classes and 100 XML articles in each class) and *ABC News* (6 classes and 200 XML news in each class). With the original datasets, we transform each XML document into probabilistic one, using the same method in [5].

6.2   Performance Evaluation

In this section we present the performance evaluation of PU-ELM, SU-ELM, the uncertain ELM proposed in [13] (hereinafter referred to as UELM) and SVM. Note that SU-ELM is based on sampling method, the performance of SU-ELM varies with the sample ratio. Therefore we first present the performance comparison among SVM, UELM and PU-ELM. The evaluation of SU-ELM will be given in separate experimental results.

One of the most important evaluation criteria of learning algorithms is *training time*, of which the comparison is presented in Table 1, from which we can see that the ELM based algorithms is much more faster than SVM. The effect of UELM pruning is not obvious.

Another important evaluation criterion is the *testing accuracy*. The comparison is presented in Table 2, which shows that ELM based algorithms outperform SVM. UELM and PU-ELM have similar accuracy.

We evaluated the *running time* of SU-ELM, which includes the sampling and training time. Figure 2(a) shows that the running time grows along with the increment of sample ratio.

Table 1. Training Time comparison among SVM, UELM, PU-ELM

Datasets	SVM	UELM		PU-ELM	
		Sigmoid	RBF	Sigmoid	RBF
Wikipedia	13.2727	2.4721	2.5213	2.5481	2.5480
IBM developWorks	8.6601	1.9604	1.9425	1.9982	1.9434
ABC News	11.9016	2.7730	2.8152	2.7920	2.8209

Table 2. Testing Accuracy comparison among SVM, UELM, PU-ELM

Datasets	SVM	UELM		PU-ELM	
		Sigmoid	RBF	Sigmoid	RBF
Wikipedia	0.6314	0.6880	0.6795	0.7005	0.6924
IBM developWorks	0.6714	0.7032	0.7188	0.7271	0.7070
ABC News	0.6801	0.7793	0.7393	0.7815	0.7515

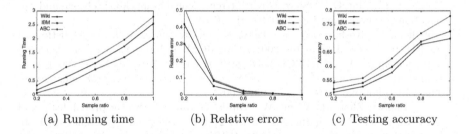

(a) Running time   (b) Relative error   (c) Testing accuracy

Fig. 2. Performance of SU-ELM

Relative error of the estimator is computed as $\delta = \dfrac{\mid R - \hat{R} \mid}{R}$. From Figure 2(b), it can be seen that the relative error decreases as sample ratio increases. Moreover, with the increment of sample ratio, the relative error plunges and converges near zero, that is, the estimator is capable of approximating the exact learning answer well. Due to the relative errors of estimator, the *testing accuracy* of SU-ELM is less than PU-ELM. Figure 2(c) demonstrates that as the sample ratio increases, the testing accuracy of SU-ELM increases.

7   Conclusion

This paper is the *first* one to address the problem of uncertain XML document classification. An *appearance probability* based algorithm PU-ELM and a *sampling* based algorithm SU-ELM are proposed. Extensive experiments are conducted to verify the effectiveness and efficiency of the proposed approaches.

Acknowledgment. This research is partially supported by the National Natural Science Foundation of China under Grant Nos. 60933001, 61272181, and 61073063; the National Basic Research Program of China under Grant No. 2011CB302200-G; the 863 Program under Grant No. 2012AA011004, and the Public Science and Technology Research Funds Projects of Ocean Grant No. 201105033.

References

1. Abiteboul, S., Kimelfeld, B., Sagiv, Y., Senellart, P.: On the expressiveness of probabilistic xml models. The VLDB Journal 18(5), 1041–1064 (2009)
2. Chang, L., Yu, J.X., Qin, L.: Query ranking in probabilistic xml data. In: Proceedings of the 12th International Conference on Extending Database Technology: Advances in Database Technology, EDBT 2009, pp. 156–167. ACM, New York (2009)
3. Huang, G.B., Zhu, Q.Y., Siew, C.K.: Extreme learning machine: a new learning scheme of feedforward neural networks. In: International Symposium on Neural Networks, vol. 2 (2004)
4. Huang, G.B., Zhu, Q.Y., Siew, C.K.: Extreme learning machine: Theory and applications. Neurocomputing 70, 489–501 (2006)
5. Kimelfeld, B., Kosharovsky, Y., Sagiv, Y.: Query efficiency in probabilistic xml models. In: Proceedings of the 2008 ACM SIGMOD International Conference on Management of Data, SIGMOD 2008, pp. 701–714. ACM, New York (2008)
6. Kimelfeld, B., Kosharovsky, Y., Sagiv, Y.: Query evaluation over probabilistic xml. The VLDB Journal 18(5), 1117–1140 (2009)
7. Li, J., Liu, C., Zhou, R., Wang, W.: Top-k keyword search over probabilistic xml data. In: Proceedings of the 2011 IEEE 27th International Conference on Data Engineering, ICDE 2011, pp. 673–684. IEEE Computer Society, Washington, DC (2011)
8. Liu, J., Ma, Z.M., Yan, L.: Querying and ranking incomplete twigs in probabilistic xml. World Wide Web 16(3), 325–353 (2013)
9. Nierman, A., Jagadish, H.V.: Protdb: Probabilistic data in xml. In: Proceedings of the 28th International Conference on Very Large Data Bases, VLDB 2002, pp. 646–657. VLDB Endowment (2002)
10. Ning, B., Liu, C., Yu, J.X., Wang, G., Li, J.: Matching top-k answers of twig patterns in probabilistic XML. In: Kitagawa, H., Ishikawa, Y., Li, Q., Watanabe, C. (eds.) DASFAA 2010. LNCS, vol. 5981, pp. 125–139. Springer, Heidelberg (2010)
11. Ning, B., Liu, C., Yu, J.: Efficient processing of top-k twig queries over probabilistic xml data. World Wide Web 16(3), 299–323 (2013)
12. Senellart, P., Abiteboul, S.: On the complexity of managing probabilistic xml data. In: Proceedings of the Twenty-Sixth ACM SIGMOD-SIGACT-SIGART Symposium on Principles of Database Systems, PODS 2007, pp. 283–292. ACM, New York (2007)
13. Sun, Y., Yuan, Y., Wang, G.: Extreme learning machine for classification over uncertain data. Neurocomputing 128, 500–506 (2014)
14. Zhao, X., Bi, X., Qiao, B.: Probability based voting extreme learning machine for multiclass xml documents classification. In: World Wide Web, pp. 1–15 (2013)

Encrypted Traffic Identification Based on Sparse Logistical Regression and Extreme Learning Machine

Juan Meng, Longqi Yang, Yuhuan Zhou, and Zhisong Pan*

Colleage of Command Information System,
PLA University of Science and Technology, Nanjing, China 210007
hotpzs@hotmail.com

Abstract. In this work, a new encrypted traffic identification algorithm using sparse logistical regression and extreme learning machine (ELM) is introduced. The proposed method is based on randomness characteristics of encrypted traffic. we utilize ℓ_1-norm regularized logistic regression to select sparse features. The identification is performed with the help of Extreme Learning Machine (ELM) because of its better identification and faster speed. In ELM, the input weights and the bias values are randomly chosen and the output weights are analytically calculated. Extensive experiments are performed using the proposed encrypted traffic identification algorithm and results are compared against state of the art techniques.

Keywords: encrypted traffic identification, randomness test, sparse logistical regression, extreme learning machine.

1 Introduction

Identification of encrypted traffic has recently received increasing attention. The accurate identification of encrypted traffic can regulate illegal data, detect network attacks and protect users' information. It has great significance to improve network security and enhance network management.

To date there mainly exist three effective fashions on identification of encrypted traffic:

1. Identification based on ports. Currently the main encryption protocols tend to have the fixed default ports. Ports can help identify the encryption type of business[1,2].
2. Identification based on content signatures. Encryption protocol data have specific content features. It achieves the purpose of identification encryption protocol traffic by matching[3,4].
3. Identification based on flow features. Encryption protocol need to establish a connection to complete version negotiation and key exchange before

---

* Corresponding author.

© Springer International Publishing Switzerland 2015   61
J. Cao et al. (eds.), *Proceedings of ELM-2014 Volume 2,*
Proceedings in Adaptation, Learning and Optimization 4, DOI: 10.1007/978-3-319-14066-7_7

transmitting encrypted data. At this stage of establishing connection, the interaction protocol packets have a relatively fixed format and content, also have the specific flow characteristics[5,6,7,8].

Currently methods of encrypted traffic identification mainly depend on the recognition of ports, content signatures and flow features which can only be implemented for a specific encryption protocol identification. The details of encryption protocol must be known. In an open network environment, the emergence of private encryption protocol makes it difficult to implement these methods. There is an urgent need to address three major research questions:1) how can we identify protocol independent encrypted traffic? 2) what is the smallest set of features most identify encrypted traffic? 3) how to improve speed and accuracy of encrypted traffic identification?

In this paper, the idea of encrypted traffic identification based on randomness estimation is inducted. Through randomness test, we can get 188-dimension randomness features. Sparse logistical regression based on ℓ_1-norm regularization is proposed as a dimensionality reduction technique. The identification of encrypted traffic is performed with the help of Extreme Learning Machine (ELM) because of its better classification and faster speed. To this end, experimental results demonstrate superiority of our proposed encrypted traffic identification algorithm over existing state of the art methods.

The main contributions of this paper include:

- Applying randomness features to differ the encrypted data and non-encrypted data.
- Proposing sparse logistical regression to build a low dimensional feature space.
- Using ELM to identify encrypted traffic.

The remainder of this paper is organized as follows. After doing randomness features extraction in Section 2, we present the idea of a ℓ_1-norm regularized logistic regression for sparse features selection in Section 3. In Section 4, we describe the extreme learning machine and we propose encrypted traffic identification algorithm in Section 5. Experimental results are presented in Section 6 and we conclude this paper in Section 7.

2   Randomness-Based Features extraction

Randomness is an important feature to differ the encrypted data and non-encrypted data. The purpose of encryption is to hide plaintext information in ciphertext output. One of the goals is trying to eliminate non-randomized characteristics of ciphertext data to ensure safety and avoid being attacked. One evaluation criterion of encryption algorithm is that the ciphertext output must pass randomness test. The output of a secure encryption algorithm must be random. Since encrypted data approximate random, we utilize the randomness characteristics to identify encrypted traffic.

At present, the test methods for randomness have been up to hundreds. The test methods or test sets include NIST test set [9], DIEHARD test set[10], Crypt-X test[11] and TestU0 test[12]. Among these test methods, there is a strong correlation between many methods. In other words, the results of the test methods have a certain correlation. It is redundant. On the other hand, many methods focus on linear correlation aspects which often based on some complex mathematical operations. So it is important to select the appropriate test method.

NIST(National Institute of Standards and Technology)statistical test suite released by U.S. National Institute of Standards and Technology in April 2010 is one of the most stringent randomness test. NIST has a good comprehensiveness, independence and rigorousness which provides 15 categories randomness test methods. The comprehensiveness is that the approach including 15 categories methods which cover the principle of existing randomness test methods. The independence is that there is a small mutual correlation between the tests and the randomness of sequences can be detected from all aspects. Rigorousness is that these methods have low false alarm rate. NIST test set return a series of statistical values of samples i.e. p-value. P-value gives a measure of confidence that the test sample is a random sequence. The larger the p-value, the greater likelihood of a test sample is a random sequence. NIST statistical test set contains 15 categories of statistical test methods which used to calculate the statistics of sequences as follows:

- a) Frequency test: test the proportion of zeroes and ones for the entire sequence.
- b) Frequency test within a Block: test the proportion of ones within M-bit blocks.
- c) Runs test: test the total number of runs in the sequence.
- d) Test for the longest run of ones in a block: test the longest run of ones within M-bit blocks.
- e) Binary matrix rank test: test the rank of disjoint sub-matrices of the entire sequence.
- f) Discrete Fourier transform test: test the peak heights in the Discrete Fourier Transform of the sequence.
- g) Non-overlapping template matching test: test the number of occurrences of pre-specified target strings.
- h) Overlapping template matching test: test the number of occurrences of pre-specified target strings.
- i) Maurer's 'Universal statistical' test: test the number of bits between matching patterns.
- j) Linear complexity test: test the length of a linear feedback shift register.
- k) Serial test: test the frequency of all possible overlapping m-bit patterns across the entire sequence.
- l) Approximate entropy test: test the frequency of all possible overlapping m-bit patterns across the entire sequence.
- m) Cumulative sums test: test the maximal excursion (from zero) of the random walk defined by the cumulative sum of adjusted $(-1, +1)$ digits in the sequence.

- n) Random excursions test: test the number of cycles having exactly K visits in a cumulative sum random walk.
- o) Random excursions variant test: test the total number of times that a particular state is visited (i.e., occurs) in a cumulative sum random walk.

According to the different input parameters, it can be regarded as 188 randomness test items and we can extract the 188-dimensional randomness features.

3 Sparse Feature Selection

Even though encrypted traffic has a large number of randomness features, only a small number of them contribute to its encryption. The underlying representations of 188-dimensional features are sparse. In order to make 188-dimensional randomness features sparse (contains multiple 0 entries), regularization term is imposed. We propose computing by solving the following optimization problem:

$$\min_x \sum_{i=1}^m w_i log(1 + exp(-y_i(x^T a_i + c))) + p\|x\|_1 \tag{1}$$

The problem in Formula (1) is the ℓ_1-norm regularized logistic regression. $\{a_i\}_{i=1}^p$ is the p-values of randomness test. Let $A = [a_1, \cdots, a_p] \in n \times p$ denotes the randomness matrix with its columns representing different features. y_i (either 1 or -1) is the response and w_i denotes the weight of the ith sample. $\|x\|_1 = \sum_{i=1}^p |x_i|$ stands for the ℓ_1-norm of a vector which equals the sum of absolute values of the vector's entries. $\rho > 0$ is a tunable parameter that controls the sparsity. The parameter ρ controls the amount of regularization applied to the estimate. Setting $\rho = 0$ reverses the problem to logistic regression which minimizes the empirical loss term[13,14]. On the other hand, a very large ρ will completely shrink x to 0 thus leading to the empty or null model. In general, moderate values of ρ will cause shrinkage of the solutions towards 0, and some coefficients may end up being exactly 0.

It is well-known the solution to this problem is sparse in the sense that many entries in x will be set to zero. Recently, many algorithms for solving the ℓ_1-norm regularized logistic regression problem have been proposed. The solution to the ℓ_1-norm regularized logistic regression can be interpreted in a Bayesian framework as the maximum a posterior probability estimate of x and c, x has a Laplacian prior distribution on $\mathbb{R}^n$ and covariance ρI, and c has the uniform prior on $\mathbb{R}$. This problem can be solved by many existing software packages, such as SLEP[15].

4 Extreme Learning Machine

Extreme learning machine was originally proposed for single hidden layer feedforward neural networks and was then extended to the generalized single hidden layer feedforward networks where the hidden layer need not be neuron alike[16,17,18]. ELM is a single hidden layer feedforward network, where the input weights are chosen randomly and the output weights are calculated analytically[19]. This algorithm tends to afford the best generalization performance at extremely fast learning speed. The structure of ELM network is shown in Fig.1. ELM contains an input layer, hidden layer and an output layer.

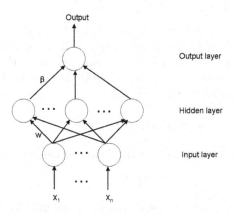

Fig. 1. Structure of ELM network

In ELM, an infinitely differentiable activation function facilitates random assignment of input weights and hidden layer biases. Consider a collection of N distinct samples (x_i, t_i), where $x_i = (x_{i1}, x_{i2}, \cdots, x_{in})^T \in \mathbb{R}^n$ and $t_i = (t_{i1}, t_{i2}, \cdots, t_{im}) \in \mathbb{R}^m$, an ELM with L hidden nodes and an activation function $g(x)$ is mathematically modeled as:

$$\sum_{i=1}^{L} \beta_i g(w_i \cdot x_k + b_i) = o_k, k = 1, \cdots, N \tag{2}$$

where $w_i = (w_{i1}, w_{i2}, \cdots, w_{in})^T$ is the weight vector connecting the ith hidden neuron and the input neurons, $\beta_i = (\beta_{i1}, \beta_{i2}, \cdots, \beta_{im})^T$ is the weight vector connecting the ith hidden neuron and the output neurons, b_i is the threshold of the ith hidden neuron and $o_k = (\beta_{k1}, \beta_{k2}, \cdots, \beta_{km})^T$ is the output vector of the SLFN. $w_i \cdot x_k$ denotes the inner product of w_i and x_k. The ELM reliably approximates N samples with minimum error:

$$\sum_{i=1}^{L} \beta_i g(w_i \cdot x_k + b_i) = t_k, k = 1, \cdots, N \tag{3}$$

The above N equations can be written compactly as:

$$H\beta = T \tag{4}$$

where

$$H = \begin{bmatrix} g(w_1 \cdot x_1 + b_1) & \cdots & g(w_L \cdot x_1 + b_L) \\ \vdots & \ddots & \vdots \\ g(w_1 \cdot x_N + b_1) & \cdots & g(w_L \cdot x_N + b_L) \end{bmatrix}_{N \times m} \tag{5}$$

$$\beta = \begin{bmatrix} \beta_1^T \\ \vdots \\ \beta_L^T \end{bmatrix}_{L \times m} \quad and \quad T = \begin{bmatrix} t_1^T \\ \vdots \\ t_N^T \end{bmatrix}_{N \times m} \tag{6}$$

Here H is called the hidden layer output matrix. If the number of neurons in the hidden layer is equal to the number of samples, then H is square and invertible. Otherwise, the system of equations needs to be solved by numerical methods, concretely by solving

$$\min_{\beta} \|H\beta - T\| \tag{7}$$

The result that minimizes the norm of this least squares equation is

$$\hat{\beta} = H^{\dagger}T \tag{8}$$

Where $H^{\dagger}$ is called Moore-Penrose generalized inverse[20]. The solution produced by ELM in 8 not only achieves the minimum square training error but also the best generalization performance on novel patterns.

5   Proposed Encrypted Traffic Identification Algorithm

The proposed encrypted traffic identification algorithm based on the randomness characteristics. we utilize ℓ_1-norm regularized logistic regression to select sparse features. Distinctive features are used to train and test an ELM classifier. The encrypted traffic identification algorithm has two parts: data preprocessing and learning. A block schematic diagram of our proposed algorithm is shown in Fig. 2.

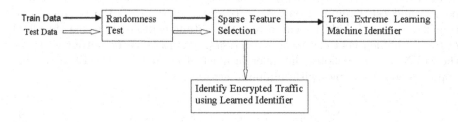

Fig. 2. Schematic diagram of proposed encrypted traffic identification algorithm

5.1   Data Preprocessing

Data preprocessing includes three main steps: packets aggregation, randomness features extraction and sparse features selection.

Data stream consists of a series packets with the same source address, destination address, source port, destination port, and same protocol. We use the quintuple (i.e., source IP, destination IP, source port, destination port, and protocol number) to aggregate packets. Bidirectional data flow is considered to be the same stream, The forward direction is defined by the first packet. The duration of the stream cannot be infinite. UDP flow termination condition is connection timeout. TCP flow termination condition is disconnection or connection timeout. Furthermore, the data flow in each direction has at least one packet, and each packet has at least one byte load.

Randomness features extraction aims for valid payload using NIST test, we can extract 188-dimensional randomness features which are totally 15 categories.

In the encrypted traffic study, we utilize ℓ_1-norm regularized logistic regression to select sparse features. Even though encrypted data stream has a large number of randomness features, only a small number of them contribute to it. The underlying representations of 188-dimensional features are sparse. In order to make 188-dimensional randomness features sparse (contains multiple 0 entries), ℓ_1 regularization term is imposed. We use sparse learning technique to improve randomness features representation.

5.2   Learning

Constructing the test set and the training set based on the selected features set. we obtain optimal parameters of ELM through experiments. ELM can be trained based on the training set and the optimal parameters, and then the test set can be identified based on the trained ELM.

Learning the Parameter. Through experiments we learn the optimal number of ELM hidden neurons. First, we set the number of hidden neurons is N_{min}. Based on the training set and the number of the current hidden neurons, we get ELM model referring to the learning method of Section 4, then identify the test set and store the identification accuracy. After learning and identification for ten data sets, the average identification accuracy can be calculated. We store the current number of neurons and the corresponding accuracy values. Increasing the number of neurons and repeating the process to get the new average accuracy, when the number of neurons reaches N_{max} or average accuracy reaches a threshold value, we stop the iteration process and choose the number of neurons corresponding to the highest average accuracy as the optimal the number of neurons.

Training. Some parameters of objective function are gained based on the training set. We train ELM based on the selected training set and the obtained parameters. We now summarize the training ELM as the follows:

1. Setting the number of neurons L;
2. Assigning random value to the input weight w_i and the bias b_i, $i = 1, \cdots, L$;

3. calculating the hidden layer output matrix H;
4. calculating the output weight $\beta = H^\dagger T$.

Identification We identify the test set based on the training ELM.

1. calculating the hidden layer output matrix H for the input test data;
2. According to the formula $f = H\beta$, calculating the identification value of the test data;
3. According to the identification results, calculating the accuracy of identification.

6   Experimental Results

This section validates the proposed encrypted traffic identification algorithm. This experimentation data consists of 10000 data stream samples. Those 10000 data stream samples are passed to NIST randomness test. The sparse features are generated from ℓ_1-norm regularized logistic regression and these features are passed to the ELM for the learning process. After learning, a data stream sample is passed to the proposed encrypted traffic identification algorithm. Then the proposed algorithm will process through its processing steps and finally it will identify whether the supplied data stream sample is encrypted or not. Thus we identify encrypted traffic based on the trained ELM.

6.1   Identifier Parameters Selection

Before performance comparison, the parameters of the encrypted traffic identifier achieved the best generalization performance are pre-estimated. For sparse features selection, The regularization parameter ρ of sparse representation is a tunable parameter that controls the sparsity. For different values of ρ, the number of features is different. As the regularization parameter ρ increases gradually, the number of features decreases. For ELM, the parameter only involves the number of hidden neurons used in a standard SLFN. Parameters of the algorithm is set as follows: the regularization parameter $\rho = [0.5, 0.2, 0.1]$, the number of neurons between 1 and 80. Table 1 shows when the optimal number of hidden neurons is above 50 and the regularization parameter ρ is set to 0.1, the algorithm can get a better performance.

Table 1. identification accuracy at varying ρ and number of neurons

ρ value	Feature numbers	neurons							
		10	20	30	40	50	60	70	80
0.5	4	0.5092	0.5533	0.5741	0.6047	0.6528	0.6020	0.5386	0.5175
0.2	9	0.5263	0.5917	0.6038	0.6502	0.7019	0.6954	0.6533	0.5972
0.1	32	0.5476	0.6294	0.7013	0.7542	0.8896	0.7215	0.7007	0.6522

6.2 Evaluation of Identification Performance

For the different ratio settings of the test set and the training set (from 0.1 to 0.9, in steps of 0.1), we calculate the accuracy for 10 times. We compare ELM method with SVDD[21](support vector data description) and GMM[22] (Gaussian mixture model). Fig.3 shows that the ELM model has a better performance compared to the other methods.

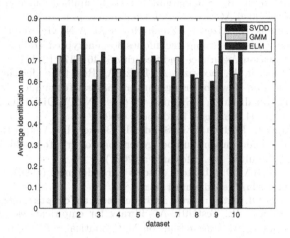

Fig. 3. Average identification rate (%) using SVDD, GMM and ELM

7 Conclusion

In this paper, an efficient encrypted traffic identification technique based on randomness features is proposed. A ℓ_1-norm regularized logistic regression is used to generate distinctive features. These features are input to an ELM to analytically learn an optimal model. Experimental results corroborate our claim that the proposed method achieves improved identification at a substantially faster rate against existing techniques. In addition, our proposed method is independent of encrypted traffic protocol. It can be taken as a general method for identification of encrypted traffic.

Next we will study whether the proposed method helps to categorize encrypted traffic.

References

1. Alshammari, R., Zincir-Heywood, A.N.: Generalization of signatures for ssh encrypted traffic identification. In: IEEE Symposium on Computational Intelligence in Cyber Security, CICS 2009, pp. 167–174. IEEE (March 2009)
2. Karagiannis, T., Broido, A., Faloutsos, M.: Transport layer identification of P2P traffic. In: Proceedings of the 4th ACM SIGCOMM Conference on Internet Measurement, pp. 121–134. ACM (October 2004)

3. Karagiannis, T., Papagiannaki, K., Faloutsos, M.: BLINC: multilevel traffic classification in the dark. ACM SIGCOMM Computer Communication Review 35(4), 229–240 (2005)
4. Kang, H.J., Kim, M.S., Hong, J.W.: Streaming media and multimedia conferencing traffic analysis using payload examination. ETRI Journal 26(3), 203–217 (2004)
5. McGregor, A., Hall, M., Lorier, P., Brunskill, J.: Flow clustering using machine learning techniques. In: Barakat, C., Pratt, I. (eds.) PAM 2004. LNCS, vol. 3015, pp. 205–214. Springer, Heidelberg (2004)
6. Bernaille, L., Teixeira, R., Salamatian, K.: Early application identification. In: Proceedings of the 2006 ACM CoNEXT Conference, p. 6. ACM (December 2006)
7. Bacquet, C., Gumus, K., Tizer, D., Zincir-Heywood, A.N., Heywood, M.I.: A comparison of unsupervised learning techniques for encrypted traffic identification. Journal of Information Assurance and Security 5, 464–472 (2010)
8. Maiolini, G., Baiocchi, A., Iacovazzi, A., Rizzi, A.: Real time identification of SSH encrypted application flows by using cluster analysis techniques. In: Fratta, L., Schulzrinne, H., Takahashi, Y., Spaniol, O. (eds.) NETWORKING 2009. LNCS, vol. 5550, pp. 182–194. Springer, Heidelberg (2009)
9. Rukhin, A., Soto, J., Nechvatal, J., Smid, M., Barker, E.: A statistical test suite for random and pseudorandom number generators for cryptographic applications. Booz-Allen and Hamilton Inc., Mclean Va (2001)
10. Marsaglia, G.: DIEHARD: a battery of tests of randomness (1996), http://stat.fsu.edu/?geo/diehard.html
11. Crypt-X Test [CP/OL], http://www.isrc.qut.edu.au/cryptx/index.html
12. L'Ecuyer, P., Simard, R.: TestU01: AC library for empirical testing of random number generators. ACM Transactions on Mathematical Software (TOMS) 33(4), 22 (2007)
13. Lee, S.I., Lee, H., Abbeel, P., Ng, A.Y.: Efficient L 1 Regularized Logistic Regression. In: Proceedings of the National Conference on Artificial Intelligence, vol. 21(1), p. 401. AAAI Press, MIT Press, Menlo Park, Cambridge (2006)
14. Torgo, L., Gama, J.: Regression by classification. In: Borges, D.L., Kaestner, C.A.A. (eds.) SBIA 1996. LNCS, vol. 1159, pp. 51–60. Springer, Heidelberg (1996)
15. Liu, J., Ji, S., Ye, J.: SLEP: Sparse learning with efficient projections. Arizona State University, 6 (2009)
16. Huang, G.B., Zhu, Q.Y., Siew, C.K.: Extreme learning machine: a new learning scheme of feedforward neural networks. In: Proceedings of the 2004 IEEE International Joint Conference on Neural Networks, vol. 2, pp. 985–990. IEEE (July 2004)
17. Huang, G.B., Siew, C.K.: Extreme learning machine: RBF network case. In: ICARCV 2004 8th Control, Automation, Robotics and Vision Conference, vol. 2, pp. 1029–1036. IEEE (2004)
18. Huang, G.B., Siew, C.K.: Extreme learning machine with randomly assigned RBF kernels. International Journal of Information Technology 11(1), 16–24 (2005)
19. Huang, G.B., Chen, L., Siew, C.K.: Universal approximation using incremental constructive feedforward networks with random hidden nodes. IEEE Transactions on Neural Networks 17(4), 879–892 (2006)
20. Serre, D.: Matrices: Theory and Applications. Springer-Verlag New York, Inc. (2002)
21. Tax, D.M., Duin, R.P.: Data domain description using support vectors. In: ESANN, vol. 99, pp. 251–256 (April 1999)
22. Reynolds, D.: Gaussian mixture models. In: Encyclopedia of Biometrics, pp. 659–663 (2009)

Network Intrusion Detection

Zhifan Ye and Yuanlong Yu*

College of Mathematics and Computer Science, Fuzhou University
Fuzhou, Fujian, 350116, China
357499398@qq.com, yu.yuanlong@fzu.edu.cn

Abstract. With the developing of Internet, network intrusion has becoming more and more common.Extreme learning machine (ELM) is an efficient learning algorithm for generalized single hidden layer feedforward networks. ELM can be used for network intrusion detection.This work introduces a method using extreme learning machine to detect network intrusion. In proposed approach, a classifier is trained and used to classify connections as one of five categories. The experiment data applied is KDD99 data, which is the benchmark data for intrusion detection. In additional, this proposed method is compared against decision tree, neural network and support vector machines .It can be seen that the proposed method which using extreme learning machine has better performance than support vector machines in terms of sensitivity.

Keywords: Network Intrusion Detection, Extreme Learning Machine, KDD 99.

1 Introduction

The information security has draw the attention from the public in recent years, because personal information and business information is more and more important today. Using the intrusion detection system (IDS) [1] is one of the good solutions to information security based on the fact that IDS can be in the form of an application or device which can detect suspicious activities in the network.

Machine Learning is one of the methods used for the IDS. There are mainly three popular machine learning algorithms named decision tree [2], neural network [3] and support vector machines [4]. Decision tree has high precision, but when applied in high dimensional data, the training speed is low and there are a lot of tuning parameters. The classical learning algorithm in neural network, e. g. backpropagation, requires predefining several parameters and might get into local minimum. Support vector machine has been widely used for classification, but its computational cost of training is very high and it has not shown great performance in the case of multiple classes.

---

* Correspondence author. This paper is supported by National Natural Science Foundation of China under Grant 61105102.

© Springer International Publishing Switzerland 2015
J. Cao et al. (eds.), *Proceedings of ELM-2014 Volume 2*,
Proceedings in Adaptation, Learning and Optimization 4, DOI: 10.1007/978-3-319-14066-7_8

Extreme Learning Machine (ELM) [5][6] is used as multiclass classifier. Normally, it is hard to adjust the classifier to achieve the best performance of each class. This paper presents an approach to identifying network connections by first designing a best classifier for each class and then link them serially as a cascaded classifier. the KDD99 data, which consists of 5 million network connection records, is used to evaluate this proposed method.

ELM is a new learning algorithm which can be used in regression and classification applications. ELM randomly generates the hidden node parameters, and then analytically determines the output weights of single hidden layer feedforward networks (SLFN) [7]. ELM only requires setting the number of hidden neurons and the activation function. It does not require adjusting the input weights and hidden layer biases during the implementation of the algorithm, and it produces only one optimal solution [8].

This paper is organized as follows: Section 2 presents the details of this proposed method, section 3 gives the experiment results of this proposed method and section 4 is the conclusion.

2   Approach

2.1   Framework

First, raw data is normalized and five best binary classifiers are trained. Then they are assembled serially. The cascaded classifier is shown in Fig.1. The rule about the arrangement of these five classifiers is based on one classifier's effect to others. If a classifier has fewer errors that it identifies samples belonging to other classes to its own class, it would be put more forward in the sequence. Each time when the data pass a binary classifier, the unlabeled data will be classified as positive or negative. The data which is classified as positive will be labeled as the corresponding class.

2.2   Five Binary Classifier

The 5-class elm classifier is not good enough especially for the last two attack categories. And when adjusting the 5-class classifier's performance of a specific class , the performance of other class will be affect. So using a 5-class elm classifier is inconvenient, using five binary classifiers instead.

In a binary classifier, positive class is a class of five categories, negative class is the rest four categories. In each set of parameters ,training is repeated for 500 times and get best result in it.

2.3   Normalization

The data value which range from -1 to 1 is has a better performance for elm algorithm, so the raw data is normalized linearly according to (1) . In formula (1), x is a value of attribute, minValue is the smallest value in that attribute, maxValue is the biggest value in that attribute.

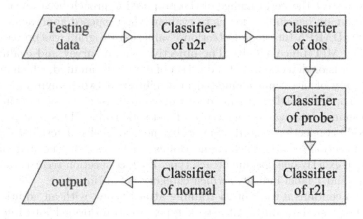

Fig. 1. The sequence of assemblage

$$x' = \frac{(x - minValue)\, 2}{(maxValue - minValue)} - 1 \tag{1}$$

2.4 Introduction of Extreme Learning Machine

Machine learning concerns the construction and study of systems that can learn from data. Representation of data instances and functions evaluated on these instances are part of all machine learning systems. Neural networks (NN) and support vector machines (SVM) play key roles in machine learning.

In fact, popular learning techniques face some challenging issues such as: intensive human intervene, slow learning speed, poor learning scalability. The learning speed of feed-forward neural networks is in general far slower than required and it has been a major bottleneck in their applications.

ELM(extreme learning machine) works for the generalized single-hidden layer feed-forward networks (SLFNs) but the hidden layer in ELM need not be tuned. ELM learning theory shows that the neurons of generalized feed-forward networks neednt be tuned and these neurons can be randomly generated. All the hidden node parameters are independent from the target functions or the training datasets.

3   Experiments Result

3.1   Experimental Setup

In experiments, a computer with Intel i7-3770 CPU , 16G memory and windows 7 operating system is used. The code is run in matlab7.0. The KDD 1999 data is selected to test the performance of the proposed approach because that data is still a common benchmark for us to evaluate the proposed approach.

The 1998 DARPA Intrusion Detection Evaluation Program was prepared and managed by MIT Lincoln Labs. The objective was to survey and evaluate research in intrusion detection. A standard set of data to be audited, which includes a wide variety of intrusions simulated in a military network environment, was provided. The 1999 KDD intrusion detection contest uses a version of this data.

The training data was seven weeks of network traffic. This was processed into 4898430 connection records. Similarly, the two weeks of test data yielded 311029 connection records. Each connection is labeled as either normal, or as an attack, with exactly one specific attack type. Each connection record consists of 41 features and 1 label indicates which category it falls in.

The data contains a total of 24 training attack types, with an additional 14 types in the test data only. All attack types are transformed into four main categories:

- probing: surveillance and other probing
- DOS: denial-of-service
- U2R: unauthorized access to local superuser (root) privileges
- R2L: unauthorized access from a remote machine

The whole training data(4898430 records) is too big for a personal computer to deal with, so the 10% training data(494021 records) which provide by KDD official website at the same time is adopted , it has similar probability distribution as the whole training data. The proportion of each category is in Tab.1.

Table 1. Proportion of each category

	normal	probe	dos	u2r	r2l
training data	0.1969	0.0083	0.7924	0.0001	0.0022
testing data	0.1948	0.0134	0.7442	0.0007	0.0469

3.2 Normalization

It can be seen from Tab.2 and table 3 that after normalizing the performance improved.

Table 2. Result before normalize

	normal	probe	dos	u2r	r2l
sensitivity	0.9801	0.6147	0.9266	0	0
precision	0.6461	0.7440	0.9945	0	0

Table 3. Result after normalize

	normal	probe	dos	u2r	r2l
sensitivity	0.9955	0.7170	0.9629	0	0
precision	0.7148	0.9387	0.9979	0	0

3.3   Getting Five Best Binary Classifier

As you can see from Fig.2 to Fig.6, different percentage of positive and negative data result in different binary classifier performance. Sensitivity measures the proportion of actual positives which are correctly identified as such. Precision measures the proportion of actual positives in the population being tested as it is about the test.

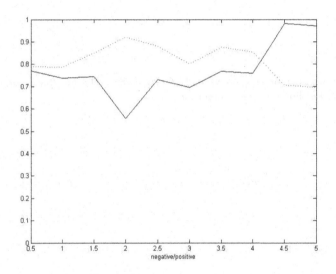

Fig. 2. Performance of binary classifier of normal class The full line represents sensitivity, the dotted line represents precision

In experiments, different numbers of hidden neurons are tried, and classifier with seven hidden neurons has the best classification performance for this data. So seven hidden neurons is used. You can see the classification performance varies with different hidden neurons in Fig.7.

Finally, best binary classifiers for five categories is got, the performance of each classifier is in Tab. 4. From Fig.2 to Fig.6, it can be found that in experiment for different proportion of data, sometimes it achieves better performance than the five best binary classifiers or it didn't achieve the performance of five best binary classifiers in Tab. 4. The reason is that ELM randomly generate hidden neurons,

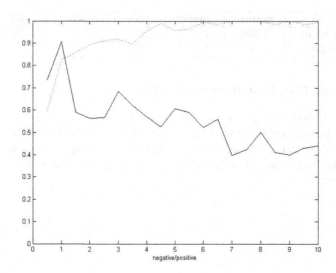

Fig. 3. Performance of binary classifier of probe class The full line represents sensitivity, the dotted line represents precision

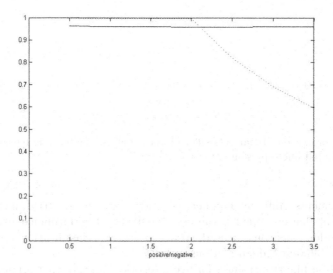

Fig. 4. Performance of binary classifier of dos class The full line represents sensitivity, the dotted line represents precision

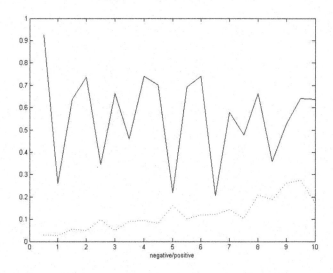

Fig. 5. Performance of binary classifier of u2r class The full line represents sensitivity, the dotted line represents precision

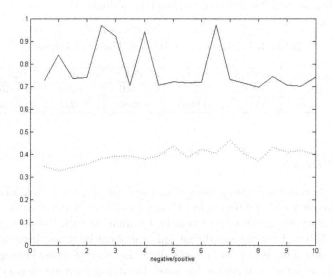

Fig. 6. Performance of binary classifier of r2l class The full line represents sensitivity, the dotted line represents precision

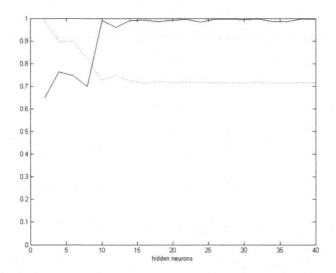

Fig. 7. Classification performance on different hidden neurons . the full line represents sensitivity, the dotted line represents precision

so when repeating the experiment with same parameter and experiment data, it will get different result. The best classifiers is got first, then the experiment for getting performance with different parameters is done.

Table 4. Performance of five best binary classifiers

	normal	probe	dos	u2r	r2l
sensitivity	0.9963	0.7931	0.9601	0.6404	0.9545
precision	0.7157	0.8889	0.9986	0.3411	0.4303

3.4 Final Result

The detection results of this proposed method are shown in Tab.5.Due to the reasonable sequence of combining five binary classifiers, it improve the classification performance of u2r and r2l obviously. Comparing with the previous 5-class classifier in Tab.6, it can be found that only the sensitivity and precision of class normal decrease a little. But in fact, the detection rate of normal category is not as important as attack categories. Because classifying normal category as attack categories won't cause serious problems.

Table 5. Performance of serial classifier

	normal	probe	dos	u2r	r2l
sensitivity	0.7872	0.7861	0.96	0.64	0.9531
precision	0.92	0.93	0.9987	0.3411	0.4392

Table 6. Performance of previous five class classifier

	normal	probe	dos	u2r	r2l
sensitivity	0.9955	0.7170	0.9629	0	0
precision	0.7148	0.9387	0.9979	0	0

3.5 Comparison

Other experiment result on KDD99 data using support vector machines by Kamularifin Abd Jalil is on Tab.7 [9]. They used 10000 records that randomly draw from KDD99 data and only test the sensitivity of their classifier.

Table 7. Performance of different approach

	probe	dos	u2r	r2l
Proposed Approach	0.7861	0.96	0.64	0.9531
SVM	0.725	0.612	0.66	0.123
Neural Network	0.812	0.56	0.601	0.168
Decision Tree	0.991	0.998	1	0.943

Comparing with support vector machines, the proposed approach using extreme learning machine has better sensitivity on probe dos and r2l categories.

4   Conclusions

This paper presents an approach that get five binary classifiers and combine them serially. It improves the classification performance of class u2r and r2l obviously compare to the original 5-class classifier. The proposed approach using extreme learning machine also has better sensitivity on most attack categories compare with support vector machines.

References

1. Zaman, S., Karray, F.: Lightweight IDS based on features selection and IDS classification scheme. In: International Conference on Computational Science and Engineering, CSE 2009, vol. 3, pp. 365–370 (2009)
2. Quinlan, J.R.: Induction of decision trees. Machine Learning 1(1), 81–106 (1986)
3. Hagan, M.T., Demuth, H.B., Beale, M.H.: Neural network design. Pws Pub., Boston (1996)

4. Burges, J.: A tutorial on support vector machines for pattern recognition. Data Mining and Knowledge Discovery 2, 121–167 (1998)
5. Rong, H.-J., Huang, G.-B., Ong, Y.-S.: Extreme learning machine for multi-categories classification applications. In: Proceedings of IEEE International Joint Conference on Neural Networks (IJCNN 2008) (IEEE World Congress on Computational Intelligence), Hong Kong, June 1-8, pp. 1709–1713 (2008)
6. Lan, Y., Soh, Y.C., Huang, G.-B.: Extreme learning machine-based bacterial protein subcellular localization prediction. In: Proceedings of IEEE International Joint Conference on Neural Networks (IJCNN 2008) (IEEE World Congress on Computational Intelligence), Hong Kong, June 1-8, pp. 1859–(1863)
7. Huang, G.-B., Chen, Y.-Q., Babri: Classification ability of single hidden layer feedforward neural networks. Neural Networks 11(3), 799–801 (2000)
8. Cheng, G.-J., Cai, L., Pan, H.-X.: Comparison of Extreme Learning Machine with Support Vector Regression for Reservoir Permeability Prediction. In: International Conference on Computational Intelligence and Security, vol. 2, pp. 173–176 (2009)
9. Jalil, K.A.: Comparison of Machine Learning Algorithms Performance in Detecting Network Intrusion. In: 2010 International Conference on Networking and Information Technology (2010)

A Study on Three-Dimensional Motion History Image and Extreme Learning Machine Oriented Body Movements Trajectory Recognition

Zheng Chang and Xiaojuan Ban

School of Computer and Communication Engineering
University of Science and Technology Beijing,
Haidian District, Beijing, China
elephant.cz@gmail.com,
banxj@ustb.edu.cn

Abstract. Based on the traditional machine vision recognition technology and traditional artificial neural networks about body movement trajectory, this paper finds out the shortcomings of the traditional recognition technology. By combining the invariant moments of the three-dimensional motion history image (computed as the eigenvector of body movements) and the extreme learning machine (constructed as the classification artificial neural network of body movements), the paper applies the method to the machine vision of the body movements trajectory. In detail, the paper gives a detailed introduction about the algorithm and realization scheme of the body movements trajectory recognition based on the three-dimensional motion history image and the extreme learning machine. Finally, by comparing with the results of the recognition experiment, we verify that the method of body movement trajectory recognition technology based on the three-dimensional motion history image and extreme learning machine has a more accurate recognition rate and a better robustness.

Keywords: Human-machine interface, Motion history image, Extreme Learning machine.

1 Introduction

With the rapid development of the natural human-computer interaction technology, the body movements trajectory tracking and recognition technology has become an important and indispensable research direction in the natural human-computer interaction technology. As we all know, since the body movement is a natural and intuitive communication mode (Daniel Weinland, 2006), therefore, beyond all doubts, the body movement recognition technology has become an useful technology to the new generation of natural human-computer interaction interface (Remi Ronfard, 2006), especially for the disabled and patients who can only use their body movements to give orders to the auxiliary equipment (such as the wheelchair, the smart television, and disabled scooter), which will bring them more convenience.

© Springer International Publishing Switzerland 2015

J. Cao et al. (eds.), *Proceedings of ELM-2014 Volume 2*,

Proceedings in Adaptation, Learning and Optimization 4, DOI: 10.1007/978-3-319-14066-7_9

The prior body movements trajectory recognition researches on the human-computer interaction mainly focus on the modelling of human skin colour and the extraction of dynamic body movements based on image attributes of the robust feature (Edmond Boyer 2003) and the BP artificial neural network, however, due to the diversity, ambiguity and disparity in time and space of the body movements, the traditional body movements trajectory recognition researches have great limitations. The paper attempts to introduce the invariant moments of the three-dimensional motion history image and the extreme learning machine into the body movement trajectory recognition, which will make the machine vision recognition of the body movement trajectory more accurate, efficient and robust.

2   Problem Description

The essence of the Body movement trajectory recognition is to classify the sample data accurately. It is a contrast between the body movements trajectory captured by the sensor and the pre-defined sample movement trajectory. Therefore, there are two steps in the body movement trajectory recognition. The first step is the extraction of dynamic body movements. The second step is the classification of the body movements trajectory captured by the sensor.

In the first extraction step, the traditional method extracts the dynamic body movements by applying the Hidden Markov Model (Guofan Huang et al., 2010). As is shown in Figure 1.

Based on the two-dimensional images and the Hidden Markov Model, the whole process of the comparison between the real trajectory from the sensor and the pre-defined sample is shown in Figure 2 (Sheng Xu et al., 2008). But based on the two-dimensional images, the recognition process still has some limitations, for example:

- Light: when the light condition is changed, the luminance information of the body will change since the images captured by the sensor are easily affected by natural light and artificial light.
- Obstruction: During the whole process, the body movement trajectory may be blocked by some objects in the environment. Since the obstruction can lead to the loss of the identification information, which will affect the reliability of the recognition greatly.
- Background: In the real recognition process, if the factors (color, texture, shape etc.) of the body movement area and the background area are similar, it will also affect the performance of the recognition (Junxue Liu et al., 2008).

Fig. 1. Body Movements Two-dimensional Image

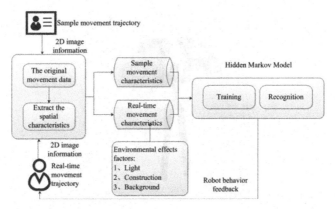

Fig. 2. The HMM Based on Two-dimensional Image

Based on the three-dimensional depth data from the structured light sensor, the three-dimensional Hidden Markov Model (3DHMM) is free from the effects of light, obstruction and background, but this method has been restricted to very few application fields because of the huge amount of calculation, the inefficiency of the training and the easy accessibility to the local optimal value etc.

In the second classification step, the traditional methods involve many machine learning algorithms (such as the average eigenvalue method, K nearest neighbour method, gradient descent-based feed-forward network learning methods). But these methods still have some limitations, as follows:

- Average eigenvalue method: In this method, we need to compute the average eigenvalue of the pre-defined sample body movement trajectory. So the average eigenvalue method is influenced greatly by some special the pre-defined sample body movement trajectory such as the wide margin action or the narrow range action.

- K nearest neighbor method: In this method, we need to compute the distance (such as Euclidean distance, Mahalanobis distance or Pearson Correlation) between the real body movements trajectory captured by the sensor and every pre-defined sample body movement trajectory. Therefore, because of the heavy computation in this method, it has been restricted to very few application fields.

- Gradient descent-based feed-forward network learning methods: The BP artificial network is one of the typical gradient descent-based feed-forward network learning methods. all the parameters of the feed forward networks need to be turned and thus there exists the dependency between different layers of parameters (weight and biases) in the BP artificial network. Therefore, the training process is very slow. And this gradient descent-based learning method may easily converge to local minima and over-fitting problem.

To solve these problems, this paper attempts to combine the three-dimensional motion history image and the extreme learning machine in order to overcome those shortcomings. In the first step, this paper attempts to combine the motion history image (MHI) with the three-dimensional depth data of the body movements in order to get the three-dimensional motion history image (3DMHI) of body movements (Figure 3).

Fig. 3. The MHI Based on Three-dimensional Depth Data

And then it calculates seven invariant moments of the three-dimensional motion history image working as the eigenvector of the body movements. In the second step, we use the extreme learning machine instead of the traditional methods (like KNN or BP). On the one hand, the new processes not only is free from the effects of the illumination, obstruction, background and other environmental factors, on the other hand, it improves the efficiency, accuracy and robustness of body movement recognition.

3   Problem Solving

3.1   Body Movements Characterized by 3D Motion History Image

To characterize the 3D motion information, the paper proposes a new method named three-dimensional motion history images approach. This method improves the application of the traditional motion history images approach based on two-dimensional image in order to combine with the three-dimensional depth data.

The motion history image approach is a branch of The Finite Difference Time Domain Method (FDTD). The mechanism of the FDTD method is to get different images from a continuous image sequences by comparing with two or three adjacent pixels in the corresponding frames, then to extract the moving regions in the image by setting the threshold. By introducing the 3D data, the paper presents the improved FDTD method named three-dimensional motion history image approach as follows:

$$\mathcal{D}(x, y, z, n) = \mathcal{J}(x, y, z, n - 1) - 2\mathcal{J}(x, y, z, n) + \mathcal{J}(x, y, z, n + 1) \qquad (1)$$

Among them, $\mathcal{J}(x, y, z, n)$ represents the pixel gray value in the position (x, y, z) in three-dimensional space, $\mathcal{D}(x, y, z, n)$ is the result of the three consecutive frames difference and also represents body movement changed area. The threshold $\mathcal{D}(x, y, z, n)$ is as follows:

$$\mathcal{B}(x, y, z, n) = \begin{cases} 1 & \mathcal{D}(x, y, z, n) > \Gamma \\ 0 & \text{otherwise} \end{cases} \qquad (2)$$

Γ is the specially selected threshold. If the value is too low, it can't effectively remove noises in images, but if the value is too high, it will impede the valuable variation of the image. So the value of the threshold should be adjustable for different experiment conditions. The experiment should be repeated many times to determine the value of the threshold. Under my experiment condition, the value of the threshold is $(10,10,15,n)$ in my follow-up work.

Three-dimensional motion history image approach of body movements is as follows:

$$\mathcal{H}_\tau(x,y,z,t) = \begin{cases} \tau & \mathcal{B}(x,y,z,t) = 1 \\ \max(0, \mathcal{H}_\tau(x,y,z,t-1)-1) & \text{otherwise} \end{cases} \quad (3)$$

Among them, $\mathcal{H}_\tau(x,y,z,t)$ represents the Pixel gray value in the position (x,y,z) and t in three-dimensional motion history image. The motion history image MHI not only reflects the external shape of the body movements, but also reflects the direction and state of them. In the motion history image, the gray value of each pixel is in proportion with the duration of the body movement in the position. The recent body gestures have the maximum gray value. Gray value changes reflect the direction of the body movements.

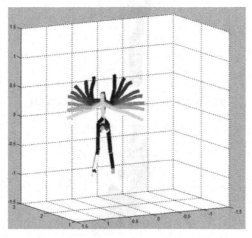

Fig. 4. Three-dimensional Motion History Image of Body Movements (MHI)

3.2   The Calculation of the Invariant Moments of the Motion History Image

Although the three-dimensional motion history image approach based on the MHI is simple and efficient, it is too sensitive to the observation position. In order to overcome this shortcoming, this paper selects the invariant moments as eigenvector of the motion history image. The method of invariant moments is a classical method to extract image feature. Its translation invariance, scaling invariance and rotation invariance properties rule out the impact on the position and angle.

To calculate the invariant moments, after getting the three-dimensional motion history image, we project it in the XY plane (Figure 5), YZ plane (Figure 6) and XZ plane (Figure 7). This method can get three views of three-dimensional motion history image with one gesture. Then we have the calculation of invariant moments for the three main views.

Fig. 5. XY Surface Projection of the MHI

Fig. 6. YZ Surface Projection of the MHI

Fig. 7. XZ Surface Projection of the MHI

For a size of $\mathcal{M} \times \mathcal{N}$ digital image $f(x, y)$, the $p + q$ order moment m_{pq} is defined as follows:

$$m_{pq} = \sum_{x=1}^{N} \sum_{y=1}^{M} f(x, y) x^p y^q \tag{4}$$

Among them, $p, q = 0,1,2, \ldots$

$p + q$ order central moment μ_{pq} is defined as follows:

$$\mu_{pq} = \sum_{x=1}^{N} \sum_{y=1}^{M} f(x,y)(x - \bar{x})^p (y - \bar{y})^q \tag{5}$$

(x,y) represents the object image point, $(\bar{x}, \bar{y})$ is the object centroid:

$$\bar{x} = \frac{m_{10}}{m_{00}}, \bar{y} = \frac{m_{01}}{m_{00}} \tag{6}$$

Among them:

$$m_{00} = \sum_{x=1}^{N} \sum_{y=1}^{M} f(x,y)$$

$$m_{10} = \sum_{x=1}^{N} \sum_{y=1}^{M} f(x,y)x^1$$

$$m_{01} = \sum_{x=1}^{N} \sum_{y=1}^{M} f(x,y)y^{1}.$$

Then through the normalizing of the central moment by the zero order central moments μ_{00} ,we can get the normalized center moment of the motion history image.

$$\eta_{pq} = \frac{\mu_{pq}}{\mu_{00}^r}, r = \frac{p + q + 2}{2}, p + q = 2,3,4, \dots \tag{7}$$

Hu·M·K get seven invariant moments based on the linear combination of two order and three order normalized central moment. The image translation, rotation and scaling are unchanged and the invariant moments are as follows: (Xiaoniu Liu et al., 2010)

$$M_1 = \eta_{20} + \eta_{02} \tag{8}$$

$$M_2 = (\eta_{20} - \eta_{02})^2 + 4\eta_{11}^2 \tag{9}$$

$$M_3 = (\eta_{30} - 3\eta_{12})^2 + (3\eta_{21} - \eta_{03})^2 \tag{10}$$

$$M_4 = (\eta_{30} + \eta_{12})^2 + (\eta_{21} + \eta_{03})^2 \tag{11}$$

$$M_5 = (\eta_{30} - 3\eta_{12}) \times (\eta_{30} + \eta_{12}) \times [(\eta_{30} + \eta_{12})^2 - 3(\eta_{21} + \eta_{03})^2] + (\eta_{03} - 3\eta_{21})(\eta_{03} + \eta_{21})[(\eta_{03} + \eta_{21})^2 - 3(\eta_{12} + \eta_{30})^2] \tag{12}$$

$$M_6 = (\eta_{20} - \eta_{02})[(\eta_{30} + \eta_{12})^2 - (\eta_{21} + \eta_{03})^2] + 4\eta_{11}(\eta_{30} + \eta_{12})(\eta_{21} + \eta_{03}). \tag{13}$$

$$M_7 = (3\eta_{21} - \eta_{03})(\eta_{30} + \eta_{21})[(\eta_{30} + \eta_{12})^2 - 3(\eta_{21} + \eta_{03})^2] - (3\eta_{12} - \eta_{30})(\eta_{03} + \eta_{21})[(\eta_{03} + \eta_{21})^2 - 3(\eta_{12} + \eta_{30})^2] \tag{14}$$

Because the invariant moments are too small, it is compressed by the absolute value of the logarithm and so the actual values need to be amended in accordance with the following formula.

$$M_k = \log|M_k|, k = 1,2,3,4,5,6,7 \tag{15}$$

The invariant moments still have a translation, rotation and scaling invariance after amendment.

Through the calculation of the projection images in three directions, we will get a 3×7 eigenvalue matrix. This eigenvalue matrix is the eigenvector for motion history volume.

3.3   The Body Movements Recognition Based on Extreme Learning Machine

In the process of recognition, we first collect the sample of body movement and build a training sample set so that we can get a better recognition performance. In order to get better recognition results, the samples of the body movements must be definite and slow, in another word, all of the training samples must comply with the criterion.

For the same body movement, different people involved shall repeat the action several times. And then we get multiple groups of three-dimensional motion history images for each movement, after that, the training sample set for each gesture will be established.

According to many precedent outstanding researches, we can draw the conclusion that the input weights and hidden layer biases of a single-hidden layer feed-forward neural network (SLFN) can be randomly assigned if the activation functions in the hidden layer are infinitely differentiable (Guang-Bin Huang et al., 2004).

Based on the conclusion, this paper proposes the extreme learning machine (ELM) for SLFN. The extreme learning machine has a faster and better generalisation performance than another traditional feed-forward artificial network (such as BP network).

For our body movements trajectory recognition process, we could construct a SLFN as follows:

There are $\mathcal{N}$ hidden nodes and $\mathcal{M}$ training samples in our SLFN. Our training sample set is (X_j, T_j), where $X_j = [X_{j1}, X_{j2}, \cdots, X_{jM},]$ is the invariant moments of training samples, $T_j = [T_{j1}, T_{j2}, \cdots, T_{jM},]^T$ is the kind of training samples. And activation function $f(x)$ is mathematically modelled as:

$$\sum_{i=1 \, j=1}^{\mathcal{N} \, \mathcal{M}} \beta_i f(W_i \cdot X_j + b_i) = R_j \tag{16}$$

Where $R_j = [R_{j1}, R_{j2}, \cdots, R_{jM},]$ is the output vector form our SLFN, $W_i = [W_{i1}, W_{i2}, \cdots, W_{iN},]^T$ is the weight vector which connect the ith hidden layer node and the input nodes, $\beta_i = [\beta_{i1}, \beta_{i2}, \cdots, \beta_{iN},]^T$ is the weight vector which connect the ith hidden layer node and the output nodes, b_i is the threshold of the ith hidden layer node.

Our goal is to make T_j close to R_j, which is equivalent to minimizing the cost function

$$C = \sum_{i=1}^{N} \sum_{j=1}^{M} \beta_i f(W_i \cdot X_j + b_i) - T_j = R_j - T_j \tag{17}$$

Previous excellent papers have rigorously proved that for any infinitely differentiable activation function SLFN with N hidden nodes can N distinct samples exactly and SLFN may require less than N hidden nodes if learning error is allowed (Guang-Bin Huang et al., 2006).

So we could obtain the equations like this:

$$\sum_{i=1}^{N} \sum_{j=1}^{M} \beta_i f(W_i \cdot X_j + b_i) = T_j \tag{18}$$

The above equations could be abbreviated as

$$BF = T \tag{19}$$

where

$$F = \begin{bmatrix} f(W_1 \cdot X_1 + b_1) & \cdots & f(W_1 \cdot X_M + b_1) \\ \vdots & \cdots & \vdots \\ f(W_N \cdot X_1 + b_N) & \cdots & f(W_N \cdot X_M + b_N) \end{bmatrix}_{N \times M}$$

$$B = [\beta_1, \beta_2, \cdots, \beta_N,]$$

$$T = [T_1, T_2, \cdots, T_M,]$$

On the both sides of the equation, we could apply the transpose operation to obtain the standard ELM model:

$$\mathcal{H}\beta = \mathcal{T} \tag{20}$$

where

$$\mathcal{H} = F^T$$

$$\beta = B^T$$

$$\mathcal{T} = T^T$$

So we could obtain the result like this:

$$\beta = \mathcal{H}^\dagger \mathcal{T} \tag{21}$$

where $\mathcal{H}^\dagger$ is the Moore-Penrose generalized inverse of matrix $\mathcal{H}$.

If $\mathcal{N} = \mathcal{M}$ and $\mathcal{H}$ has an inverse, the solution is existing and unique. And the answer is, evidently, $\mathcal{H}^\dagger = \mathcal{H}^{-1}$. So we could obtain the result like this:

$$\beta = \mathcal{H}^{-1}\mathcal{T} \tag{22}$$

In other cases, we should consider to use other method to solve the problem. There are several methods which could be used to calculate the Moore-Penrose generalized inverse of matrix $\mathcal{H}$. In this paper, we apply the Spectral Theorem and Tikhonov's regularization method to calculate $\mathcal{H}^\dagger$ Moore-Penrose generalized inverse of matrix $\mathcal{H}$ (J. C. A. Barata et al., 2012).

$$\mathcal{H}^\dagger = \sum_{\substack{b=1 \\ \alpha_b \neq 0}}^{s} \frac{1}{\alpha_b} \left[\prod_{\substack{l=1 \\ l \neq b}}^{s} (\alpha_b - \alpha_l)^{-1} \right] \left[\prod_{\substack{l=1 \\ l \neq b}}^{s} (\mathcal{H}^*\mathcal{H} - \alpha_l E_n) \right] \mathcal{H}^* \tag{23}$$

Where $\mathcal{H}^*$ represents the adjoint matrix of $\mathcal{H}$ and $\alpha_b, b = 1,2,\dots,s$ are the eigenvalues of $\mathcal{H}^*\mathcal{H}$ (or the singular values of $\mathcal{H}$).

According to the training sample set, we could obtain the weight vector β and then complete the training of ELM. As presented in the previous outstanding paper, ELM has many important properties, as follows:

- Smallest norm of weights;
- Minimum approximation error;
- The minimum norm least-squares solution of $\mathcal{H}\beta = \mathcal{T}$ is unique;

After the training process, we could put the invariant moments of real body movement captured by the structured light sensor into the ELM. Although the training process of the ELM would cost some time, the recognition process of the real body movement is very fast.

4   Results of the Experiment

4.1   Data Pre-processing

This trajectory recognition experiments are done in normal laboratory environment. In the experiment, people should keep the body forward, perpendicular to the horizontal plane and be about 1.2 to 2 meters to the structured light sensor. In this paper, we debounce the physical movements monitored and record the centre position data of the prior frame to compare with the centre position data of the current frame. If the deviation is within the threshold range, we should adopt the position data of the prior frame to ignore the jitter of the current frame.

When using the real movements trajectory, invalid frames will appear at the beginning and the end of the movement. In order to remove the invalid part and get the middle part, we debounce the physical movements to make sure that the motion part displacement will decrease and all the frames can be used.

In the experiment, we ask four people to do four kinds of body movements, as is shown in Figure 8, Figure 9, Figure 10 and Figure 11. Each kind of body movements is repeated 10 times and generates 40 samples for each body movement. Every movement lasts five to fifteen seconds with the image size of1200 × 900.

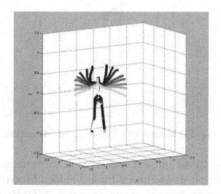

Fig. 8. Motion History Image for Movement A

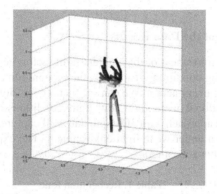

Fig. 9. Motion History Image for Movement B

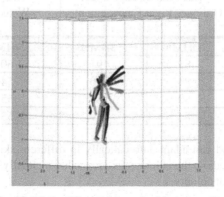

Fig. 10. Motion History Image for Movement C

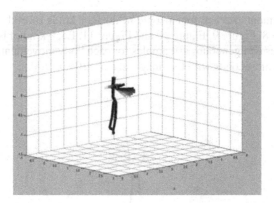

Fig. 11. Motion History Image for Movement D

We choose the first 20 samples of each kind of movements for training to get the standard movement templates and the traditional hidden Markov model (3DHMM) and three-dimensional motion history image approach (3DMHI) to test the other 20 remaining samples.

4.2   The Analysis of the Data Results

In order to verify the robustness of the ambient light factor, here we do the experiment under different lighting conditions. Table 1 is the recognition accuracy in normal light and the weak light for every action. We can see that recognition accuracy rate declined sharply by the traditional method in low light conditions. The three-dimensional motion history image approach with 3D depth data will capture the trajectory of human motion very well even under weak light environment. The experiment shows that the new method has a better robustness.

Table 1. Recognition Rate

Movement Types	Recognition Rate under Ordinary Light Environment (%)		Recognition Rate under Weak Light Environment (%)	
	3DHMM	3DMHI	3DHMM	3DMHI
A	85	91	60	89
B	84	90	54	88
C	80	89	55	86
D	83	89	53	87

The above method can adapt automatically to different skin colours and clothes. There is no need to adjust the colour reference value manually according to the target body's clothes or skin colours.

In order to verify the efficiency and accuracy of the body movement recognition process based on the extreme learning machine, here we do the experiment under two different recognition algorithms (ELM and BP).

To keep the same experiment condition, we apply the same number of the hidden nodes (7 hidden nodes), and adopt the same activation function (standard Sigmoid Function).

$$f(x) = \frac{1}{1 + e^{-x}} \tag{24}$$

Table 2 is the recognition accuracy and elapsed time in two different recognition algorithms for each action. We can see that there is no significant difference between the two algorithms, but the extreme learning machine costs less time than the BP network in the training process.

Table 2. Recognition Rate

Movement Types	ELM		BP	
	Accuracy(%)	Time(s)	Accuracy(%)	Time(s)
A	83	0.021	85	4.219
B	87	0.016	89	3.820
C	88	0.013	87	3.761
D	84	0.019	81	4.172

From the above two aspects, we could come to the conclusion that the new process in this paper has a better accuracy, efficiency and robustness.

5   Conclusion

Combined with the three-dimensional motion history image and the extreme learning machine, this paper overcomes the shortcomings of the traditional body movements trajectory recognition method, such as the light, the obstruction, the background, the influence of some specific data, the heavy calculation and the slow recognition process. The new process in this paper realizes the recognition of body movements trajectory. Finally through the two different experiments, we verify that the new process in the paper has a better robustness, efficiency and accuracy.

Acknowledgement. This work was supported by National Nature Science Foundation of P.R. China (No.61272357, 61300074), the Fundamental Research Funds for the Central Universities (FRF-TP-09-016B), the new century personnel plan for the Ministry of Education (NCET-10-0221) and by China Post-doctoral Science Foundation (No. 20100480199).

References

Weinland, D.: Automatic discovery of action taxonomies from multiple views. In: IEEE Computer Society Conference on Computer Vision and Pattern Recognition (2006)

Boyer, E.: Motion history volumes for free viewpoint action recognition. In: IEEE International Workshop on Modeling People and Human Interaction (2005)

Huang, G.-B., Zhu, Q.-Y., Siew, C.-K.: Extreme learning machine: A New Learning Scheme of Feed-forward Neural Networks. In: Proceedings of International Joint Conference on Neural Networks (IJCNN 2004), Budapest, Hungary, July 25-29 (2004)

Huang, G.-B., Zhu, Q.-Y., Siew, C.-K.: Extreme learning machine: Theory and applications. Neurocomputing 70, 489–501 (2006)

Huang, G., Cheng, X.: An automatic recognition approach of human gestures. Journal of Southwest China Normal University 35(4), 136–140 (2010)

Barata, J.C.A., Hussein, M.S.: The Moore-Penrose Pseudoinverse. A Tutorial Review of the Theory. Brazilian Journal of Physics 42(1-2), 146–165 (2012)

Liu, J., Qu, Z.: Real-time detecting and tracking of multiple moving object based on improved motion history image. Computer Applications 28(6), 198–201 (2008)

Ronfard, R.: Free viewpoint action recognition using motion history volumes. Computer Vision and Image Understanding 104, 249–257 (2006)

Xu, S., Peng, Q.: Three-dimensional object recognition based combined moment invariants and neural network. Computer Engineering and Applications 44(31), 78–80 (2008)

Liu, X., Yuan, K.: Weight moment method based on hu invariant moments and applications. Journal of Dalian Nationaloties University 12(5), 470–472 (2010)

An Improved ELM Algorithm
Based on PCA Technique*

Haigang Zhang, Yixin Yin, Sen Zhang**, and Changyin Sun

School of Automation and Electrical Engineering,
University of Science and Technology Beijing, Beijing 100083, China
zhangsen@ustb.edu.cn

Abstract. This paper proposes a modified ELM algorithm named P-ELM subject to how to select the number of hidden nodes and how to get rid of the multicollinear problem in calculation based on PCA technique. By reducing the dimension of hidden layer output matrix(H) without loss of any information through PCA theory, the proposed P-ELM algorithm can not only ensure the full column rank of newly generated hidden layer output matrix(H'), but also reduce the number of hidden nodes resulting the improvement in training speed. In order to verify the effectiveness of P-ELM algorithm, the comparative simulations are performed. The simulation results illustrate the better generalization performance and the stability of the proposed P-ELM algorithm.

Keywords: Extreme learning machine, Principle component analysis, Multicollinear problem, Dimensionality reduction.

1 Introduction

Extreme learning machine(ELM) is a novel single hidden layer feedforward neural network proposed by Huang in 2006[1]. In ELM, the input weights and the bias of hidden nodes are generated randomly without human tuning and the output weights are determined based on the method of least squares[1,2,7]. Unlike the traditional feedforward neural network learning algorithm, ELM has the fast training speed and gets rid of the opportunity to converge to local minima[1]. Nowadays, because of its good generalization, ELM algorithm has been applied in many aspects like image segmentation[3], fault diagnosis[4,5], human action recognition, human computer interface[8] and so on.

So far, ELM algorithm has drawn more and more attention. However, two open problems often occur in many real industrial applications: (1)how to select the number of hidden nodes and (2)how to deal with the multicollinear problem faced in the calculation. For the first problem, inappropriate selection of the

---

* This work has been supported by the National Natural Science Foundation of China (NSFC grant No. 61333002) and Beijing Natural Science Foundation (grant No. 4132065).

** Corresponding author.

hidden node number will lead to overfitting or underfitting results[6]. According to the theory of Huang et. al., ELM with N hidden nodes can exactly learn N distinct observations[1]. However, in many case, the number of hidden nodes is often lower within the tolerable range in order to improve the calculating speed. [9] proposed a modified ELM algorithm to select the number of hidden nodes from a small size to a proper one based on the minimal error. Other techniques like P-ELM(Haijun Rong)[12] and OP-ELM[10] can also choose the appropriate number at different angles. In this note, a novel attempt to select the number of hidden nodes is proposed based on PCA technique. One chooses an appropriate number of hidden nodes according to operation experience and then reduce the dimension to a lower one, which represents all or most information of hidden layer output matrix. In addition, the multicollinear problem in ELM algorithm deteriorates its generalization performance in many complex industrial process. When calculating the output weights, Huang just adds a threshold to every element of matrix $H^T H(\hat{\beta} = \left(H^T H + {1}/{c}\right)^{-1} H^T T)$[1], which lacks the rigorous mathematical proof and seems irrational in applications. In our previous work, LDL^T decomposition to $H^T H$ is employed to overcome the multicollinear problem[11]. One just needs to set a proper threshold to some singular elements of matrix D after the decomposition. However, any change to the hidden layer output matrix will affect the performance indexes(like Expectation, Variance, Mean Square Error) of the final results. Here, following the novel approach to select lower number of hidden nodes, the improved ELM algorithm can also get rid of the multicollinear problem.

In this paper, a modified ELM algorithm named P-ELM is proposed based on PCA technique. Compared with the previous methods, the proposed P-ELM algorithm can not only get rid of the puzzle of multicollinear restriction but also reduce the number of hidden nodes to improve the training speed. The improved P-ELM algorithm shows a better performance in dealing with interfered data obtained from complex industrial process. In order to verify the robustness of P-ELM, we add a certain amount of white noise to the data artificially in the simulation part. The experiment results are satisfactory. It is worth mentioning that [13] also employed PCA theory to the ordinary ELM algorithm. However, in this manuscript, PCA technique is used to deal with the hidden layer output matrix rather than the original sampling data.

The following sections are organized as follows: Section 2 is the review of ordinary ELM algorithm. The improved P-ELM algorithm will be presented in section 3 and simulation results are given in section 4. Section 5 summarizes the conclusion of this paper.

2   The Theory of Ordinary ELM Algorithm

In the last few years, ELM algorithm has received very wide range of applications and development because of its fast train speed and good generalization performance. In this section, a review of the ordinary ELM algorithm is presented and some properties of the solution will be discussed.

Suppose there are N arbitrary samples (x_i, t_i), where $x_i = [x_{i1}, x_{i2}, \cdots, x_{in}]^T$ $\in R^n$ denotes an n-dimensional feature of the ith sample and $t_i = [t_{i1}, t_{i2}, \cdots, t_{im}] \in R^m$ denotes the target vector. The mathematical model of SLFNs with $\tilde{N}$ hidden nodes is as follows:

$$\sum_{i=1}^{\tilde{N}} \beta_i g_i(x_k) = \sum_{i=1}^{\tilde{N}} \beta_i g(w_i \cdot x_k + b_i) = o_k, \qquad k = 1, 2, \cdots, N \qquad (1)$$

where w is $\tilde{N} \times n$ input weight matrix connecting the hidden and the input nodes, β is $\tilde{N} \times m$ output weight matrix connecting the hidden and the output nodes, b is $\tilde{N} \times 1$ bias of hidden layer nodes. And $w_i \cdot x_k$ denotes the inner product of w_i and x_k. o is $N \times 1$ output vector under this model with activation function $g(x)$.

The above $\tilde{N}$ equations can be written in matrix form as

$$H\beta = T \qquad (2)$$

where $H(W, B, X) = \begin{pmatrix} g(w_1 \cdot x_1 + b_1) & \cdots & g(w_{\tilde{N}} \cdot x_1 + b_{\tilde{N}}) \\ \vdots & \ddots & \vdots \\ g(w_1 \cdot x_N + b_1) & \cdots & g(w_{\tilde{N}} \cdot x_N + b_{\tilde{N}}) \end{pmatrix}_{N \times \tilde{N}}$ called the

hidden layer output matrix. According to the theory of Least Square, the output weight β can be estimated as

$$\hat{\beta} = H^+ T \qquad (3)$$

where H^+ is the Moore-Penrose gnenralized inverse of H.

There are several methods to calculate the Moore-Penrose gnenralized inverse. Here singular value decomposition(SVD) method is widely used, where $H^+ = (H^T H)^{-1} H^T$[19]. So

$$\hat{\beta} = (H^T H)^{-1} H^T T \qquad (4)$$

It is apparent from (4) that one can not obtain correct and satisfactory result if the matrix $H^T H$ is singular despite that some mathematical softwares like Matlab have the corresponding methods to deal with the inverse of singular matrix. Next some properties of the solution of ELM algorithm will be discussed. In a complex industrial production environment, the data will often be interfaced by external noise. Then the model (2) should be modified as

$$H\beta + \xi = T \qquad (5)$$

where white noise is considered here and $\xi \in N(0, \sigma^2)$.

After adding the interface of external noise, the solution of ELM will also be modified as

$$\hat{\beta} = \left(H^T H\right)^{-1} H^T T = \frac{\sum\limits_{i=1}^{\tilde{N}} H_i^T \left(H_i \beta_i + \xi_i\right)}{\sum\limits_{i=1}^{\tilde{N}} H_i^T H_i} = \beta + \frac{\sum\limits_{i=1}^{\tilde{N}} H_i^T \xi_i}{\sum\limits_{i=1}^{\tilde{N}} H_i^T H_i} \tag{6}$$

Next we analyze the result from three aspects: expectation(E), variance(V) and mean square error(MSE). Then one can get

$$\bullet\, E\left(\hat{\beta}\right) = \beta$$

$$\bullet\, V\left(\hat{\beta}\right) = E\left(\hat{\beta}^2\right) - E\left(\hat{\beta}\right)^2 = \frac{\sigma^2}{\sum\limits_{i=1}^{\tilde{N}} H_i^T H_i} = \sigma^2 \sum\limits_{i=1}^{\tilde{N}} \frac{1}{\lambda_i} \tag{7}$$

$$\bullet\, MSE\left(\hat{\beta}\right) = \frac{1}{N} E\left[\left(\hat{\beta} - \beta\right)^T \left(\hat{\beta} - \beta\right)\right] = \frac{1}{N} \cdot \frac{\sigma^2}{\sum\limits_{i=1}^{\tilde{N}} H_i^T H_i} = \frac{\sigma^2}{N} \sum\limits_{i=1}^{\tilde{N}} \frac{1}{\lambda_i}$$

where λ_i is the ith eigenvalue of $H^T H$[11].

As we all know, the input weights and the bias of hidden nodes in the model of ELM algorithm are generated randomly without human tuning. In most case, the number of hidden nodes is far less than that of the samples($\tilde{N} \ll N$). So H is usually not a square matrix and $H^T H$ may not always be nonsingular. That is to say when $H^T H$ is multicollinear, some eigenvalues will tend to zero, while $V\left(\hat{\beta}\right)$ and $MSE\left(\hat{\beta}\right)$ will become larger. Therefore, the solution is not convincing.

3   The Introduction for Improved P-ELM Algorithm

This section is the main part of the note. Here the improved P-ELM algorithm is proposed to overcome the multicollinear problem using principle component analysis(PCA) technique. PCA is a useful statistical technique that has found application in fields such as face recognition and image compression, and is a common technique for finding patterns in high dimension data[14, 16]. For a series of data with high dimension, PCA is a powerful tool for identifying patterns and expressing the data in such a way as to highlight their similarities and differences by reducing the number of dimensions[15].

Fig. 1 presents the distribution of sample data with two dimensions where we can see that most of the data distribute along the direction of w_1. Here w_1 is called the first principle component direction which can character the main information of data distribution. In addition, w_2 stands for another direction of the second principle component which means less important information like external disturbance. The main purpose of the theory of PCA is to represent the

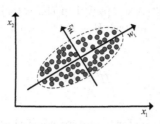

Fig. 1. The distribution of sample data

sample data using the main principle components. In the theory of mathematical statistics, the first principle component direction is the one with maximum variance of the data distribution[16]. So the sort of principle components is based on the size of variances of sample data in different directions. Next a brief theoretical derivation of PCA is presented[17].

For a set of real data $H \in R^{m \times n}$ where with m samples and n indicators, we want to find some principle components to present the information of sample data as more as possible. So

$$T = HW \tag{8}$$

where T is the principle matrix after reducing the dimension of ordinary H and W is the unit orthogonal transformation matrix.

Considering the theory of maximum variance, the problem can be transformed into

$$\begin{aligned} \max \quad & \text{cov}(T) \\ s.t. \quad & W^T W = I \end{aligned} \tag{9}$$

Then it equals to

$$\begin{aligned} \max \quad & \langle HW, HW \rangle \\ s.t. \quad & W^T W = I \end{aligned} \tag{10}$$

Applying Lagrangian algorithm, one can get

$$L = W^T H^T H W - \lambda(W^T * W - I) = 0 \tag{11}$$

where λ is Lagrangian parameter.

Then

$$\begin{cases} \frac{\partial L}{\partial W} = 2H^T H W - 2\lambda W = 0 \\ \frac{\partial L}{\partial \lambda} = W^T * W - 1 = 0 \end{cases} \tag{12}$$

So one can get

$$H^T H W = \lambda W \tag{13}$$

where we can see that λ is a diagonal matrix composed of the eigenvalues of $H^T H$ and W is the transform matrix consisting of the corresponding eigenvectors and the column of W are the principle components of the sample data. Additionally, the sort of importance of principle components are based on the values of eigenvalues of $H^T H$.

Next the technique of PCA is applied in the calculation of ELM algorithm. For the hidden layer output matrix $H \in R^{N \times \tilde{N}}$ with $\tilde{N}$ hidden nodes

$$H' = HG \tag{14}$$

where $G : R^{\tilde{N} \times p}$ is the transform matrix consisting with the corresponding eigenvectors of nonzero eigenvalues of $H^T H$ and $H' : R^{N \times p}$ is the new hidden layer output matrix with lower dimension. p is the new dimension and also equals to the number of nonzero eigenvalues. After the transform, one can get

$$H'^T H' = G^T H^T HG = \begin{bmatrix} \lambda_1 & & \\ & \ddots & \\ & & \lambda_p \end{bmatrix}_{p \times p} \tag{15}$$

where $\lambda_i \neq 0, i = 1, 2, \cdots, p$. It is appear that the improved ELM algorithm can get rid of the multicollinear problem after the PCA transform and the calculation of inverse of H'^T is reasonable and less time-consuming. Then the output estimated weight $\hat{\beta}$ will be modified as

$$\hat{\beta} = \left(H'^T H' \right)^{-1} H'^T T \tag{16}$$

In fact, the improved P-ELM algorithm is equals to adding an additional layer to the model of ordinary ELM like in Fig. 2. PCA technique is employed to transform H to an orthogonal matrix H' with lower dimension (lower number of hidden nodes). According to the mathematical statistical theory, it means no less of information in hidden layer matrix if one chooses the transformation matric G consisting with the eigenvectors of all the nonzero eigenvalues. Additionally, in most case, one can select the number of principle components based on the following cumulative contribution rate(CCR) equation

$$CCR = \frac{\sum\limits_{i=1}^{p} \lambda_i}{\sum\limits_{i=1}^{\tilde{N}} \lambda_i} \geq \varepsilon \tag{17}$$

where ε is the artificial threshold like 90% and p is the number of selected principle components[18].

Then recalculate the performance indicators under the situation of selecting all the principle components:

$$\bullet E\left(\hat{\beta}\right) = \beta$$

$$\bullet V\left(\hat{\beta}\right) = E\left(\hat{\beta}^2\right) - E\left(\hat{\beta}\right)^2 = \frac{\sigma^2}{\sum\limits_{i=1}^{p} H_i^T H_i} = \sigma^2 \sum\limits_{i=1}^{p} \frac{1}{\lambda_i} \le \frac{p\sigma^2}{\lambda_{\min}}$$

$$\bullet MSE\left(\hat{\beta}\right) = \frac{1}{p} E\left[\left(\hat{\beta} - \beta\right)^T \left(\hat{\beta} - \beta\right)\right] = \frac{1}{p} \cdot \frac{\sigma^2}{\sum\limits_{i=1}^{p} H_i^T H_i} \qquad (18)$$

$$= \frac{\sigma^2}{p} \sum\limits_{i=1}^{p} \frac{1}{\lambda_i} \le \frac{\sigma^2}{\lambda_{\min}}$$

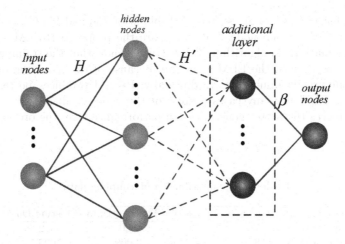

Fig. 2. The model of P-ELM algorithm

As we can see from the above three indicators, P-ELM algorithm can deal with the multicollinear problem under the premise of unbiased estimation. Compared with our previous work in [11], this improvement is more reasonable and accurate. In addition, the reducing of hidden node number can guarantee the raise in training speed. Then the improved P-ELM algorithm can be summarized as following

Step 1: Assign arbitrary input weights W and the bias of hidden nodes B.

Step 2: Calculate the hidden layer output matrix H.

Step 3: Select the number of principle components for H and calculate the transformation matrix G.

Step 4: Calculate the additional hidden layer output matrix H'.

Step 5: Calculate the output weights: $\hat{\beta} = \left(H'^T H'\right)^{-1} H'^T T$.

4 Simulation Results

This section presents the simulation results subject to the proposed P-ELM algorithm applied in four databases including two regression applications and two classification applications. The benchmark problems are described in Table 1. Additionally, all the experiments have been conducted in Matlab 7.8.0 running on a desktop PC with AMD Athlon II X2 250 processor, 3.00 GHz CPU and 2G RAM.

All the regression data are normalized into the range $[0, 1]$, while we normalize the classification attributes into the range $[-1, 1]$. For simplicity, there is two type of activation function taken into consideration: the sigmoidal additive activation function $G(a, b, x) = 1/(1 + \exp(-(a \cdot x + b)))$. and the Gaussian RBF activation function $G(a, b, x) = \exp\left(-b\|x - a\|^2\right)$ All the experiments are carried out 50 trials, and all the performance indicators are calculated by the average of 50 conclusions.

The simulation results are stated in Table 2(for sigmoid hidden nodes) and Table 3(for RBF hidden nodes). We compare the proposed P-ELM algorithm with the ordinary ELM algorithm. Table 2 and 3 show that the training time in P-ELM is less than in the ELM algorithm because of the reduction of hidden layer output matrix dimension. Additionally, P-ELM produces better testing results than ELM in many cases despite of the wine database. Consequently, P-ELM achieves the better generalization performance than the ordinary ELM algorithm.

Table 1. The distribution of sampling data

Dataset	#Attributes	#Classes	#Training Data	#Testing Data
Auto-MPG	7	-	320	72
Abalone	8	-	3,000	1,177
Zoo	17	7	70	31
Wine	13	3	120	58

Table 2. Simulation results(sigmoid hidden nodes)

Dataset	Algorithm	#nodes	Training time(s)	Testing time(s)	RMSE or accuracy Training	Testing
Auto-MPG	ELM	25	0.0034	0.0025	0.0558	0.1290
	P-ELM	14	0.0044	0.0012	0.0609	0.1224
Abalone	ELM	50	0.0964	0.0119	0.0751	0.0730
	P-ELM	40	0.0880	0.0128	0.0756	0.0723
Zoo	ELM	35	0.0044	0.0013	100%	89.01%
	P-ELM	20	0.0025	6.24×10^{-4}	100%	91.53%
Wine	ELM	30	0.0025	6.24×10^{-4}	99.98%	95.69%
	P-ELM	18	0.0019	0.0016	98.47%	95.10%

Table 3. Simulation results(RBF hidden nodes)

Dataset	Algorithm	#nodes	Training time(s)	Testing time(s)	RMSE or accuracy Training	RMSE or accuracy Testing
Auto-MPG	ELM	25	0.0044	0.0012	0.0572	0.1273
	P-ELM	14	0.0031	0.0022	0.0631	0.1306
Abalone	ELM	50	0.0908	0.0131	0.0752	0.0724
	P-ELM	40	0.0836	0.0100	0.0757	0.0719
Zoo	ELM	35	0.0022	0.0019	99.98%	86.27%
	P-ELM	20	0.0050	0.0001	97.74%	90.73%
Wine	ELM	30	0.0031	0.0019	99.43%	95.48%
	P-ELM	18	0.0022	0.0011	98.10%	94.79%

Table 4. Simulation results for auto-MPG database with white noise

White noise	Algorithm	#nodes	RMSE Training	RMSE Testing	SD Training	SD Testing	CV
5%	ELM	25	0.0586	0.1276	0.0015	0.0106	0.1325
	P-ELM	20	0.0504	0.1234	0.0014	0.0098	0.1225
15%	ELM	40	0.0602	0.1448	0.0021	0.0133	0.1511
	P-ELM	30	0.0594	0.1391	0.0020	0.0129	0.1499

In order to verify the robustness of P-ELM, two scalars of white noise contaminate the auto-MPG database. In addition, we introduce two performance indicators: standard deviation(SD) and coefficient of variation(CV) in this simulation. Table. 4 presents the results which show that the proposed P-ELM algorithm produces smaller SD and CV, meaning P-ELM is more stable and robust than the ordinary ELM algorithm in dealing with the contaminated data.

5 Conclusions

In practical applications, ELM algorithm encounters some constrains: (1)inappropriate number of hidden nodes and (2)multicollinear problem in calculation of output weight matrix. In this paper, an improved ELM algorithm named P-ELM is proposed subject to the above two limits based on PCA theory. Through reducing the dimension of hidden layer output matrix(H) without loss of any statistic information, the newly generated hidden layer output matrix(H') has full rank of the columns. This approach can not only solve the multicollinear problem but also improve training speed compared with the ordinary ELM algorithm. The experimental results show that P-ELM algorithm has more stability in handling the contaminated industrial data and produces the better generalization performance.

References

1. Huang, G.B., Zhu, Q.Y., Siew, C.K.: Extreme learning machine: Theory and applications. Neurocomputing 70(1-3), 489–501 (2006)
2. Liang, N.Y., Huang, G.B., Saratchandran, P., Sundararajan, N.: A Fast and Accurate Online Sequential Learning Algorithm for Feedforward Networks. IEEE Trans. Neural Networks 17(6), 1411–1423 (2006)
3. Pan, C., Park, D.S., Yang, Y., Yoo, H.M.: Leukocyte image segmentation by visual attention and extreme learning machine. Neural Computing and Applications 21(6), 1217–1227 (2012)
4. Wang, P.K., Yang, Z.X., Vong, C.M., Zhong, J.H.: Real-time fault diagnosis for gas turbine generator systems using extreme learning machine. Neurocomputing 128, 249–257 (2014)
5. Hu, X.F., Zhao, Z., Wang, S., Wang, F.L., He, D.K., Wu, S.K.: Multi-stage extreme learning machine for fault diagnosis on hydraulic tube tester. Neural Computing and Applications 17(4), 399–403 (2008)
6. Fu, Y., Chai, T.Y.: Nonlinear multivariable adaptive control using multiple models and neural networks. Automatica 43(6), 1101–1110 (2007)
7. Zhang, H., Zhang, S., Yin, Y.: An Improved ELM Algorithm Based on EM-ELM and Ridge Regression. In: Sun, C., Fang, F., Zhou, Z.-H., Yang, W., Liu, Z.-Y. (eds.) IScIDE 2013. LNCS, vol. 8261, pp. 756–763. Springer, Heidelberg (2013)
8. Huang, G.B., Zhou, H., Ding, X., Zhang, R.: Extreme Learning Machine for Regression and Multiclass Classification. IEEE Transactions on Systems, Man, and Cybernetics - Part B: Cybernetics 42(2), 513–529 (2012)
9. Feng, G.R., Huang, G.B., Lin, Q.P., Gay, R.: Error minimized extreme learning machine with growth of hidden nodes and incremental learning. IEEE Trans. Neural Networks 20(8), 1352–1357 (2009)
10. Miche, Y., Sorjamaa, A., Bas, P., Simula, O., Jutten, C., Lendasse, A.: OP-ELM: Optimally Pruned Extreme Learning Machine. IEEE Trans. Neural Networks 21(1), 158–162 (2010)
11. Zhang, H.G., Zhang, S., Yin, Y.X.: A Novel Improved ELM Algorithm for a Real Industrial Application. Mathematical Problems in Engineering (2014), doi:10.1155/2014/824765
12. Rong, H.J., Ong, Y.S., Tan, A.H., Zhu, Z.X.: A fast pruned-extreme learning machine for classification problem. Neurocomputing 72(1-3), 359–366 (2008)
13. Castao, A., Fernandez-Navarro, F., Hervas-Martínez, C.: PCA-ELM: A Robust and Pruned Extreme Learning Machine Approach Based on Principal Component Analysis. Neural Processing Letters 37(3), 377–392 (2013)
14. Kirby, M., Sirovich, L.: Application of the Karhunen-Loeve Procedure for the Characterization of Human Faces. IEEE Trans. Pattern Analysis and Machine Intelligence 12(1), 103–108 (1990)
15. Good, R.P., Kost, D., Cherry, G.A.: Introducing a Unified PCA Algorithm for Model Size Reduction. IEEE Trans. Semiconductor Manufacturing 23(2), 201–209 (2010)
16. Yan, L.: A PCA-based PCM data analyzing method for diagnosing process failures. IEEE Trans. Semicond. Manuf. 19(4), 404–410 (2006)
17. Jolliffe, I.T.: Principal Component Analysis, 2nd edn. Springer (2002) ISBN: 978-0-387-22440-4
18. Tang, X.L., Zhang, L., Hu, X.D.: The support vector regression based on the chaos particle swarm optimization algorithm for the prediction of silicon content in hot metal. Control Theory & Application 26(8), 838–842 (2009)
19. Huang, G.B.: An insight into Extreme Learning Machines: Random Neurons, Random Features and Kernels. Cogn. Comput. (2014), doi:10.1007/s12559-014-9255-2

Wi-Fi and Motion Sensors Based Indoor Localization Combining ELM and Particle Filter

Xinlong Jiang[1,2,*], Yiqiang Chen[1], Junfa Liu[1], Yang Gu[1,2], and Zhenyu Chen[1,2]

[1] Institute of Computing Technology, Chinese Academy of Sciences, Beijing, China, 100190
[2] University of Chinese Academy of Sciences, Beijing, China, 100190
jiangxinlong@ict.ac.cn

Abstract. Indoor localization and tracking become more and more popular in our daily life. Due to the time consuming of Wi-Fi based location and error accumulation of motion sensor based location. A timely, enduring location system with high accuracy is difficult to be made. In this paper we present an effective combined model to handle this problem: we use ELM regression algorithm to predict position based on emotion sensor, and then combine the Wi-Fi location result to motion sensor based location result using particle filter. The experiments show that we can get higher accurate location result in every 200ms. And the trajectory is smoother as the real one than traditional Wi-Fi fingerprinting method.

Keywords: Wi-Fi Indoor location, Sensor Data, ELM, Particle Filter.

1 Introduction

Nowadays, Indoor Location and tracking are very important in our daily life. Given that many buildings are equipped with WLAN APs (Access Points). It becomes easy to locate in these indoor environments without any other infrastructure investment. A common idea of Wi-Fi based indoor location is finger print method [1]. It has two phases including offline training and online localization.

Although Wi-Fi signal can generate good prediction model for indoor localization, it cannot offer continuous location service as taking time to scan Wi-Fi signal. And the power consumption of Wi-Fi and communication module is several dozen times higher than the IMU (Inertial Measurement Unit) consists of gyroscopes and accelerometers [2].

In order to reduce the shortcoming of Wi-Fi based location, we can take use of sensors integrated in smartphone to measure people's motion state, including walking speed and orientation. But based on motion sensor alone, we can only estimate the position with high accuracy in a short time, it is difficult to offer localization independently for a long time period.

---

* Corresponding author.

© Springer International Publishing Switzerland 2015
J. Cao et al. (eds.), *Proceedings of ELM-2014 Volume 2*,
Proceedings in Adaptation, Learning and Optimization 4, DOI: 10.1007/978-3-319-14066-7_11

Therefore, in this paper we combine the advantages of both two methods and propose a Wi-Fi and motion sensors based indoor localization combining ELM and particle filter.

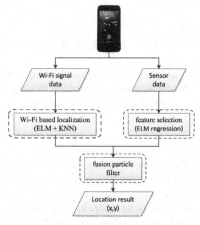

Fig. 1. Method flow chart

The architecture of our method is shown in figure 1. We design the method composed of three main parts: 1) sensor data based short-time localization with ELM (Extreme Learning Machine) regression. 2) Wi-Fi based localization with ELM classification and KNN (K-nearest neighbor). 3) Fusion particle filter module to eliminate the cumulative error caused by motion sensor based localization with Wi-Fi location result and offer a smooth location trajectory.

2   ELM Algorithm

As ELM model is competitively fast in offline learning and online prediction, we adopt ELM [3] to train an effective model which can offer real-time prediction with high precision depending on the motion sensor data. Our lab has done a lot of researches based on ELM and got good result [4] [5] [6].

ELM is an ANN (Artificial Neutral Network), especially an SLFN (Single Layer Feedforward Networks), which developed by Huang et al. Given N arbitrary distinct samples $(\mathbf{x}_i, \mathbf{t}_i) \in R^n \times R^m, i = 1, 2, \cdots, N$. Here, $\mathbf{x}_i$ is a $n \times 1$ input vector $\mathbf{x}_i = [x_{i1}, x_{i2}, \cdots, x_{in}]^T$ and $\mathbf{t}_i$ is a $m \times 1$ target vector $\mathbf{t}_i = [t_{i1}, t_{i2}, \cdots, t_{im}]^T$.

If an SLFN with $\tilde{N}$ hidden nodes can approximate these N samples with zero error, is then implies that there exist $\beta_i, \mathbf{a}_i$ and b_i such that

$$f_{\tilde{N}}(x_j) = \sum_{i=1}^{\tilde{N}} \beta_i G(a_i, b_i, x_j), \quad j = 1, \cdots, N \tag{1}$$

Equation (1) can be summarized as

$$H\beta = T \tag{2}$$

Where

$$H(\mathbf{a}_1, \cdots, \mathbf{a}_{\tilde{N}}, b_1, \cdots, b_{\tilde{N}}, \mathbf{x}_1, \cdots, \mathbf{x}_{\tilde{N}}) = \begin{bmatrix} G(\mathbf{a}_1, b_1, \mathbf{x}_1) & \cdots & G(\mathbf{a}_{\tilde{N}}, b_{\tilde{N}}, \mathbf{x}_1) \\ \vdots & \ddots & \vdots \\ G(\mathbf{a}_1, b_1, \mathbf{x}_N) & \cdots & G(\mathbf{a}_{\tilde{N}}, b_{\tilde{N}}, \mathbf{x}_N) \end{bmatrix} \tag{3}$$

$$\beta = \begin{bmatrix} \beta_1^T \\ \vdots \\ \beta_{\tilde{N}}^T \end{bmatrix}_{\tilde{N} \times m} \quad \text{and} \quad T = \begin{bmatrix} \mathbf{t}_1^T \\ \vdots \\ \mathbf{t}_N^T \end{bmatrix}_{N \times m} \tag{4}$$

According to [3], the hidden node parameters $\mathbf{a}_i$ and b_i (input weights and biases or centers and impact factors) of SLFNs need not be tuned during training and may simply be assigned with random values. Equation (2) then become a linear system and the output weights β are estimated as

$$\hat{\beta} = H^{\dagger} T = (H^T H)^{-1} H^T T \tag{5}$$

where $H^{\dagger}$ is the Moore-Penrose generalizes inverse of hidden layer output matrix H. equation (2) can be seen as a linear system, the least-squares solution of equation (5) is the answer of ELM.

From all the information above, we can see that ELM model can be used in regression and classification problems. In our application, the velocity and orientation are all continuous variable. So we use ELM as a regression model.

- **ELM model for velocity:**

  For velocity, the input vector is a vector with six features:
  $$\mathbf{x}_i = [acc_i.\mathrm{x}, acc_i.\mathrm{y}, acc_i.\mathrm{z}, gry_i.\mathrm{x}, gry_i.\mathrm{y}, gry_i.\mathrm{z}]^T$$
  and, the output vector is only a scalar value $\mathbf{t}_i = [v_i]^T$.

- **ELM model for orientation:**

  For orientation, the input is a vector with four features:
  $$\mathbf{x}_i = [ori_i.\mathrm{x}, ori_i.\mathrm{y}, ori_i.\mathrm{z}, mag_i.\mathrm{x}, .mag_i.\mathrm{y}, mag_i.\mathrm{z}]^T$$
  and, the output vector in also a scalar value $\mathbf{t}_i = [o_i]^T$.

  We have offline training and online prediction phases. During offline, we collect the sensors' data and the trajectory to train these two models mentioned above. During the online phases, we use sensors' data and models to determine v and o. Then we can use equation (6) to calculate the current position.

$$\begin{bmatrix} x_t \\ y_t \end{bmatrix} = \begin{cases} \begin{bmatrix} x_{t-1} \\ y_{t-1} \end{bmatrix} + \begin{bmatrix} T_w & 0 \\ 0 & T_w \end{bmatrix} * \begin{bmatrix} v_t & 0 \\ 0 & v_t \end{bmatrix} * \begin{bmatrix} \sin(o_i) \\ \cos(o_i) \end{bmatrix} & t > 1 \\ \begin{bmatrix} x_0 \\ y_0 \end{bmatrix} & t = 0 \end{cases} \tag{6}$$

By now, we can already track the trace. Given the initial position, in every T_w second, we collect the sensors' data and use these two ELM models to determine v and o. So that, the current position will be calculate by equation (6). But unfortunately, sensors on smart phone are not accurate enough to offer a data with high confidence for a long time.

3   Particle Filter for Sensor Based PDR

To solve this problem, we will use particle filter with some other regulation. Particle filter is an on-line posterior density estimation algorithm that estimating the posterior density of the state-space by directly implementing the Bayesian recursion equations. For a lot of position and navigation problems, a common motion model based filters can be applied. Models that are linear in the state dynamics and non-linear in the measurements are considered [7].

State transition equation:

$$\mathbf{x}_{t+1} = \mathbf{A}\mathbf{x}_t + \mathbf{B}_u\mathbf{u}_t + w_t \tag{7}$$

Observation equation:

$$\mathbf{y}_t = h(\mathbf{x}_t) + e_t \tag{8}$$

Where, $\mathbf{x}_t$ is state vector; $\mathbf{u}_t$ is measured inputs; w_t is process noise; $\mathbf{y}_t$ is measurements; and e_t is measurement error.

In our application, we change the model, equation (7) and (8), into a specific one, as shown in follows equations (9) and (10).

$$\begin{bmatrix} x_t \\ y_t \end{bmatrix} = \begin{bmatrix} 1 & 0 \\ 0 & 1 \end{bmatrix}\begin{bmatrix} x_{t-1} \\ y_{t-1} \end{bmatrix} + \begin{bmatrix} T_w & 0 \\ 0 & T_w \end{bmatrix}\begin{bmatrix} v_{x_t} \\ v_{y_t} \end{bmatrix} + Q \tag{9}$$

$$\begin{bmatrix} X_t \\ Y_t \end{bmatrix} = \begin{bmatrix} 1 & 0 \\ 0 & 1 \end{bmatrix}\begin{bmatrix} x_t \\ y_y \end{bmatrix} + R \tag{10}$$

In equation (9), $[x_t, y_t]^T$ denotes the state vector associated to each particle (position). T_w is the elapsed time between the $(t-1)^{th}$ and the tth measurements. Q is the process noise. Here the used process is a zero mean Gaussian noise with a $0.1m/s^2$ variance which is a realistic model of pedestrian movement.

Equation (10) is observation equation, which denotes the relation between the hidden states $[x_t, y_t]^T$ and the observation states $[X_t, Y_t]$. Because we directly observe the coordinate point from ELM regression algorithm, there is no measurement error in this process, so we set R to be 0.

A particle filter approximates the position at time t by a set of N particles. Each particle contains a position z_k^i and a probability $P_r[x_t|z_t^i]$. $P_r[x_t|z_t^i]$ indicates the importance of the i-th particle. In the case of an indoor movement, the following law has been retained:

$$P_r[x_t|z_t^i] = \frac{1}{\sqrt{2\pi}\sigma} \exp\left[-\frac{\left(X_{x_t} - X_{z_t^i}\right)^2}{2\cdot\sigma^2}\right] \tag{11}$$

With X_{x_t} is the position returned by the ELM regression algorithm, $X_{z_t^i}$ is the position of the ith particle at time t and σ is the measurement confidence. Here, $\sigma = 50$ was chosen.

According to the process of particle filter algorithm, there are four steps in a circle: prediction, correction, particle update and resampling. At each T_w second, we will determine a new position by the following steps:

Feature selection:

1. We collect sensors' data from mobile, and calculate the feature vector.
2. Then we use two ELM regression models to calculate the orientation o_i and velocity v_i.

Then we go to particle filter phase:

3. Prediction:
   During this step, every particle will propagate to a new one based on state transition equation (9). After that, all the particles move to new places.
4. Correction:
   Now, we have all the particles and we must give them the weight, according to o_i, v_i and location of last moment, we can calculate the observed value by equation (10). Then we can use the equation (11) to update the probability.
5. Particle Update:
   The weight update equation is given in [8]:

$$w_t^i = w_{t-1}^i \cdot P_r[x_t | z_t^i] \qquad (12)$$

   To obtain the posterior density function, we have to normalize those weights. After a few iterations, only a few particles can still alive due to others have a wrong direction. So we should go to resampling step to avoid having few remaining particle.
6. Resampling:
   The resampling step is the critical point for the PF. The fundamental idea is to remove the particles which have too low weight. Various resampling algorithms can be chose. We choose the SIR (Sequential Importance Resampling) for the reason that it is a simple one which is wildly used in SLAM (Simultaneous Location and Mapping) [9]. Take N samples with replacement from the set $\{x_t^i\}_{i=1}^N$, where the probability to take ith sample is w_t^i, Let $w_t^i = 1/N$.
7. Finally, we can calculate the new location result by $\hat{x}_t = \sum_{i=1}^N w_t^i x_t^i$. And then go to step 1) until stop.

4   Combined Model of Wi-Fi Fingerprint and Motion Particle Filter

By now, we have only used the sensors' data to locate, as the accuracy of sensors is limited, so they cannot offer long time location result with high accuracy. Thus, we present a confusion particle filter, which can combine the advantages of motion sensor based location and Wi-Fi based localization.

Wi-Fi fingerprint based localization method includes offline learning and online prediction. During the offline learning we collect fingerprints consisted of coordinate $\{x, y\}$ and Wi-Fi RSSI (Received Signal Strength Indication) $\{r_1, r_2, ..., r_k\}$ measured at $\{x, y\}$. During the online phase, we use ELM classification and KNN algorithm to locate with finger print information.

Using ELM, we can find the K nearest neighbor points, but we cannot know the distance because of the ELM will offer the nearest points but not the weight. So we use the Euclidean distance of signal strength to determine the weight according to equation (13) and (14).

$$dis_k = \sqrt{\sum_{i=1}^{n}\left(rss_{k_i} - rss_i'\right)^2} \tag{13}$$

$$w_k = \frac{\sum_{i=1}^{K} dis_i - dis_k}{(n-1)\sum_{i=1}^{K} dis_i} \tag{14}$$

When we use smart phone to do Wi-Fi based indoor location, it takes a little while to collect Wi-Fi data. So we can get a location result in every T_{wifi} second. To eliminate the cumulative error caused by sensor location, we add the Wi-Fi location result to the particle filter, as shown in figure 2. When go to the correction step, we should inquiry whether there comes a Wi-Fi location result. If yes, we can use the Wi-Fi location point as the observed point to calculate all particles' weight. If not, we still use the result predicted by motion sensor data to calculate. After the confusion, our system can offer a highly accurate location result with limited time.

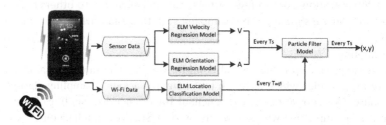

Fig. 2. Flow chart of the total system

5   Experiments and Performance Evaluation

Now, we will evaluate the performance of our method through all the experiments running on smart phones and a computer with following configuration:

Smart Mobile Phones:

No.	Version	OS	CPU(Hz)	RAM	ROM
1	HTC Desire S	Android, v2.3	1G	768MB	1GB
2	Samsung Galaxy S3	Android, v4.0	1.4G	1GB	16GB

Computer:
Operation System: Windows XP Professional SP3; CPU: Intel Pentium(R) 4 CPU; Main Frequency: 3.2 GHz; RAM: 2G

5.1 Experimental Design and Data Preparing

In order to evaluate the performance, we build a test platform in our work space. It is about $100m^2$ (8.5m * 12m). For the reason we apply ELM algorithm in both of Wi-Fi based localization and senor based localization, we will collect data to train the model during offline phase.

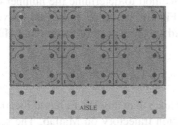 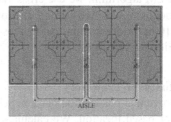

Fig. 3. (a) Finger-print points is in yellow color; (b) Pre-setting route

For the Wi-Fi fingerprinting based location, we select 36 points to build a signal distribution map, shown in Figure 3(a). We register four times at each point, and collect Wi-Fi data for three times then use the mean value.

Concerning sensor data, we plan a fixed route including straights and curves, shown in Figure 3(b). People walk along this trajectory with any speed: slow, middle, fast, and even run. After collecting the original sensor data, we use Cubic Spline Interpolation to smooth the trajectory. Then we use sliding window to segment the trajectory with window length T_w.

During the online phase, people walk along the trajectory, shown in Figure 3(b). Then we will use our method to predict the path people walked, and evaluate the performance by comparing the true path with the default trajectory.

5.2 Experimental Performance

Determine the Time Window Length T_w

Firstly, we will determine the window length T_w. For the reason that the sensor frequency is $f_s \approx 65Hz$. In order to void the influence caused by noise, we use the mean value to act as the features. So we should ensure that there are more than 10 samples in each window. T_w should be bigger than 160 millisecond.

We use MSE (Mean Square Error) [10] to measure the accuracy in ELM regression algorithms. The results are shown in table 1. We can see that, the lowest MSE in both orientation and velocity can be got when T_w is 200 ms.

Table 1. MSE of orientation and velocity in different window length

Window length	200m s	400m s	600m s	800m s	1000 ms
MSE of o(orientation)	25.05 29	27.85 70	26.87 27	26.57 59	26.45 03
MSE of v(velocity)	42.01 71	56.27 87	58.21 29	64.98 13	62.09 68

Parameter of ELM Regression Models:
For ELM algorithm, we use cross validation method to determine the number of hidden nodes $\tilde{N}$ and Activation Function $G(x)$. Finally, we set $\tilde{N}$ to be **500** for velocity and **1000** for orientation. And choose **Sigmoid** function for orientation and **Hardlim** function for velocity.

Evaluate the Trajectory:
To prove that our method has a better performance than any Wi-Fi based location and Sensor based location. We calculate the location trajectory on Matlab. The result of sensor based location, Wi-Fi only location and our method are shown in figure 4.

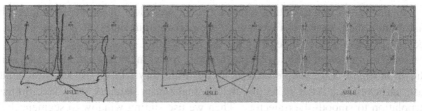

Fig. 4. Trajectory of different data: (a) Trajectory of sensor based location; (b) Trajectory of Wi-Fi based location; (c) Trajectory of our system

We can see that trajectory of sensor based localization is not so good. As time goes on, the trajectory is getting more and more far away from the pre-setting one. The Wi-Fi based location result is better than sensor based, but the location time consumption is too much that only 16 separate location results can be get for the whole walking process. Combining these two methods, we can see our system's result is really better than those methods mentioned above. It can offer a timely and highly accurate location result.

6   Conclusion

In this paper, we have presented a Wi-Fi and motion sensors based indoor localization combining ELM and particle filter. The reason we raised this problem is that: as the development of smart phone, a lot of sensors are integrated in it. So we can easily get sensor data and Wi-Fi signals which can be taken full use to do indoor localization.

But any single method cannot get satisfactory result. So we combine Wi-Fi based location and sensor based inertial navigation. And we also use particle filter and ELM algorithm to improve the location accuracy. From the experiments result, we can see that our method have achieved a good performance.

Acknowledgments. This work is supported by NSFC General Program under grant No. 61173066, and No.41201410, Guangdong Province Funds Supported Project for the Development of Strategic Emerging Industry No.2011912030.

References

1. Bahl, P., et al.: RADAR: An in-building RF-based user location and tracking system. In: Proceedings of the IEEE Nineteenth Annual Joint Conference of the IEEE Computer and Communications Societies, INFOCOM 2000, vol. 2, pp. 775–784. IEEE (2000)
2. Carroll, et al. An analysis of power consumption in a smartphone. In: Proceedings of the 2010 USENIX Conference on USENIX Annual Technical Conference (2010)
3. Huang, G.-B., et al.: Extreme learning machine: a new learning scheme of feedforward neural networks. In: Proceedings of the 2004 IEEE International Joint Conference on Neural Networks, vol. 2, pp. 985–990. IEEE (2004)
4. Liu, J., Chen, Y., et al.: SELM: Semi-supervised ELM with application in sparse calibrated location estimation. Neurocomputing 74(16), 2566–2572 (2011)
5. Liu, J., Guyang, et al.: Incremental Localization in WLAN Environment with Timeless Management. Chinese Journal of Computers 36(7), 1448–1455 (2013)
6. Chen, Z., et al.: Online Sequential ELM based Transfer Learning for Transportation Mode Recognition. In: The 6th IEEE International Conference on Cybernetics and Intelligent Systems (CIS 2013), pp. 78–83 (2013)
7. Gustafsson, F., et al.: Particle filters for positioning, navigation, and tracking. IEEE Transactions on Signal Processing 50(2), 425–437 (2002)
8. Arulampalam, M.S., Maskell, S., Gordon, N., Clapp, T.: A tutorial on particle filters for online nonlinear/non-Gaussian Bayesian tracking. IEEE Transactions on Signal Processing 50(2), 174–188 (2002)
9. Thrun, S.: Simultaneous localization and mapping. In: Jefferies, M.E., Yeap, W.-K. (eds.) Robot. & Cogn. Approach. to Spat. Map. STAR, vol. 38, pp. 13–41. Springer, Heidelberg (2008)
10. Tuchler, M., Singer, A.C., Koetter, R.: Minimum mean squared error equalization using a priori information. IEEE Transactions on Signal Processing 50(3), 673–683 (2002)

Online Sequential Extreme Learning Machine for Watermarking

Ram Pal Singh[1], Neelam Dabas[2], Vikash Chaudhary[3], and Nagendra[3]

[1] Department of Computer Science,DDUC, University of Delhi, Delhi, India
[2] Department of Computer Science, University of Delhi, Delhi, India
[3] Department of Computer Science,BNC, University of Delhi, Delhi, India
rprana@gmail.com

Abstract. Protecting and securing an information of digital media is very crucial due to illegal reproduction and modification of media has become an acute problem for copyright protection now a day. A Discrete Wavelet Transform (DWT) domain based robust watermarking scheme with online sequential extreme learning machine (OSELM) has been implemented on different images. The proposed scheme which combine DWT domain with OSELM and watermark is embedded as an ownership information. Experimental results demonstrate that the proposed watermarking scheme is imperceptible and robust against image processing and attacks such as blurring, cropping, noise addition, rotation, scaling, scaling-cropping, sharpening etc. Performance and efficacy of algorithm on watermarking scheme is determined and calibrated results are compared with other machine learning methods.

Keywords: BER,ELM, OSELM, PSNR, Watermarking.

1 Introduction

In recent years, there is a rapid development of multimedia including images, audio and video, are reprinted, duplicated and easily redistributed over internet. Duplication and redistribution of digital media has become very serious problem thereby needs to protect intellectual property right (IPR) of digital media. Therefore, digital watermarking has received considerable attention for finding unauthorized use of digital media. In digital watermarking,a watermark or a trademark or a sequence is embedded into image for copyright protection and embedded watermark may be extracted from watermarked media in order to prove ownership. An image is transformed in frequency domain by using DWT for digital watermarking where the coefficients of image are modulated by adding an additional information. In image authentication techniques, in which visually recognizable pattern is embedded as watermark in low frequency sub-band in DWT domain, gives a trade off between imperceptibility and robustness. A digital watermarking scheme must have following requirement (i) imperceptible or transparency (ii) difficult to extract without affecting quality of an image (iii) should have robustness against image processing,conventional and geometrical attacks. Therefore, developing a computational algorithm which exhibits these requirements is not so easy job. The number of information hiding schemes have been used and reported in literature [4]. Recently, computational intelligence techniques have been widely used and out of these, neural networks and support vector machine (SVMs)[5] playing very dominating role but are prone to unavoidable drawbacks.

Most of the drawbacks in neural networks based back propagation and support vector machine, ELM learning algorithm tries to address them. ELM [8]-[14] uses a very generalized type of *single layer feed forward networks* (SLFNs) has become attractive among the researchers working in different research areas.In this paper, we explore the possibility of ELM and OSELM for information hiding in natural images where every pixel has high relevance to its neighbours, it can be predicted by its neighbours [6] using ELM and OSELM and this pixels relevance help in extraction of watermark. In ELM, the input weights and bias of hidden layer nodes are randomly generated and fixed while updating the output weight between hidden layer and output layer. In OSELM, learning is performed by writing data one-by-one or trunk by trunk with fixed or varying size and the input weights and bias of hidden nodes are randomly generated and fixed but output weights between hidden layer and output layer is updated or changed only when new data arrives rather than all data[15].Further,we propose an implementation of watermarking scheme based on ELM and OSELM in DWT domain. The experimental results show better imperceptibility and good robustness against different attacks.

Rest of this paper is organized as follows. We briefly given theories about ELM and OSELM in Section 2. Watermarking schemes through ELM and OSELM is described in Section 3.Experimental results are given in Section 4. Finally, conclusions are given in last Section 5.

2   Brief Theories of ELM and OSELM

2.1   ELM

Over last decades, batch learning algorithms for machine learning have wide range of application in different kind of research areas such as watermarking and information hiding [6],[16] etc. Huang et al.[8]-[14] has proposed new machine learning algorithm, extreme learning machine for single hidden layer feedforward neural networks(SLFNs).In this algorithm, input layer weights do not need to be tuned iteratively and may be generated randomly, however, the output weights are determined analytically using least squares method. This algorithm has fast learning speed and high learning accuracy with good generalization ability. ELM has been upgraded to online sequential extreme leaning machine (OSELM)[15] which is in detail discussed in subsequent section of this paper.

ELM is a batch learning type of algorithm having single hidden layer feedforward neural networks(SLFNs). Given N arbitrary distinct data samples $\{x_i, t_i\}_i^N$, the output function of SLFNs with $\hat{L}$ number of hidden nodes can approximate

N input samples with zero error then β_i, a_i and b_i hold such that

$$f_L(x) = \sum_{i=1}^{\hat{L}} \beta_i g_i(x) = \sum_{i=1}^{\hat{L}} \beta_i G(a_i, b_i, x), a_i \in \Re^n, b_i \in \Re, \beta_i \in \Re^m \quad (1)$$

where $g_i = G(a_i, b_i, x)$ output function of ith hidden node and a_i is weight vector connecting i^{th} hidden neuron and input neuron and b_i is threshold of ith hidden neuron, are learning parameters of hidden nodes. β_i the weight vector connecting i^{th} hidden node to output neuron. For additive nodes, the activation function $g(x) : R \to R$ for i^{th} hidden node, g_i is defined as

$$g_i = G(a_i, b_i, x) = g(a_i \cdot x + b_i), a_i \in \Re^n, b_i \in \Re \quad (2)$$

The above two equations can be written in matrix form as

$$H\beta = T \quad (3)$$

where H is called the hidden layer output matrix of the SLFNs[10]and T is the target vector.

As described in [13], the parameter of hidden layer node neuron as weights a_i and bias b_i need not to be adjusted again and again but these are randomly assigned and fixed. Therefore, for known values of hidden layer output matrix H and output matrix T, the solution of output parameter β can be obtained as

$$\hat{\beta} = H^{\dagger}T \quad (4)$$

where $H^{\dagger}$ is the Moore-Penrose generalized pseudoinverse [17] of the hidden layer output.

2.2 OSELM

As discussed in [18],by OSELM data may be learned one-by-one or trunk-by trunk or a block of data and not using data which has already been used for training of machine model.Batch learning based algorithm ELM may be reformulated for this case so as to make it online sequential.The output weight matrix is $\hat{\beta}$ is a least-squares solution of (4).The orthogonal projection method can be used for determining Moore-Penrose generalized inverse as $H^{\dagger} = (H^T H)^{-1} H^T$ provided $H^T H$ is nonsingular in nature. Substituting the value $H^{\dagger} = (H^T H)^{-1} H^T$ into (4), sequential implementation of the least-squares solution is possible in OSELM. Therefore, (4) becomes as

$$\hat{\beta} = (H^T H)^{-1} H^T T \quad (5)$$

For OSELM, consider initial training data sample (x_i, t_i), $\forall i = 1, \ldots, L_0$ and $L_0 \geq \hat{L}$, for batch learning ELM, the solution is obtained by minimizing $\|H_0\beta - T_0\|$ as

$$\beta^0 = C_0^{-1} H_0^T T_0 \quad (6)$$

where $C_0 = H_0^T H_0$. As detail analysis of OSELM is given in [18],on new chunk of data samples arrive as $(x_i, t_i), \forall i = L_0 + 1, \ldots, L_0 + L_1$, then

$$C_1 = \begin{pmatrix} H_0 \\ H_1 \end{pmatrix}^T \begin{pmatrix} H_0 \\ H_1 \end{pmatrix} = (H_0^T \ H_1^T) \begin{pmatrix} H_0 \\ H_1 \end{pmatrix} = C_0 + H_1^T H_1 \qquad (7)$$

Therefore, the new model updated parameter due to arrival of new chunk of data sample

$$\beta^{(1)} = C_1^{-1} \begin{pmatrix} H_0 \\ H_1 \end{pmatrix}^T \begin{pmatrix} T_0 \\ T_1 \end{pmatrix} = C_1^{-1}(C_1\beta^{(0)} - H_1^T H_1\beta^{(0)} + H_1^T T_1) \qquad (8)$$

$$= \beta^{(0)} + C_1^{-1}H_1^T(T_1 - H_1\beta^{(0)})$$

With generalization and recursive approach, as new data sample arrives, a recursive least-squares algorithm solution can be written as

$$C_{k+1} = C_k + H_{k+1}^T H_{k+1} \qquad (9)$$

$$\beta^{(k+1)} = \beta^{(k)} + C_{k+1}^{-1}H_{k+1}^T(T_{k+1} - H_{k+1}\beta^{(k)}) \qquad (10)$$

From (10), it is clear that C_{k+1}^{-1} is used for computation of $\beta^{(k+1)}$ from $\beta^{(k)}$ in (10), using Woodbury formula, an updated formula for $K_{k+1} = C_{k+1}^{-1}$ for machine learning for new arrival data sample after simplifying can be written as

$$K_{k+1} = K_k + K_k H_{k+1}^T(I + H_{k+1}K_k H_{k+1}^T)^{-1} \times H_{k+1}^T K_k \qquad (11)$$

$$\beta^{(k+1)} = \beta^{(k)} + K_{k+1}H_{k+1}^T(T_{k+1} - H_{k+1}\beta^{(k)}) \qquad (12)$$

The value of an updated weight $\beta^{(k+1)}$ for new data arrival at $(k+1)^{th}$ trunk can be determined iteratively with recursive formula (13) and accordingly output estimation function (1) may be determined for ELM for updated weight $\beta^{(k+1)}$ at $(k+1)^{th}$ trunk,an output estimation function for OSELM is given by

$$f_{(OSELM)}(x) = \sum_{i=1}^{\hat{L}} \beta_i^{(k+1)} G(a_i, b_i, x), a_i \in \Re^n, b_i \in \Re, \beta_i \in \Re^m \qquad (13)$$

$$= \sum_{i=1}^{\hat{L}} \beta_i^{(k)} + K_{k+1}H_{k+1}^T(T_{k+1} - H_{k+1}\beta^{(k)})g(a_i \cdot x, b_i)$$

3   Watermarking Scheme in DWT Domain

In natural image analysis, it is found that every pixel has relationship to its neighbour, therefore, it can be predicted by its neighbours[2].ELM and OSELM may be used for prediction of this relationship used for watermark embedding and extraction.

3.1   ELM and OSELM Based Watermark Embedding

Let us assume there is a color image $I = [R, G, B]$ with size $m \times n$, where R,G and B are the red, green and blue channel component of color image, respectively.An image to be embedded as watermark,w, is a binary image with the size $p \times q$. Here watermark logo is embedded into blue channel as human vision system (HSV)[19] is insensitive to blue channel. The embedding algorithm is:

1. The logo watermark is permuted and reshaped into line w of size $p \times q$, as $w = \{w_i\}_{i=1,...,p \times q}$.
2. The blue channel components, B, of the color image, I, to be watermarked transferred through (DWT) and its low frequency sub-band denoted as B.
3. For training ELM and OSELM for regression and reference pixel position $p_t = \{i_t, j_t\}$, the training data set may be extracted from every 3×3 active window obtained after applying DWT and training data set is obtained as:

$$S = d_t | d_t = [B(i_{t-1}, j_{t-1}), B(i_{t-1}, j_t), B(i_{t-1}, j_{t+1}), B(i_t, j_{t-1}), \quad (14)$$
$$B(i_t, j_{t+1}), B(i_{t+1}, j_{t-1}), B(i_{t+1}, j_t), B(i_{t+1}, j_{t+1})]$$

where $B(\cdot, \cdot)$ is the pixel value in the B-channel of color image. And the pixels set $B(i, j)$ are the training objective of ELM and OSELM and their respective regression output is given by (20) and (21)

$$y = f_{ELM}(d_t) = \sum_{i \in m \times n} \beta_i g(a_i \cdot d_t, b_i), a_i \in \Re^n, b_i \in \Re, \beta_i \in \Re^m \quad (15)$$

$$y = f_{OSELM}(d_t) = \sum_{i \in m \times n} \beta_i^{(k+1)} G(a_i, b_i, d_t), a_i \in \Re^n, b_i \in \Re, \beta_i \in \Re^m \quad (16)$$
$$= \sum_{i \in m \times n} \beta_i^{(k)} + K_{k+1} H_{k+1}^T (T_{k+1} - H_{k+1}\beta^{(k)}) g(a_i \cdot d_t, b_i)$$

where $\beta^{(k+1)}$ is output weight vector in $(k + 1)^{th}$ trunk of new data for OSELM.

4. For each embedding position (i_t, j_t), eight pixels in 3×3 active window are collected to form training data set in blue channel color image as

$$S' = d_t' | d_t' = [B(i_{t-1}, j_{t-1}), B(i_{t-1}, j_t), B(i_{t-1}, j_{t+1}), B(i_t, j_{t-1}), \quad (17)$$
$$B(i_t, j_{t+1}), B(i_{t+1}, j_{t-1}), B(i_{t+1}, j_t), B(i_{t+1}, j_{t+1})]$$

we obtain predicting pixel at each embedding position (i_t, j_t),where $t = 1,...,p \times q$ and trained output from ELM and OSELM

$$y_t' = f_{(ELM)}(d_t') = \sum_{i=1}^{\hat{L}} \beta_i g(a_i \cdot d_t', b_i), a_i \in \Re^n, b_i \in \Re, \beta_i \in \Re^m \quad (18)$$

$$y_t' = f_{(OSELM)}(d_t') = \sum_{i \in m \times n} \beta_i^{(k+1)} G(a_i, b_i, d_t), a_i \in \Re^n, b_i \in \Re, \beta_i \in \Re^m \quad (19)$$

$$= \sum_{i \in m \times n} \beta_i^{(k)} + K_{k+1} H_{k+1}^T (T_{k+1} - H_{k+1} \beta^{(k)}) g(a_i \cdot d_t', b_i)$$

Now after comparing predicted value $y_t = f(d_t')$ and actual value $B(i_t, j_t)$, one can embed watermark according to following equation (21) as

$$B^*(i_t, j_t) = \begin{pmatrix} max((B(i_t, j_t), f^*(d_t')(1 + \eta_t)) \\ if \mapsto w_t = 1 \\ min((B(i_t, j_t), f^*(d_t')(1 - \eta_t)) \\ if \mapsto w_t = 0 \end{pmatrix} \quad (20)$$

where $f^*(d_t') = f_{(ELM)}$ for ELM, $f^*(d_t') = f_{(OSELM)}$ for OSELM and η_t is the embedding strength and $B^*(i_t, j_t)$ is the modulated wavelet coefficient in low frequency sub-band domain at position (i_t, j_t).

5. After embedding blue channel with desire watermark, B-channel is combined with R and G color channels and resultant image is a final watermarked image.

3.2 Watermark Detection

For watermark detection, the original watermark image is not necessary and extraction algorithm is

1. The blue channel B of watermarked color image is decomposed through DWT and low frequency component is designated by B'.
2. For each pixel reference position $p_t = \{i_t, j_t\}$, dataset can be constructed by collecting 3×3 active window. The formed data set $\tilde{S}$ is based on reference pixel $\tilde{B}(i_t, j_t)$, is

$$\tilde{S} = \tilde{d}_t | \tilde{d}_t = [\tilde{B}(i_{t-1}, j_{t-1}), \tilde{B}(i_{t-1}, j_t), \tilde{B}(i_{t-1}, j_{t+1}), \tilde{B}(i_t, j_{t-1}), \quad (21)$$
$$\tilde{B}(i_t, j_{t+1}), \tilde{B}(i_{t+1}, j_{t-1}), \tilde{B}(i_{t+1}, j_t), \tilde{B}(i_{t+1}, j_{t+1})]$$

3. For each embedding position, by using well-trained based ELM and OSELM, we can determined corresponding output y' as:

$$y' = f_{(ELM)}(\tilde{d}_t) = \sum_{i=1}^{\hat{L}} \beta_i g(a_i \cdot \tilde{d}_t, b_i), a_i \in \Re^n, b_i \in \Re, \beta_i \in \Re^m \quad (22)$$

$$y' = f_{(OSELM)}(\tilde{d}_t) = \sum_{i \in m \times n} \beta_i^{(k+1)} G(a_i, b_i, \tilde{d}_t), a_i \in \Re^n, b_i \in \Re, \beta_i \in \Re^m \quad (23)$$

$$= \sum_{i \in m \times n} \beta_i^{(k)} + K_{k+1} H_{k+1}^T (T_{k+1} - H_{k+1} \beta^{(k)}) g(a_i \cdot \tilde{d}_t, b_i)$$

The watermark can be extracted by comparing for $f^*(\tilde{d}_t) = f_{(ELM)}$ for ELM, $f^*(\tilde{d}_t) = f_{(OSELM)}$ for OSELM between predicted pixel value blue channel B by trained ELM and OSELM output and actual pixel value as:

$$w'_t = \begin{pmatrix} 1, if \mapsto \tilde{B}(i_t, j_t) > f^*(\tilde{d}_t) \\ 0, \quad else \end{pmatrix} \tag{24}$$

4. Finally, one-dimensional watermark bit sequence $w_1, w_2, \ldots, w_{p \times q}$ is converted into a two-dimensional watermark logo image w'. The detection result may be verified by parameter BER (bit error rate) and PSNR (peak signal to noise ratio) between image $I(x)$ and watermarked image $I'(x)$ as

$$BER = \sum_{t=1}^{p \times q} (w_t \oplus w'_t)/p \times q \tag{25}$$

$$PSNR = 10 \log_{10} \frac{255 \times 255}{\sum_{i=0}^{m-1} \sum_{j=0}^{n-1} (x_{ij} - x'_{ij})^2}/(m \times n) \tag{26}$$

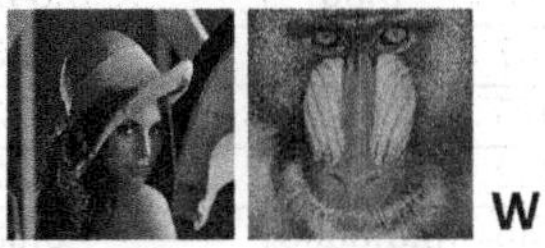

Fig. 1. Lena and Baboon images used for watermarking and watermark image

Fig. 2. Lena and Baboon watermarked images with no attack with (PNSR=53.06dB, 48.5dB) at activation function RBF

Table 1. The comparison results PSNR (dB) values under different attacks on *Lena* watermarked image with ELM and OSELM

Attacks	ELM(RBF))	OSELM	Chang's method[2]	Chen's method[3]
JPEG	32.77	34.68	30.89	27.25
Blurring	30.73	31.22	29.36	27.75
Cropping	18.20	18.94	11.68	17.27
Noise	29.90	30.95	30.29	28.78
Rotation	24.77	24.89	14.95	15.80
Scaling	43.02	43.08	29.80	27.13
Scaling-Cropping	19.54	19.98	15.93	15.29
Sharpening	36.02	36.32	27.83	33.12

Table 2. The comparison results PSNR (dB) values under different attacks on *Baboon* watermarked image with ELM and OSELM

Attacks	ELM(RBF))	OSELM	Chang's method[2]	Chen's method[3]
JPEG	30.33	14.61	23.46	31.75
Blurring	29.98	23.54	21.40	25.50
Cropping	15.90	15.10	11.35	15.74
Noise	32.79	24.91	30.32	33.10
Rotation	20.58	20.58	13.83	14.48
Scaling	30.25	30.34	21.14	28.32
Scaling-Cropping	17.60	17.60	15.12	15.07
Sharpening	19.54	19.49	17.54	24.60

Table 3. The comparison results BER values under different attacks on *Lena* watermarked image with ELM and OSELM

	ELM(RBF))	OSELM	Shen[6]	Kutter's Method[7]	Yu's Method[1]
JPEG	0.176	0.154	0.36	0.3425	0.3512
Blurring	0.0321	0.0157	0.0	0.0537	0.0113
Noise	0.0179	0.0075	0.0350	0.0737	0.0058
Rotation	0.0407	0.0332	0.0963	0.1225	0.1021
Scaling	0.0407	0.0245	0.0.0605	0.08	0.07

Table 4. Attacked Lena images and corresponding PSNR values (dB) and extracted binary ownership as a watermark from attacked watermarked Lena image with ELM

Attacks:	Blurring	Cropping	JPEG	Noise	Rotation	Scaling	Scale+Cropp	Sharp
Image								
PSNR	30.73	18.20	32.77	29.90	24.77	43.02	19.54	36.02
Attacks	Blurring	Cropping	JPEG	Noise	Rotation	Scaling	Scale+Cropp	Sharp
Ownership	W	W	W	W	W	W	W	W
BER	0.0321	0.104	0.176	0.0179	0.0407	0.0021	0.0498	0.0876

Table 5. Attacked Lena images and corresponding PSNR values (dB) and extracted binary ownership as a watermark from attacked watermarked Lena image with OSELM

Attacks:	Blurring	Cropping	JPEG	Noise	Rotation	Scaling	Scale-Cropp	Sharp
Image								
PSNR	31.22	18.94	34.68	30.95	24.89	43.08	19.98	36.32
Attacks	Blurring	Cropping	JPEG	Noise	Rotation	Scaling	Scale+Cropp	Sharp
Ownership	W	W	W	W	W	W	W	W
BER	0.0301	0.101	0.147	0.0112	0.0307	0.0020	0.03465	0.0667

4   Experimental Results

The experiment is performed on original color images *Lena* and *Baboon* of size 512×512. An Images Lena and Peppers representing low complexity as having smooth regions while Baboon has high complexity because it contains region of more complex texture. A watermark image is shown in Fig. 1 of size 32×32 has been used as a watermark used for embedding purpose in color images used in experimental results. All images used here for experimental purpose are decomposed using two-level wavelet transformation in frequency domain having sub-image LL_2 of size 128×128 used for watermark embedding. Since in this watermarking scheme color images are used, watermark logo is embedded into blue channel as human vision system (HSV)[19] is insensitive to blue channel which lead to improve imperceptibility of scheme.

In order to evaluate the robustness of watermarking scheme where ELM and OSELM algorithm are used as learning machine, the number of attack are conducted on watermarked image Lena. After subjecting image with number of attacks, embedded watermark has been extracted with the help of ELM and OSELM by learning neighbouring relationship among pixels. It is found that neighbouring relationship among pixels remains unchanged even though watermarked images are subjected to attacks. Therefore, this pixels relationship used for prediction of pixels relationship and thereby embedded watermark is extracted through these machine learning techniques.We conducted experiment and values of PSNR and BER are calculated and results are compared with already existed mathods[1] [2][6][7][10] as shown in Tables 1-4. It was found that calculated parameters show the efficacy of our used machine learning method in the form of robustness and imperceptibility. Experimental results show that our implemented watermarking schemes based on ELM and OSELM outperformed Yu's method [1], Chang's method[2] and Kutter's method[7] and against different attacks including blurring, JPEG, rotation, cropping, scaling, scaling-cropping, sharpening etc.

5   Conclusion

A robust and imperceptible watermarking scheme based on ELM and OSELM proposed for coptright protection and authentication of ownership. The proposed schemes are composed of DWT domain with ELM and OSELM in low frequency band LL_2 used in two-level wavelet transformation. The watermark is embedded into low frequency band of wavelet domain based on ELM and OSELM regression training. The concept of high correlation with its neighbour pixels is used for prediction. This relationship can be predicted by training machine learning for regression and used for watermarking scheme. When watermarked image is subject to attacks, it is found this relationship among pixels do remain unchanged and it can be learned by training process of ELM and OSELM for regression. This process can be used for watermark extracted from watermarked image. As machine learning methods used have high generalization ability, there is always possibility of correct extraction of watermark unless the watermarked image is severely attacked. Experimental results show our implemented methods outperformed when compared to other learning methods.

References

1. Yu, P.T., Tsai, H.-H., Sun, D.-W.: Digital watermarking based on neural networks for color images. Signal Processing 81(3), 663–671 (2001)
2. Chang, C.-Y., Wang, H.-J., Pan, S.-W.: A robust DWT-based copyright verification scheme with Fuzzy ART. The Journal of Systems and Software 82, 1906–1915 (2009)
3. Chen, T.H., Horng, G., Lee, W.B.: A publicaly verifyable copyright-proving scheme resistant to malicious attacks. IEEE Trans. on Industrial Electronics 52(2), 327–334 (2005)
4. Mohanty, S.P.: Digital Watermarking: A tutorial review (1999)
5. Vapnik, V.N.: Statistical Learning Theory. John Wiley & Sons, New York (1998)
6. Shen, R.-M., Fu, Y.G., Lu, H.T.: A novel image watermarking scheme based on support vector regression. The Journal of Systems and Software 78, 1–8 (2005)
7. Kutter, M., Jordan, F., Bossen, F.: Digital signature of color images using amplitude modulation. Journal of Electronic Imaging 7(2), 326–332 (1998)
8. Huang, G.B., Zhu, Q.Y., Siew, C.K.: Extreme learning machine: theory and applications. Neurocomputing 70, 489–501 (2006)
9. Zhu, Q.-Y., Qin, A., Suganthan, P., Huang, G.B.: Evolutionary extreme Learning Machine. Neurocomputing 70, 1759–1763 (2005)
10. Huang, G.-B., Chen, L., Siew, C.-K.: Universal approximation using incremental constructive Feed forward networks with random hidden nodes. IEEE Trans. Neural Networks 17(4), 879–892 (2006)
11. Huang, G., Chen, L.: Enhanced random search based incremental extreme learning machine. Neurocomputing 71(16-18), 3460–3468 (2008)
12. Huang, G., Chen, L.: Convex inceremental extreme learning machine. Neurocomputing 70, 3056–3062 (2007)
13. Huang, G.-B., Wang, D.H., Lan, Y.: Extreme learning machines: a survey. International Journal of Machine Learning & Cybernetic 2, 1107–1122 (2011)
14. Feng, G., Huang, G.-B., Lin, Q., Ray, R.: Error minimized extreme learning machine with growth of hidden nodes and incremental learning. IEEE Trans. Neural Networks 20(8), 1352–1357 (2009)
15. Ye, Y., Squartini, S., Piazza, F.: Online sequential extreme learning machine in nonstationary environments. Neurocomputing 116, 94–101 (2013)
16. Tsai, H.-H., Sun, D.-W.: Color image watermark extractionbased on support vector machine. Information Science 177, 550–569 (2007)
17. Rao, C.R., Mitra, S.K.: Generalized Inverse of Matrices and its Applications. Wiley, New York (1971)
18. Liang, N.-Y., Huang, G.-B., Saratchnadran, P., Sundarajan, N.: A fast and accurate online sequential learning algorithm for Feedforward Networks. IEEE Trans. Neural Networks 17(6), 1411–1423 (2006)
19. Liu, C.C., Tsai, H.-H.: Wavelet-based image watermarking with visibility range estimation based on HSV and neural networks. Pattern Recognition (2010), doi:10.1016.

Adaptive Neural Control of a Quadrotor Helicopter with Extreme Learning Machine

Yu Zhang[1,*], Bin Xu[2], and Hongbo Li[3]

[1] School of Aeronautics and Astronautics, Zhejiang University,
Zhejiang, Hangzhou 310027, China
[2] School of Automation, Northwestern Polytechnical University,
Shaanxi, Xi'an 710072, China
[3] Department of Computer Science and Technology, Tsinghua University,
Beijing 100084, China
zhangyu80@zju.edu.cn,smileface.binxu@gmail.com,hbli@mail.tsinghua.edu.cn

Abstract. A new controller design method for the quadrotor helicopter based on the extreme learning machine (ELM) is proposed. ELM based neural controller and sliding mode controller are combined to stabilize the attitude systems of quadrotors including roll, pitch and yaw. A single hidden layer feedforward network based on ELM with fast learning speed is used to approximate the unmodeled nonlinear attitude dynamics while the sliding mode controller is employed to eliminate the external disturbances. In this way, precise dynamic model and prior information of disturbances are not needed. The simulation study is presented to show the effectiveness of the proposed control algorithm.

Keywords: Quadrotor helicopter, Extreme learning machine, Flight control, Neural networks.

1 Introduction

In recent years, a special kind of rotorcraft called quadrotor which has a compact form is becoming more and more popular than conventional rotorcrafts as they are mechanically and dynamically simpler and easier to control. In spite of this, the quadrotor is still a dynamically unstable system and control of quadrotor is also challenging work because of the inherent system characteristics such as nonlinearities, cross couplings due to the gyroscopic effects and underactuation[1]. Besides, most applications require a vehicle with stable and accurate performance of motion. Therefore, how to design a high quality controller for quadrotor is an important and meaningful problem.

Many different control theories and methods are employed to design the attitude stabilizer or motion controller for quadrotors[11]. In the early stage, Bouabdallah et al. first present the application of two different control techniques PID and Linear Quadratic (LQ) to a micro Quadrotor called OS4[1]. These two kinds

---

* Corresponding author.

© Springer International Publishing Switzerland 2015
J. Cao et al. (eds.), *Proceedings of ELM-2014 Volume 2,*
Proceedings in Adaptation, Learning and Optimization 4, DOI: 10.1007/978-3-319-14066-7_13

of controller are easy to design and apply while the disadvantage is that they cannot handle the unmolded dynamics and external disturbances. Because of the nonlinearity, feedback linearization technology is adopted to design the controller for quadrotors[15]. However, precise models of quadrotors are difficult to obtain. One potential way to solve this problem is intelligent control methods such as fuzzy logic control[4], neural networks control[3] and learning based control[10], etc. It should be mentioned that the main challenge we are facing in neural control and learning based control is the convergence. Slow convergence may cause failure in real time control system.

In this paper, a novel computational intelligence technique named extreme learning machine (ELM)[7] is induced to controller system. It is mainly used to compensate the dynamic uncertainties. ELM has successfully used in several control systems[12,13,14,16]. Actually, it is a learning policy for generalized single-hidden layer feedforward networks (SLFNs). Compared with backpropagation (BP) method and support vector machines (SVMs), ELM provides better generalization performance at a much faster learning speed and with least human intervene[5,6,8]. This is the exact reason why ELM can be used to estimate and compensate the unmodeled uncertainties in real time.

This paper is organized as follows: In Section 2, a basic mathematical description of ELM is given. Dynamic model of quadrotors is described in Section 3. In Section 4, the details of designing an ELM based quadrotor controller are presented. Simulation results are given in Section 5 to show the performance of the proposed controller. Finally, the paper is concluded in Section 6.

2   Description of Extreme Learning Machine

In this section, the basic idea of ELM is briefly reviewed. ELM is a special SLFN whose learning speed can be much faster than conventional feedforward network learning algorithm such as BP algorithm while obtaining better generalization performance[9]. The essence of ELM is that the input weights and the parameters of the hidden layer do not need to adjust during the learning procedure. Here, we take a SLFN with L hidden nodes as an example. The output of the SLFN can be modeled as

$$f_L(\mathbf{x}) = \sum_{i=1}^{L} \beta_i G(\mathbf{x}, \mathbf{c}_i, a_i), \ x \in \mathbf{R}^n, \mathbf{c}_i \in \mathbf{R}^n \ , \tag{1}$$

where β_i is the output weight connecting the ith hidden node to the output node, $G(\mathbf{x}, \mathbf{c}_i, a_i)$ is the activation function of the ith hidden node, $\mathbf{c}_i$ and a_i are the parameters of the activation function which are randomly generated and fixed.

Then, N sample pairs $(\mathbf{x}_k, \mathbf{y}_k) \in \mathbf{R}^n \times \mathbf{R}^m (k = 1, ..., N)$ are employed to train the SLFN. If this networks can approximate N samples with zero error, there must exist $\beta_i^*, \mathbf{c}_i$ and a_i such that

$$\sum_{i=1}^{L} \beta_i^* G(\mathbf{x}_k, \mathbf{c}_i, a_i) = \mathbf{y}_k, k = 1, \cdots, N \ . \tag{2}$$

ELM aims to minimize not only the training error but also the norm of output weights, which would yield a better generalization performance[8].

3 Dynamic Model of Quadrotors

Many works on quadrotors modeling are reported, so here we directly give the state space model[2] of a quadrotor for controller design.

The state vector is defined as $\mathbf{X} = \begin{bmatrix} \phi \ \dot{\phi} \ \theta \ \dot{\theta} \ \psi \ \dot{\psi} \ z \ \dot{z} \ x \ \dot{x} \ y \ \dot{y} \end{bmatrix}^T$, where ϕ, θ and ψ is the attitude angles which represent roll, pitch and yaw. (x, y, z) is the quadrotor's position in the inertia frame. if we define the speeds of the four rotors as $\Omega_1, \Omega_2, \Omega_3$ and Ω_4, the control input vector can be further defined as $\mathbf{U} = \begin{bmatrix} U_1 \ U_2 \ U_3 \ U_4 \end{bmatrix}^T$, which is mapped by:

$$\begin{cases} U_1 = k_F(\Omega_1^2 + \Omega_2^2 + \Omega_3^2 + \Omega_4^2) \\ U_2 = k_F(-\Omega_2^2 + \Omega_4^2) \\ U_3 = k_F(\Omega_1^2 - \Omega_3^2) \\ U_4 = k_M(\Omega_1^2 - \Omega_2^2 + \Omega_3^2 - \Omega_4^2) \end{cases} , \tag{3}$$

where k_F and k_M are the aerodynamic force and moment constants respectively.

Then, the compact form of the state space model of the quadrotor is given by

$$\dot{\mathbf{X}} = \begin{pmatrix} \dot{\phi} \\ \dot{\theta}\dot{\psi}(I_{yy} - I_{zz})/I_{xx} + \dot{\theta}\Omega_r J_r/I_{xx} + U_2 l/I_{xx} \\ \dot{\theta} \\ \dot{\phi}\dot{\psi}(I_{zz} - I_{xx})/I_{yy} - \dot{\phi}\Omega_r J_r/I_{yy} + U_3 l/I_{yy} \\ \dot{\psi} \\ \dot{\theta}\dot{\phi}(I_{xx} - I_{yy})/I_{zz} + U_4/I_{zz} \\ \dot{z} \\ g - (\cos\phi\cos\theta)U_1/m \\ \dot{x} \\ (\cos\phi\sin\theta\cos\psi + \sin\phi\sin\psi)U_1/m \\ \dot{y} \\ (\cos\phi\sin\theta\sin\psi - \sin\phi\cos\psi)U_1/m \end{pmatrix}, \tag{4}$$

where I_{xx}, I_{yy} and I_{zz} are the moment of inertia around X, Y and Z axis, respectively. J_r is the propeller inertia coefficient. l is the arm length of the quadrotor.

4 Controller Design with ELM

By investigating the relationship between the state variables in equation(4), the attitude angles of the quadrotor do not depend on the translational motion. Meanwhile, the translation of the quadrotor depends on the attitude angles. Thus, the whole system can be decoupled into two subsystems which are attitude loop and position loop. The structure of the control system is described in Fig.1.

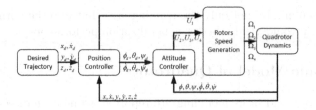

Fig. 1. The structure of the quadrotor control system

, where $(x_d, \dot{x}_d, y_d, \dot{y}_d, z_d, \dot{z}_d)$ is the desired trajectory and $(\phi_d, \theta_d, \psi_d, \dot{\phi}_d, \dot{\theta}_d, \dot{\psi}_d)$ is the desired attitude angles and their accelerations generated by the position controller. U_1 is produced by the position controller related to the height of the vehicle while the U_2, U_3 and U_4 are calculated by the attitude controller.

4.1 Position Control Loop

The outer loop is a position control system which has slower dynamics compared to the attitude control loop. In addition, the desired roll, pitch and yaw angles are normally small that makes the position subsystem an approximate linear system near the equilibrium points. Because of this, a PD controller is employed to control the quadrotor's position.

For horizontal movement, the desired accelerations $\ddot{x}_d$ and $\ddot{y}_d$ can be calculated by the PD controller based on the desired trajectory:

$$\begin{cases} \ddot{x}_d = k_{px}(x_d - x) + k_{dx}(\dot{x}_d - \dot{x}) \\ \ddot{y}_d = k_{py}(y_d - y) + k_{dy}(\dot{y}_d - \dot{y}) \end{cases}, \tag{5}$$

where (k_{px}, k_{dx}) and (k_{py}, k_{dy}) are the proportional and differential control gains in X and Y directions. Control gains to stabilize the position control system can be acquired using pole placement design method. Substitute $\ddot{x}_d$ and $\ddot{y}_d$ into equation(4) obtains:

$$\begin{cases} (\cos\phi_d \sin\theta_d \cos\psi + \sin\phi_d \sin\psi)U_1/m = \ddot{x}_d \\ (\cos\phi_d \sin\theta_d \sin\psi - \sin\phi_d \cos\psi)U_1/m = \ddot{y}_d \end{cases}. \tag{6}$$

Using the small angle assumption around the equilibrium position, the reference roll and pitch angles can be solved by

$$\begin{bmatrix} \phi_d \\ \theta_d \end{bmatrix} = \frac{m}{U_1} \begin{bmatrix} \sin\psi & \cos\psi \\ -\cos\psi & \sin\psi \end{bmatrix}^{-1} \begin{bmatrix} \ddot{x}_d \\ \ddot{y}_d \end{bmatrix}. \tag{7}$$

For vertical movements, the altitude control law can be calculated by

$$U_1 = k_{pz}(z_d - z) + k_{dz}(\dot{z}_d - \dot{z}) + mg, \tag{8}$$

where k_{pz} and k_{dz} are the proportional and differential control gains in Z direction, and mg is the term to compensate the gravity.

4.2 Attitude Control Loop

The quadrotor's attitude system is unstable and sensitive to the disturbances. It is also a nonlinear system with unmodeled dynamic uncertainties compared to the real quadrotor, hence an ELM based SLFN with very fast learning speed is employed to estimate the model of the quadrotor's attitude system in real-time.

As the roll, pitch and yaw subsystems have the same form of expression, without loss of generality, we take roll subsystem as an example to introduce the design procedure.

Model Based Control Law. According to the equation(4), the roll subsystem is expressed in the following form:

$$\begin{cases} \dot{x}_1 = x_2 \\ \dot{x}_2 = f(\mathbf{x}) + ul/I_{xx} + d \\ y = x_1 \end{cases} , \tag{9}$$

where $f(\mathbf{x})$ is assumed to be unknown but bounded. d is the unknown disturbances which is bounded and satisfies $|d| \leq d_{max}$, where d_{max} is the upper bound of the disturbances. Tracking error $\mathbf{E} = [e, \dot{e}]^T$, where $e = y_d - x_1$ and $\dot{e} = \dot{y}_d - x_2$.

Suppose all the functions and parameters in equation (9) are precisely known, the perfect control law u^* can be obtained using feedback linearization method as follows:

$$u^* = \frac{I_{xx}}{l}(\ddot{y}_d - f(\mathbf{x}) - d + \mathbf{K}^T\mathbf{E}) , \tag{10}$$

where $\mathbf{K} = [K_1, K_2]^T$ is a real number vector. Substitute equation (10) into (9), the following error equation is obtained:

$$\ddot{e} + K_2\dot{e} + K_1 = 0 , \tag{11}$$

where K_1 and K_2 can be determined when all the roots of the polynomial $s^2 + K_2s + K_1 = 0$ are in the open left half plane, which indicates that the tracking error will converge to zero.

Adaptive Neural Control Law. Since $f(\mathbf{x})$ is coupled with other variables and can not be accurately modeled, d is also an unknown disturbances, control law of equation (10) can not be directly implemented. Here, we present an ELM based neural controller to solve this problem.

The inner loop controller is mainly composed of two parts. One is the neural controller u_n and another is the sliding mode contoller u_s, so the overall control law

$$u = u_n + u_s . \tag{12}$$

The neural controller is used to estimate the unknown function $f(\mathbf{x})$ while the sliding mode controller is employed to eliminate the external disturbances d.

For the neural controller, according to the equation (10), the optimal neural control law is expected as

$$u_n^* = \frac{I_{xx}}{l}(\ddot{y}_d - f(\mathbf{x}) + \mathbf{K}^T \mathbf{E}) . \tag{13}$$

Here, a SLFN whose parameters are determined based on the ELM is employed to approximate above desired neural control law. Then, the actual neural control law u_n is given by

$$u_n = \mathbf{H}(\mathbf{r}, \mathbf{c}, \mathbf{a})\boldsymbol{\beta} , \tag{14}$$

where $\mathbf{r} = [\ddot{y}_d; \mathbf{x}; \mathbf{E}]^T$ that is the input vector. As we mentioned in Section 2, $\mathbf{c}$ and $\mathbf{a}$ are hidden node parameters which are generated randomly and fixed. Training a SLFN with ELM is simply equivalent to finding a least-square solution of the output weights $\boldsymbol{\beta}^*$, such that

$$u_n^* = \mathbf{H}(\mathbf{r}, \mathbf{c}, \mathbf{a})\boldsymbol{\beta}^* + \varepsilon(\mathbf{r}) , \tag{15}$$

where $\varepsilon(\mathbf{r})$ is the approximation error. It is bounded with the constant ε_N[14].

For the sliding mode controller, the standard sliding mode control law is given by

$$u_s = G_s(\mathbf{x})\mathrm{sgn}(f(\mathbf{E})) , \tag{16}$$

where $G_s(\mathbf{x})$ is the control gain, and $f(\mathbf{E})$ is the sliding mode surface function. Suppose the Lyapunov function is chosen as:

$$V = \frac{1}{2}\mathbf{E}^T\mathbf{PE} + \frac{1}{2\eta}\tilde{\boldsymbol{\beta}}^T\tilde{\boldsymbol{\beta}} . \tag{17}$$

The tuning rule of the output weight of the ELM and the parameters of the sliding mode controller can be acquired according to the Lyapunov stable theory, such that

$$\dot{\boldsymbol{\beta}}^T = \eta\mathbf{E}^T\mathbf{PBH} . \tag{18}$$

$$u_s = \left(\frac{d_{\max}I_{xx}}{l} + \varepsilon_N\right)\mathrm{sgn}(\mathbf{E}^T\mathbf{PB}) . \tag{19}$$

Finally, the overall control laws for quadrotor's attitude system are given by

$$\begin{cases} U_2 = \mathbf{H}_\phi(\mathbf{r}_\phi, \mathbf{c}_\phi, \mathbf{a}_\phi)\beta_\phi + \left(\frac{d_{\max}I_{xx}}{l} + \varepsilon_N\right)\mathrm{sgn}(\mathbf{E}_\phi{}^T\mathbf{P}_\phi\mathbf{B}_\phi) \\ U_3 = \mathbf{H}_\theta(\mathbf{r}_\theta, \mathbf{c}_\theta, \mathbf{a}_\theta)\beta_\theta + \left(\frac{d_{\max}I_{yy}}{l} + \varepsilon_N\right)\mathrm{sgn}(\mathbf{E}_\theta{}^T\mathbf{P}_\theta\mathbf{B}_\theta) \\ U_4 = \mathbf{H}_\psi(\mathbf{r}_\psi, \mathbf{c}_\psi, \mathbf{a}_\psi)\beta_\psi + (d_{\max}I_{zz} + \varepsilon_N)\mathrm{sgn}(\mathbf{E}_\psi{}^T\mathbf{P}_\psi\mathbf{B}_\psi) \end{cases} , \tag{20}$$

where the subscript ϕ, θ and ψ represent the roll, pitch and yaw subsystem, respectively. $\mathbf{B}_\phi = [0\ l/I_{xx}]^T$. $\mathbf{B}_\theta = [0\ l/I_{yy}]^T$. $\mathbf{B}_\psi = [0\ 1/I_{zz}]^T$. To avoid chattering problem, the sign function can be replaced by the saturation function.

5   Simulations

In this section, attitude and position control simulations are both carried out on a nonlinear quadrotor model to evaluate the performance of the proposed ELM based controller. The parameters of the quadrotor for simulation are measured from a real platform which are listed as follows: $m = 2.2kg$, $I_{xx}=0.01676kg.m^2$, $I_{yy}=0.01676kg.m^2$, $I_{zz}=0.02314kg.m^2$, $J_r = 0.1kg.m^2$, $l = 0.18m$.

For attitude control simulations, the task is to control the quadrotor helicopter from an initial attitude to a target attitude. The initial values are set as $\phi = \pi/8rad$, $\theta = \pi/8rad$ and $\psi = 0rad$ while the objective values are that $\phi = -\pi/16rad$, $\theta = -\pi/16rad$ and $\psi = 0rad$. In the simulation, the model uncertainty d is given by $0.5 * \sin(\pi t)$ and $d_{max} = 0.5$. The controller parameters are chosen as follows: $K_1 = 8$, $K_2 = 128$, $\eta = 0.05$, $\varepsilon_N = 12$. The number of hidden nodes $L = 5$. RBF nodes are selected as the hidden nodes whose parameters c are generated in the interval $[-5\pi/16, 5\pi/16]$ and $a = 5$.

To show the good performance of the proposed controller, a PD controller is employed to carry out the same simulation for comparison. The parameters of the PD controller are chosen as follows: $k_{p\phi} = 5$, $k_{d\phi} = 1.2$, $k_{p\theta} = 4.5$, $k_{d\theta} = 1$, $k_{p\psi} = 5$ and $k_{d\psi} = 1$. A test with only sliding mode controller is also conducted to validate the effectiveness of the ELM.

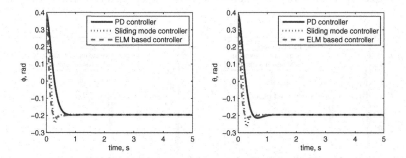

Fig. 2. Roll and Pitch Angle of the Quadrotor

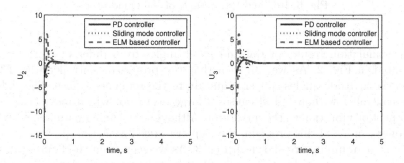

Fig. 3. Control Input U_2 and U_3

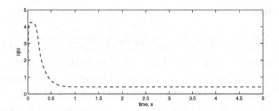

Fig. 4. Learning Curve of $\|\beta\|$ in Attitude Control

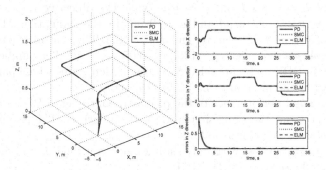

Fig. 5. Position Tracking Curves and Errors in X, Y and Z Directions

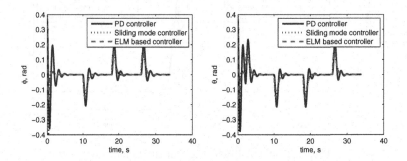

Fig. 6. Roll and Pitch Angle of the Quadrotor

The simulation results of roll and pitch angles from different methods are illustrated in Fig. 2. As we can seen that the proposed controller with ELM makes the attitude angles converge quickly to the reference values. The sliding mode controller without ELM causes a large overshoot which may result in a big impact or vibration to the quadrotor. Although, by tuning parameters of PD controller can avoid this problem, the response speed is sacrificed.

The control inputs are shown in Fig. 3. As we seen from the figures, all the control inputs are bounded and applicable. Inputs in our method are high at first but dropped quickly. This is the reason why the response speed and accuracy

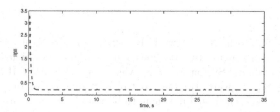

Fig. 7. Learning Curve of $\|\beta\|$ in Position Control

are acquired simultaneously. It also indicates that the output weights β trained by ELM can converge very fast as shown in Fig. 4.

To further validate the performance, position tracking control simulation is also carried out. The specific task is to make the quadrotor track a trajectory in 3D space after its taking off from ground. The tracking results are shown in Fig. 5. In this simulation, our proposed scheme as a inner loop attitude controller is also compared with the PD and sliding mode controller. These inner loop controllers both cooperate with the same PD outer loop controller to complete the position tracking task. We can see that, especially during the vertical taking off stage, the controller with ELM has better performance than the other two controllers in terms of stability and tracking accuracy. The variation of the attitude angles such as roll and pitch during the tracking process are shown in Fig. 6. It is obvious that ELM based controller provides more stable attitude variation. The training process of β is also given in Fig. 7. It shows that the output weights converge quickly and remain unchanged afterwards.

6   Conclusions

In this paper, An ELM based adaptive neural controller combined with a sliding mode controller is designed for control of the quadrotor. A single hidden layer feedforward network whose input weights and hidden node parameters are generated randomly and fixed is used to approximate the attitude dynamic model and internal uncertainties. Different from the standard ELM algorithm, the output weights of this neural network is updated based on the Lyapunov second method to guarantee the stability of the attitude control system. A sliding mode controller is employed to compensate the modelling error of the SLFN and eliminate the external disturbances. The simulations on quadrotor's attitude regulation and position tracking are carried out to validate the effectiveness of the proposed control approach. The simulation results shows that the proposed controller has better performance on response speed and control accuracy than simple PD controller and sliding mode controller.

Acknowledgments. This work was supported by the National Natural Science Foundation of China (Grant No. 61005085) and Fundamental Research Funds for the Central Univer-sities (2012QNA4024).

References

1. Bouabdallah, S., Noth, A., Siegwart, R.: PID vs LQ control techniques applied to an indoor micro quadrotor. In: 2004 IEEE/RSJ International Conference on Intelligent Robots and Systems (IROS), vol. 3, pp. 2451–2456. IEEE (2004)

2. Bouabdallah, S., Siegwart, R.: Full control of a quadrotor. In: 2007 IEEE/RSJ International Conference on Intelligent Robots and Systems, pp. 153–158. IEEE (October 2007)

3. Dierks, T., Jagannathan, S.: Output feedback control of a quadrotor UAV using neural networks. IEEE Transactions on Neural Networks 21(1), 50–66 (2010)

4. Gautam, D., Ha, C.: Control of a Quadrotor Using a Smart Self-Tuning Fuzzy PID Controller. International Journal of Advanced Robotic Systems 10, 1 (2013)

5. Huang, G.B., Chen, L.: Convex incremental extreme learning machine. Neurocomputing 70(16-18), 3056–3062 (2007)

6. Huang, G.B., Ding, X., Zhou, H.: Optimization method based extreme learning machine for classification. Neurocomputing 74(1-3), 155–163 (2010)

7. Huang, G.B., Wang, D.H., Lan, Y.: Extreme learning machines: a survey. International Journal of Machine Learning and Cybernetics 2(2), 107–122 (2011)

8. Huang, G.B., Zhu, Q.Y., Siew, C.K.: Extreme learning machine: Theory and applications. Neurocomputing 70(1-3), 489–501 (2006)

9. Huang, G.B., Zhu, Q.Y., Siew, C.K.: Real-time learning capability of neural networks. IEEE Transactions on Neural Networks 17(4), 863–878 (2006)

10. Jafari, M., Shahri, A.M., Shouraki, S.B.: Attitude control of a Quadrotor using Brain Emotional Learning Based Intelligent Controller. In: 2013 13th Iranian Conference on Fuzzy Systems (IFSC), pp. 1–5. IEEE (August 2013)

11. Li, Y., Song, S.: A survey of control algorithms for Quadrotor Unmanned Helicopter. In: 2012 IEEE Fifth International Conference on Advanced Computational Intelligence (ICACI), pp. 365–369. IEEE (October 2012)

12. Rong, H.J., Jia, Y.X., Zhao, G.S.: Aircraft recognition using modular extreme learning machine. Neurocomputing 128, 166–174 (2014)

13. Rong, H.J., Suresh, S., Zhao, G.S.: Stable indirect adaptive neural controller for a class of nonlinear system. Neurocomputing 74(16), 2582–2590 (2011)

14. Rong, H.J., Zhao, G.S.: Direct adaptive neural control of nonlinear systems with extreme learning machine. Neural Computing and Applications 22(3-4), 577–586 (2012)

15. Shakev, N., Topalov, A.: Comparative Results on Stabilization of the Quad-rotor Rotorcraft Using Bounded Feedback Controllers. Journal of Intelligent Robotic Systems (2012)

16. Xu, B., Pan, Y., Wang, D., Sun, F.: Discrete-time hypersonic flight control based on extreme learning machine. Neurocomputing 128, 232–241 (2014)

Keyword Search on Probabilistic XML Data Based on ELM

Yue Zhao*, Guoren Wang, and Ye Yuan

College of Information Science and Engineering, Northeastern University,
Liaoning, Shenyang 110819
zhaoy0927@163.com

Abstract. This paper describe a keyword search measure on probabilistic XML data based on ELM (Extreme Learning Machine). We use this method to carry out keyword search on probabilistic XML data. A probabilistic XML document differs from a traditional XML document to realize keyword search in the consideration of possible world semantics. A probabilistic XML data can be seen as a set of nodes consisting of ordinary nodes and distributional nodes. ELM has good performance in text classification applications. As the typical semi-structured data, the label of XML data possesses the function of definition self. Label and keyword which has been contained in the node can be seen as the text data of the node. ELM offers significant advantages such as fast learning speed, ease of implementation and classification nodes effectively. Keyword search on the set after it classified by using ELM can pick up the speed of query. This paper uses ELM to classify nodes and carry keyword search on the set which has been classified. The experiments can show that the speed of query can receive significant improvement.

Keywords: Extreme learning machine, Node classification, Probabilistic XML data.

1 Introduction

Traditional databases only manage deterministic information, but many applications that use databases to involve uncertain data such as information extraction, information integration, web data mining, etc. Because of the flexibility of XML data model, it can easily allow a natural representation of uncertain data. Now, many probabilistic XML models are designed and analyzed[1-4]. This paper select a popular probabilistic XML model $PrXML^{\{ind,mux\}}$[5], which is discussed in [6]. In this model, a probabilistic XML document (called a p-document) is considered as a labeled tree which has two types of nodes, *ordinary* nodes and

---

* Yue Zhao, Ye Yuan and Guoren Wang were supported by the NSFC (Grant No.61025007, 61328202 and 61100024), National Basic Research Program of China (973, Grant No.2011CB302200-G), National High Technology Research and Development 863 Program of China (Grant No.2012AA011004), and the Fundamental Research Funds for the Central Universities (Grant No. N130504006).

distributional nodes. Ordinary node is used to represent the actual data and distributional node is used to represent the probability distribution on the child nodes. There are two types of nodes in distributional nodes, IND and MUX. If a node is an IND node, its children nodes are *independent* of each other, while the children of a MUX node are *mutually − exclusive*. A real number from (0,1] is attached on each edge in the XML tree, indicating the conditional probability that the child node will appear under the parent node given the existence of the parent node. Keyword search has been widely applied on XML data. Users don't need know the knowledge of the underlying data structures and complex query language beforehand. So, keyword search is an easy method for ordinary users. In the past years, the definition of common ancestor node has several choices, such as LCA (Lowest Common Ancestor), SLCA (Smallest LCA) and so on. This paper select SLCA as the root node of result subtree.

ELM[7-10] has good performance on classification applications, and can be used to classify nodes before query XML data. Classification is considered as an important cognitive computation task[11-14]. A probabilistic XML data tree can be seen as a set of all the nodes including root node (only one), connected nodes, leaves nodes and distributional nodes. So, the classification need to consider two kinds of information, and they are keyword information and probability distributional information.

This paper is organized as follows: Section 2 introduces the probabilistic XML model and the formal semantics of keyword search result on probabilistic XML data. Section 3 shows that how to classify nodes. In section 4, we propose an algorithm to query keyword on probabilistic XML data by using ELM to classify nodes. The experimental and performance evaluation are presented in section 5. Section 6 gives the conclusion and future works.

2   Problem Definitions

2.1   Probabilistic XML Data

A probabilistic XML document (p-document) can be seen as a set of many deterministic XML documents. Each deterministic document is called a possible world. Ordinary nodes are prime XML nodes and they are appearing on both deterministic XML data and probabilistic XML data. Distributional nodes are only used to define the probabilistic process of generating deterministic documents, while those nodes do not occur on deterministic XML data. This paper adopts $PrXML^{\{ind,mux\}}$ as the probabilistic XML model. For example, figure 1(a) shows a p-document T.

Given a p-document T, we can traverse T in a top-down fashion. When we visit a distributional node, there are two situations according to the different typies. One situation is that if a node is an IND node with m children nodes, we generate 2^m copies. Another situation is that if a node is a MUX node with m children nodes, we generate m or $m + 1$ copies. For example, figure 1(b) shows the copies of a p-document with their probabilities. Figure 1(b) select node b as the only child node of node a, and the probability is $0.7 * (1 - 0.6) = 0.28$.

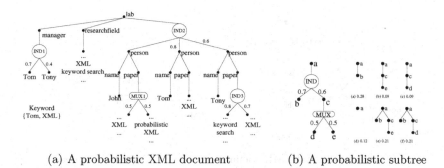

(a) A probabilistic XML document   (b) A probabilistic subtree

Fig. 1. Probabilistic XML data

If there is not any node selected as the children nodes of a, the probability of this copy is $(1 - 0.7) * (1 - 0.6) = 0.12$. Node a selects nodes b and c as its children nodes, and node c selects node d as its child node. The probability is $0.7 * 0.6 * 0.5 = 0.21$. The probabilities of the other copies (possible worlds) are easy to calculate from the above procedure.

2.2 Keyword Query

Usually, we model an XML tree as a labeled ordered tree, where nodes represent elements, and edges represent direct nesting relationship between nodes. Recently, keyword search has been studied in XML documents more and more. Given a set of keywords and a XML document, most work took LCA and SLCA of the matched nodes as the results. The function $lca(v_1, v_2, ..., v_k)$ computes the Lowest Common Ancestor of nodes $v_1, v_2, ..., v_k$. Given k keywords and the inverted lists $\{S_1, S_2, ..., S_k\}$ of them. The LCA of these keywords on T is defined as:

$$lca(v_1, v_2, ..., v_k) = lca(S_1, S_2, ..., S_k)$$
$$= \{lca(n_1, n_2, ..., n_k) \mid n_1 \in S_1, ..., n_k \in S_k\} \tag{1}$$

$child(v, n_i)$ denote the children nodes of node v on the path from v to n_i. The SLCA is defined as follows:

$$slca(\{v_1\}, S_2, ..., S_k) =$$
$$\{v \mid v \in lca(\{v_1\}, S_2, ..., S_k), \forall v' \in lca(\{v_1\}, S_2, ..., S_k)(v \nprec_a v')\} \tag{2}$$

For example, figure 2(a) give a traditional XML tree which is generated by figure 1(a) and a query $Q = \{Tom, XML\}$. The result of query is shown in figure 2(b). This paper selects SLCA as the result for the keyword search on probabilistic XML data. Because SLCA is the smallest set, every SLCA node should be seen as the suitable for the users.

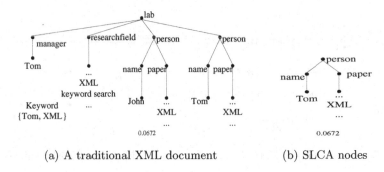

(a) A traditional XML document   (b) SLCA nodes

Fig. 2. SLCA nodes on XML data

A keyword search on p-document consists of a p-document T, a query $Q = \{k_1, k_2, ..., k_n\}$. We define the answers for a keyword search on T as ordinary nodes on T to be SLCAs in the possible worlds generated by T. The probability of a node v being an SLCA in the possible worlds is denoted as $Pr^T_{slca}(v)$. The formal definition is shown as follows:

$$Pr^T_{slca}(v) = \sum_{i=1}^{m}\{Pr(w_i) \mid slca(v, w_i) = true\} \tag{3}$$

where $Pr(w_i)$ is the existence probability of the possible world w_i. $\{w_1, w_2, ..., w_n\}$ denotes the possible worlds generated by T. $slca(v, w_i) = true$ indicates that v is an SLCA in the possible world w_i.

Definition 1: (SLCA on probabilistic XML data) *Given a query Q in a probabilistic XML tree T, an SLCA query finds the SLCA nodes v in all possible worlds with the probability of all the probabilities of the possible worlds in which the node v is an SLCA node.*

3   Classification of Nodes

3.1   Classification of Ordinary Nodes

From section 2, we can see that if we can find keyword nodes tree, the set intersection operation for keyword nodes tree should achieve SLCA nodes quickly. Figure 3(a) and 3(b) shows the keyword nodes tree. When we use set intersection operation to obtain the common ancestor nodes tree such as shown in figure 4(c). So, the important section is how to receive the keyword nodes tree.

To receive the keyword nodes tree, we need add dummy node for actual node which contains more than one keyword. If the subtree rooted at the node v contains two keywords, we should add one dummy node as the sibling node of node v. For example, node $\{lab\}$ and $\{person\}$ in figure 4 has its dummy node. These dummy nodes don't exist in the actual tree. The aim of adding dummy nodes is to classify nodes effectively.

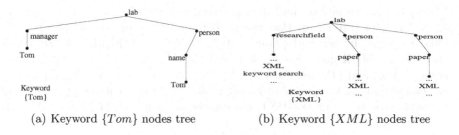

(a) Keyword {*Tom*} nodes tree   (b) Keyword {*XML*} nodes tree

Fig. 3. Keyword nodes tree

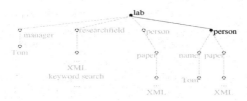

Fig. 4. A common nodes tree

3.2   Classification of Distributional Nodes

Distributional nodes can represent the probability distribution of the children nodes. A p-document defines a probability distribution over a space of deterministic XML documents. According to the different types of the distributional node, the number of copies is different. If a node is an IND node, and it has n children nodes, the number of copies is 2^n. Otherwise, if a node is a MUX node, the number is n or $n+1$. Each copy has its probability value. Some of the copies will contain the keyword, and the copies which contain the keyword are important for our probabilistic keyword search.

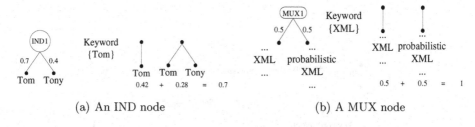

(a) An IND node   (b) A MUX node

Fig. 5. Distributional nodes

Figure 5(a) shows an example of an IND node. For the keyword {*Tom*}, IND1 is a parent node of nodes {*Tom*} and {*Tony*}. The copy with the existing of

$\{Tom\}$ has two situations, and their probabilities are $0.7 \times (1 - 0.4) = 0.42$ and $0.7 \times 0.4 = 0.28$. It means that the probability of the subtree rooted at node IND1 contains the keyword $\{Tom\}$ is $0.42 + 0.28 = 0.7$. Figure 5(b) shows an example of a MUX node. For the keyword $\{XML\}$, MUX1 is a parent node of node $\{XML\}$ and $\{ProbabilisticXML\}$. The copy with the existing of $\{XML\}$ has two situations, the probabilities are all 0.5. It means that the probability of the subtree rooted at node MUX1 contains the keyword $\{XML\}$ is $0.5+0.5 = 1$.

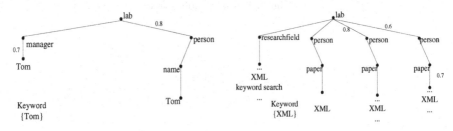

(a) Keyword $\{Tom\}$ probabilistic tree   (b) Keyword $\{XML\}$ probabilistic tree

Fig. 6. Keyword nodes probabilistic tree

For each keyword, all its ancestor nodes and itself nodes will constitute a tree. This tree contains ordinary nodes and distributional nodes. To present the probability contribution situation of this tree which contains keyword, we will delete distributional nodes and connect its children nodes to its parent node with the existence probability of containing the keyword of the subtree rooted at its parent node according to the type of distributional node. For example, figure 6(a) shows a tree contained keyword $\{Tom\}$. Node $\{manager\}$ is a parent node of node $\{IND1\}$. The probability of a subtree rooted at node $\{IND1\}$ which contains keyword $\{Tom\}$ is 0.7. As shown in figure 6(b), it is the situation of the probabilistic tree which contains keyword $\{XML\}$.

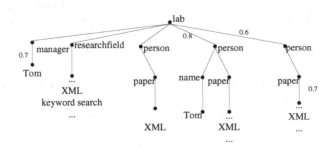

Fig. 7. A keyword nodes probabilistic tree

We merge all the keywords probabilistic trees together. A keyword nodes probabilistic tree can be generated. We need calculate SLCA nodes on this tree with the probability and delete the subtree rooted at SLCA nodes. Next, we need continue to calculate SLCA results on remaining nodes tree. So, if we repeat such operation, all the SLCA results will generate. For example, figure 7 is a keyword nodes probabilistic tree. This tree retains all the probabilities of the subtree which contains keyword. If we calculate SLCA results on this tree, the node $\{person\}$ is the only result. So, we need delete the subtree which is rooted at node $\{person\}$, and the remaining other nodes tree. To repeat calculated operation on the remaining nodes tree, we can see that the node $\{lab\}$ is another result. The calculation of the probabilities of all the SLCA nodes can be shown in next section.

4   Keyword Search on Probabilistic XML Data Based on ELM

Keyword search on probabilistic XML data based on classification mainly include four steps, they are shown as following: 1) Adding dummy nodes according to the number of keywords. 2) To classify nodes with ELM based on keyword according to the type of the nodes. 3) Use the set merging operation to structure the common ancestor nodes probabilistic tree. 4) Repeat the operation of calculating SLCA and deleting the subtree, all the SLCA results will generate. The key of the keyword search on p-document is how to calculate the probability of the SLCA results. Step 2 and step 4 all contain the computation of the probabilities.

Each node contains two kinds of information, they are code and keyword it contained. If a node is a distributional node, there are the third information in this node, that is probability. Code is used to judge the relationships between nodes, such as finding the common ancestor nodes. The keyword which is contained in a node is the key of keyword search. When we use ELM to classify nodes, keyword can be used as the label of the classification set. Every set represent one keyword. For given the query, we will find all the sets of keywords which is given by users, and operate set merging to obtain a keyword nodes tree.

Next, we introduce four steps of the keyword search algorithm used ELM to classify nodes on probabilistic XML data one by one.

First, adding dummy nodes according to the number of keywords. We can see that the probabilistic XML tree in figure 1(a) contains two keywords $\{Tom, XML\}$. The algorithm use Dewey code to encode the XML tree. So, the first step is adding the dummy nodes for the node v which contains keywords in the subtree rooted at the node v. If the subtree which is rooted at the node v has n keywords, we will add $n - 1$ dummy nodes.

Second, to classify nodes with ELM based on keyword according to the type of the nodes. From the dummy nodes tree, all the nodes and the distributional nodes are consisted of the classified nodes. ELM can classify the nodes to two sets such as shown in figure 8(a). The first set represent keyword Tom, and the second set represent keyword XML. Each distributional node has a probability which represents the keyword probability of the subtree which is rooted at the distributional node. For example, the node $IND1$ has the probability with 0.7, that means the probability of containing keyword $\{Tom\}$ of the subtree which is rooted at $IND1$ is 0.7.

Then, we need delete all the distributional nodes and connect all their children nodes to their parent node. The probability of distributional node will be moved to its child node. For example, in figure 8(b), the node $\{Tom\}$ accept the probability 0.7 from its parent node $\{IND1\}$.

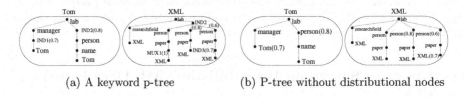

(a) A keyword p-tree   (b) P-tree without distributional nodes

Fig. 8. Probabilistic tree

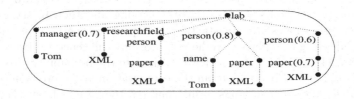

Fig. 9. A common nodes p-tree

Third, use the set merging operation to structure the common ancestor nodes probabilistic tree. The intersection of the two sets is the set which includes node $\{lab\}$ and $\{person\}$. Figure 9 shows the union set of two keyword sets.

Finally, repeat the operation of calculating SLCA and deleting the subtree, all the SLCA results will generate. Let's calculate SLCA result on the tree which is shown in figure 9. The node $\{person\}$ with the SLCA probability of 0.8 is generated. So, the subtree which is rooted at $\{person\}$ will be deleted, and the probability $1 - 0.8 = 0.2$ will be leaved to the other nodes tree. Next, the node $\{lab\}$ is another result. The probability of this result is $0.2 \times 0.7 = 0.14$. Because, the extensive probability of the node Tom is 0.7, and the extensive probability of node $\{XML\}$ is 1, the SLCA probability of $\{lab\}$ is 0.14.

5   Performance Verification

The dataset we used is shown in table 1. In this paper, the algorithm select two datasets XMARK and DBLP. For each XML dataset used, we generate the

Table 1. properties of probabilistic XML data

ID	name	size	Ordinary	IND	MUX
DOC 1	XMARK 1	10M	159,307	14,630	15,471
DOC 2	XMARK 2	20M	364,199	41,251	38,100
DOC 3	XMARK 3	40M	689,470	77,228	61,535
DOC 4	XMARK 4	80M	1,497,433	161,277	159,495
DOC 5	DBLP 1	20M	361,370	68,345	70,790
DOC 6	DBLP 2	40M	731,561	238,450	227,540
DOC 7	DBLP 3	80M	1,477,345	440,007	405,857
DOC 8	DBLP 4	160M	3,260,109	788,367	771,320

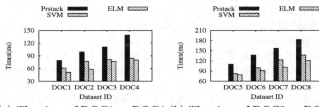

(a) The time of DOC1 to DOC4 (b) The time of DOC5 to DOC8

Fig. 10. Vary query over DOC1 to DOC8

corresponding probabilistic XML tree, using the same method as used in [5]. We visit the nodes in the original XML tree in pre-order way. For each node v visited, we randomly generate some distributional nodes as children of v. For the original children of v, we select them as the children of the new generated distributional nodes and assign them random probability distributions. We need restrict that the sum of children nodes for a MUX node is no more than 1. The keyword has 8 situations. The number of keywords is 2 to 3.

We compare the query times of three situations about keyword search in probabilistic XML data. The first situation is using the method in [15] to retrieval the SLCA nodes, and the second situation is using SVM to classify nodes for the keyword search on p-document. The third situation classify nodes by using ELM. The speed of classification is shown in figure 10.

From the figure 10, we can see that ELM has advantages of speed compared with Prstack and SVM. Prstack will compute all the nodes probabilities of all the ancestor nodes of the keyword node and it will record all the situation of the node which contains the keyword. ELM can classify nodes according to the code and keywords by retrieve all the nodes once. So, the algorithm has high speed by using ELM to classify.

6   Conclusions

This paper use ELM to classify for keyword search on probabilistic XML data. Keyword search on probabilistic XML data has been received much attention in the literature. Finding efficient query processing method for keyword search on probabilistic XML data is an important topic in this area. In this paper, SLCA is selected as the results. Classification for nodes is important among all the operations. ELM can increase retrieval speed for the classification. So, ELM can support keyword search on probabilistic XML data.

References

1. Guo, L., Shao, F., Botev, C., Shanmugasundaram, J.: Xrank: Ranked keyword search over xml documents. In: SIGMOD (2003)
2. Xu, Y., Papakonstantinou, Y.: Efficient keyword search for smallest lcas in xml databases. In: SIGMOD (2005)
3. Sun, C., Chan, C.Y., Goenka, A.K.: Multiway slca-based keyword search in xml data. In: WWW (2007)
4. Zhou, J., Bao, Z., Wang, W., Ling, T.W., Chen, Z., Lin, X., Guo, J.: Fast SLCA and ELCA Computation for XML Keyword Queries based on Set Intersection. In: ICDE (2012)
5. Kimelfeld, B., Kosharovsky, Y., Sagiv, Y.: Query efficiency in probabilistic xml models. In: SIGMOD (2008)
6. Nierman, A., Jagadish, H.V.: ProTDB: Probabilistic data in xml. In: VLDB (2002)
7. Huang, G.-B.: Learning capability and storage capacity of two-hidden-layer feedforward networks. IEEE Transactions on Neural Networks (2003)
8. Huang, G.-B., Siew, C.-K.: Extreme learning machine with randomly assigned RBF kernels. International Journal of Information Technology (2005)
9. Huang, G.-B., Chen, L.: Enhanced random search based incremental extreme learning machine. Neurocomputing (2008)
10. Huang, G.-B., Chen, L.: Convex incremental extreme learning machine. Neurocomputing (2007)
11. Taylor, J.G.: Cognitive computation. Cognitive Computation (2009)
12. Wöllmer, M., Eyben, F., Graves, A., Schuller, B., Rigoll, G.: Bidirectional lstm networks for context-sensitive keyword detection in a cognitive virtual agent framework. Cognitive Computation (2010)
13. Mital, P.K., Smith, T.J., Hill, R.L., Henderson, J.M.: Clustering of gaze during dynamic scene viewing is predicted by motion. Cognitive Computation (2011)
14. Cambria, E., Hussain, A.: Sentic computing: Techniques, tools, and applications (2012)
15. Li, J., Liu, C., Zhou, R., Wang, W.: Top-k Keyword Search over Probabilistic XML Data. In: CICDE (2011)

A Novel HVS Based Gray Scale Image Watermarking Scheme Using Fast Fuzzy-ELM Hybrid Architecture

Anurag Mishra[1,*] and Amita Goel[2]

[1] Department of Electronics, Deendayal Upadhyay College
University of Delhi, New Delhi, India
anurag_cse2003@yahoo.com
[2] Department of Computer Science,
Teerthanker Mahaveer University Moradabad, India
goelamita@rediffmail.com

Abstract. In this paper, a novel image watermarking scheme using HVS characteristics is proposed in transform domain. The HVS is modeled using 10 rules constituting the Fuzzy Inference System (FIS) by taking into account luminance sensitivity, edge sensitivity and contrast sensitivity. The output of the FIS is a single output known as weighing factor (WF). A dataset is prepared using the image coefficients and the WF. A fast neural network known as extreme learning machine (ELM) is trained using this dataset. The ELM produces a column vector of size 1024 x 1 used as normalized watermark embedded in low frequency coefficients of gray scale images. The signed images are subject to quality assessment before and after executing image processing attacks to examine the issue of robustness vis-a-vis imperceptibility. It is concluded that the proposed Fuzzy-ELM architecture for image watermarking yields good results both for visual quality of signed / attacked images and robustness. The order of time complexity of this scheme is suitable to produce commercial watermarking applications on a real time scale.

Keywords: Grayscale Image watermarking, Human Visual System, Fuzzy Inference System, Extreme Learning Machine, Hybrid Architecture, Robustness.

1   Introduction

Several image processing applications such as digital watermarking of images and video require execution of code within specified time constraints. Moreover, the embedding and extraction of external data should be such that it must fulfill two mutually exclusive objectives - robustness and visual imperceptibility. The third parameter i.e., the watermark embedding capacity is assumed to be constant as the embedded content is very small in comparison to the size of the host signal. Therefore, the embedding and extraction processes need to be optimized and should not result in much loss of visual quality after watermark embedding and execution of attacks over signed images to examine the issue of robustness. Moreover, all this is required to be

---

[*] Corresponding author.

© Springer International Publishing Switzerland 2015
J. Cao et al. (eds.), *Proceedings of ELM-2014 Volume 2,*
Proceedings in Adaptation, Learning and Optimization 4, DOI: 10.1007/978-3-319-14066-7_15

completed with minimum time complexity. Various soft computing techniques are used to optimize these processes to develop a robust watermarking package [1-6]. On the other hand, it still remains an interesting avenue to meet required timelines for these processes without compromising visual quality and robustness. Such an algorithm may be successfully applied to video at a later stage. Two different categories of ANNs are successfully employed to embed and extract watermarks. These are:- (1) a slow gradient descent based network such as BPN [1] and (2) the radial basis function neural network (RBFNN) which is comparatively faster than the BPN [2]. Being adaptive in nature, these techniques produce good results both in terms of visual imperceptibility and robustness. More specifically among these, training of BPN based schemes is slow due to inherent mathematics involved. Several groups have implemented fuzzy inference system (FIS) based image watermarking with or without taking Human Visual System (HVS) characteristics into account [3-4]. These pure FIS based schemes, although not adaptive to implement, yet they produce good results as far as the aforesaid twin issues of image watermarking are concerned. They are also not found to be fast enough to be implemented on real time scale. Machine learning based algorithms are also used for this task [5-6]. Different categories of SVMs in regression mode such as ordinary Support Vector Regression (SVR), Finite Newton Support Vector Regression (FNSVR) and Least Square Support Vector Regression (LSSVR) are more common among these. Recently, we reported the use of LSSVR to embed and extract binary image as a watermark in three different grayscale images [5]. However, as the FNSVR algorithm finishes its training in a few iterations only, it is believed to be the fastest among them. This probably yields a comparatively fast watermark embedding and extraction scheme among this category of algorithms [6].

Although, all these algorithms have been used to implement image watermarking and its related issues such as minimizing trade-off between visual imperceptibility and robustness, there is hardly any paper reported by any group that would have focused on the issue of time complexity of this problem to the best of our information. Therefore, it remains an interesting avenue to examine the embedding and extraction time complexities to develop a faster watermarking scheme which is capable to complete the task on a real time scale. This may further be utilized for video also. E. G. B. Huang et al. have developed a fast algorithm for training of ANNs with a completely different architecture in place, popularly known as the extreme learning machine (ELM). The benefits of this approach are that it has only one tunable parameter, namely, the number of hidden neurons, and its training process consists of only a single step, thereby reducing the training time up to a large extent. The training of this machine is found to be extremely fast and on regular image databases used by authors, it is reported to have been finished within milliseconds with a reasonably good accuracy [7-9]. Mishra et. al [10] have recently proposed a novel digital image watermarking algorithm based on Extreme Learning Machine (ELM) for two gray scale images. As mentioned, the ELM algorithm is very fast and completes its training in milliseconds unlike its other counterparts such as BPN or FIS based systems They use the ELM output as the watermark to be embedded within the host image using Cox's formula [11] to obtain the signed image. The authors report high PSNR values which indicate that the quality of signed images is good. The computed high value of $SIM(X, X*)$ establishes that the extraction process is successful and according to them the algorithm finds good practical applications, especially in situations that warrant meeting time constraints. Hans-Arno Jacobsen [12] discussed a generic architecture for hybrid intelligent systems in his paper. He has emphasized over integration of intelligent systems that aims at overcoming the limitations of individual techniques through hybridization or fusion of various techniques. According to author, it is difficult to identify merits and demerits of different individual approaches. He argues that neural networks are suited for learning and adaptation but it is like a black box and hence it is not interpretable. On the other hand, the fuzzy knowledge based systems show complementary characteristics. In this case, the incorporation and interpretation of knowledge is straight forward whereas learning and adaptation constitute major drawbacks in case of fuzzy systems. Due to these reasons ANNs and Fuzzy system are complementary to each other and suffer from their inherent drawbacks. The integration of these intelligent systems helps out in overcoming limitations of individual techniques. We, therefore, expect that fusing two different techniques – Fuzzy Inference System (FIS) and a fast neural network (ELM) should yield better results in terms of optimization of processes involved in this problem.

2   Review of Extreme Learning Machine Model

The Extreme Learning Machine (ELM) [7-9] is a Single hidden Layer Feed forward Neural Network (SLFN) architecture. Unlike traditional approaches such as Back Propagation (BP) algorithms which may face difficulties in manual tuning control parameters and local minima, the results obtained after ELM computation are extremely fast, have good accuracy and finally has a solution as that of a system of linear equations. For a given network architecture, ELM does not have any control parameters like stopping criteria, learning rate, learning epochs etc., and therefore, the implementation of this network is very simple and easy. The main concept behind this algorithm is that the input weights (linking the input layer to the hidden layer) and the hidden layer biases are randomly chosen based on some continuous probability distribution function such as uniform probability distribution in our simulation model and the output weights (linking the hidden layer to the output layer) are then analytically calculated using a simple generalized inverse method known as Moore – Penrose generalized pseudo inverse [8].

2.1   Mathematics of ELM Model

Given a series of training samples $(x_i, y_i)_{i=1,2..N}$ and $\tilde{N}$ the number of hidden neurons where $x_i = (x_{i1}, ..., x_{in}) \in \Re^n$ and $y = (y_{i1}, ..., y_{im}) \in \Re^m$, the actual outputs of the single-hidden-layer feed forward neural network (SLFN) with activation tion $g(x)$ for these N training data is mathematically modeled as

$$\sum_{k=1}^{\tilde{N}} \beta_k\, g(\langle w_k, x_i \rangle + b_k) = o_{i,} \forall i = 1, ..., N \tag{1}$$

where $w_k = (w_{k1}, ..., w_{kn})$ is a weight vector connecting the kth hidden neuron, $\beta_k = (\beta_{k1}, ..., \beta_{km})$ is the weight vector connecting the kth hidden neuron and output neurons and b_k is the threshold bias of the kth hidden neuron. The weight vectors w_k are randomly chosen. The term $\langle w_k, x_i \rangle$ denotes the inner product of the vectors w_k and x_i and g is the activation function. The above N equations can be written as

$$H\beta = 0 \tag{2}$$

and in practical applications $\hat{N}$ is usually much less than the number N of training samples and $H\beta \neq Y$,where

$$H = \begin{bmatrix} g(\langle w_1,x_1 \rangle + b_1) & \cdots & g(\langle w_{\hat{N}},x_1 \rangle + b_{\hat{N}}) \\ \vdots & \cdots & \vdots \\ g(\langle w_1,x_N \rangle + b_1) & \cdots & g(\langle w_{\hat{N}},x_N \rangle + b_{\hat{N}}) \end{bmatrix}_{N \times \hat{N}}$$

$$\beta = \begin{bmatrix} \beta_1 \\ \vdots \\ \beta_{\hat{N}} \end{bmatrix}_{\hat{N} \times m}, \quad O = \begin{bmatrix} O_1 \\ \vdots \\ O_N \end{bmatrix}_{N \times m} \quad \text{and} \quad Y = \begin{bmatrix} Y_1 \\ \vdots \\ Y_N \end{bmatrix}_{N \times m} \tag{3}$$

The matrix H is called the hidden layer output matrix. For fixed input weights, $w_k = (w_{k1}, \ldots, w_{kn})$ and hidden layer biases b_k, we get the least-squares solution $\hat{\beta}$ of the linear system of equation $H\beta = Y$ with minimum norm of output weights β, which gives a good generalization performance. The resulting $\hat{\beta}$ is given by $\hat{\beta} = H^+Y$ where matrix H^+ is the Moore-Penrose generalized inverse of matrix H [8]. The above algorithm may be summarized as follows:

2.2   The ELM Algorithm

Given a training set $S = \{(x_i,y_i) \in \mathfrak{R}^{m+n}, y_i \in \mathfrak{R}^m\}_{i=1}^N \Sigma$, for activation function g(x) and the number of hidden neurons $\hat{N}$;

Step1: For $k = 1, \ldots \hat{N}$ randomly assign the input weight vector $w_k \in \mathfrak{R}^n$ and as$b_k \in \mathfrak{R}$

Step2: Determine the hidden layer output matrix H

Step3: Calculate H^+

Step4: Calculate the output weights matrix $\hat{\beta}$ by using $\hat{\beta} = H^+T$

Many activation functions can be used for ELM computation. In the present case, Sigmoid activation function is used to train the ELM

2.3   Computing the Moore-Penrose Generalized Inverse of a Matrix

A matrix G of order $\hat{N} \times N$ is the Moore-Penrose generalized inverse of real matrix A of order if $N \times \hat{N}$ $AGA = A, GAG = G$ and AG, GA are symmetric matrices.

Several methods, for example orthogonal projection, orthogonalization method, iterative methods and singular value decomposition (SVD) methods exist to calculate the Moore-Penrose generalized inverse of a real matrix. In ELM algorithm, the SVD method is used to calculate the Moore-Penrose generalized inverse of H. Unlike other learning methods, ELM is very well suited for both differential and non – differential activation functions. As stated above, in the present work, computations are done using "Sigmoid" activation function.

3   Experimental Details

This research paper is classified under following categories.

3.1   Computation of HVS Parameters

The Luminance Sensitivity: The average of the DC coefficients of the 8 x 8 DCT blocks of the host image in spatial domain is used as luminance sensitivity according to following formula:

$$L = X_{DCi} / X_{DCM} \tag{4}$$

where L is the luminance sensitivity, X_{DCi} denotes the DC coefficient of the ith block and X_{DCM} is the mean value of the DC coefficients of all the blocks put together.

The Edge Sensitivity: As the edge is detected within the image using threshold operation, edge sensitivity can be quantified as a natural corollary to the computation of the block threshold T. The Matlab image processing toolbox implements graythresh ( ) which computes the block threshold using histogram – based Otsu's method [13]. The implementation of this routine is as follows:

$$T = graythresh(f) \tag{5}$$

where f is the host sub-image (block) in question and T is the computed threshold value.

The Contrast Sensitivity: An important way to describe a region is to quantify its texture content or contrast sensitivity. The computed variance value of an image block is the direct metric to quantify this parameter. A Matlab routine proposed by Gonzalez et al. [13] to compute the block variance is used for this purpose. The implementation of this routine is as follows

$$t = statxture(f) \tag{6}$$

where f is the input image or the sub-image (block) and t is the output in the form of a 7 – element row vector, one of which is the variance of the block in question. These three parameters are now fed as input to the Mamdani type Fuzzy Inference System (FIS) as available in the Fuzzy toolbox of Matlab software. Fig. 1 depicts the block diagram of this FIS.

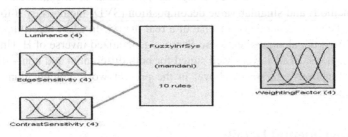

System FuzzyInfSys: 3 inputs, 1 outputs, 10 rules

Fig. 1. Block Diagram of the Fuzzy Inference System

3.2   The Design of Fuzzy Inference System

The Luminance sensitivity of the eye varies from highest dark to highest bright. It has four levels as per the inference rules used in this work. Fig. 2 depicts four levels of luminance sensitivity used in this experiment.

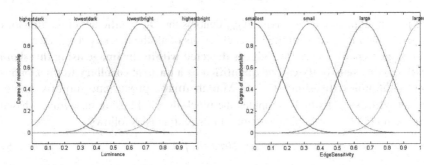

Fig. 2. Luminance Sensitivity Inference Scheme   **Fig. 3.** Edge Sensitivity Inference Scheme

The edge sensitivity of the eye varies from smallest to largest and is also having four levels as per the inference rules used in this work. Fig. 3 depicts four levels of edge sensitivity used in this experiment. The contrast sensitivity of the eye varies from highest smooth to highest rough and is also having four levels according to the inference rules used in this work. Fig. 4 depicts contrast sensitivity used in this experiment. Fig. 5 depicts the variation of the weighting factor (WF) obtained as output variable after executing the FIS as described above. The output varies from very small to very large in this experiment.

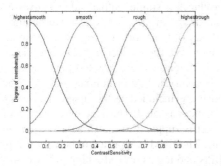 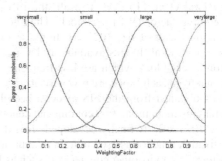

Fig. 4. Contrast Sensitivity Inference Scheme   **Fig. 5.** Inference scheme for the weighting factor as Fuzzy Output

In this work, 10 fuzzy inference rules are used which are listed below:

1. If (Luminance is highestdark) and (EdgeSensitivity is smallest) and (ContrastSensitivity is highestsmooth) then (WeightingFactor is verysmall)
2. If (Luminance is highestdark) and (EdgeSensitivity is large) and (ContrastSensitivity is highestsmooth) then (WeightingFactor is small)
3. If(Luminance is highestdark) and (EdgeSensitivity is smallest) and (ContrastSensitivity is smooth) then (WeightingFactor is small)
4. If (Luminance is highestdark) and (EdgeSensitivity is large) and (ContrastSensitivity is rough) then (WeightingFactor is small)
5. If (Luminance is lowestbright) and (EdgeSensitivity is small) and (ContrastSensitivity is rough) then (WeightingFactor is small)
6. If (Luminance is lowestdark) and (EdgeSensitivity is small) and (ContrastSensitivity is highestrough) then (WeightingFactor is large)
7. If (Luminance is lowestbright) and (EdgeSensitivity is small) and (ContrastSensitivity is highestrough) then (WeightingFactor is small)
8. If (Luminance is lowestbright) and (EdgeSensitivity is largest) and (ContrastSensitivity is highestrough) then (WeightingFactor is large)
9. If (Luminance is highestbright) and (EdgeSensitivity is largest) and (ContrastSensitivity is rough) then (WeightingFactor is large)
10. If (Luminance is heighestbright) and (EdgeSensitivity is largest) and (ContrastSensitivity is highestrough) then (WeightingFactor is verylarge)

3.3   Generating and Embedding the Watermark

In this work, three standard gray scale images – Lena, Baboon and Cameraman of size 256 x 256 are used for embedding the watermark in transform domain. The images are first divided into 8 x 8 blocks in spatial domain which are then converted into transform domain by using DCT method. Three different HVS parameters namely - Luminance Sensitivity, Edge Sensitivity and Contrast Sensitivity are computed for every block individually. These block-wise computed values are subsequently fuzzified using same 10 fuzzy rules initially proposed by Lou et al. [14] and later used by

Agarwal et al. [15]. As a result, the FIS produces a single output value known as weighing factor (WF) per block. Thus, for an image of size 256 x 256, we obtain a column vector of size 1024 x 1. In the second part of this work, the zig - zag scanning is applied to all blocks available in DCT domain. We then select first 21 coefficients from every block which are low frequency coefficients. These low frequency coefficients from each block are row wise arranged in columns 2-22 in an image data set. The WFs obtained from the FIS as described above are included in the image dataset by arranging them in the first column and denoting them as label. Thus, an image dataset of size 1024 x 22 is finally developed. This dataset is supplied to the ELM and Sigmoid activation function is used to train this machine. The ELM is reported to be very fast in comparison to its other counterparts such as BPN and it gets trained within milliseconds time frame for an image of this size. We, therefore, expect that the faster training of this machine would realize the watermark embedding in images on a real time scale. Note that a faster embedding is an important requirement of video watermarking also. As mentioned earlier, the ELM after its training produces a single output per block (or row). Thus, for an image of size 256 x 256, divided into 8 x 8 sized blocks, the ELM finally gives a column vector of size (1024 x 1). The coefficients of this column vector are found to be normalized according to mean 0 and variance 1 or N (0, 1) and are used as watermark to be embedded in the host image using Cox's formula given in Eqn. 7.

$$v_i' = v_i(1.0 + \alpha x_i) \tag{7}$$

where x_i is output obtained from ELM after training, v_i are DCT coefficients and v_i' are coefficients of the signed image. The parameter α is known as scaling coefficient or embedding strength and is optimized to be 0.15 for all our practical calculations. A detailed description of taking $\alpha = 0.15$ is given in section 4.1. The computed watermark is embedded into the largest coefficient of the identified 21 coefficients of each block after zig - zag scan. Finally, inverse DCT of each block is computed to retrieve the signed image in spatial domain. Listing 1 gives watermark embedding algorithm.

Listing 1: Watermark Embedding Algorithm

1. Divide the host image into 8 x 8 size blocks in spatial domain
2. Block wise convert the image into transform domain by using DCT method
3. Apply zig - zag scan to all AC coefficients of each block and select first 21 coefficients from each block barring the DC coefficient and thus develop a dataset of size 1024 x 21 using these coefficients
4. Compute Luminance Sensitivity, Threshold and Variance of all blocks of host image and supply them to the Mamdani Fuzzy Inference System (FIS) built on 10 fuzzy rules. Obtain Weighting Factor (WF) for all blocks from this FIS and develop a column vector of size 1024 x 1
5. Use this column vector as label in the first column of the dataset obtained in step 3 and thus obtain a dataset of size 1024 x 22
6. Supply this dataset to the Extreme Learning Machine in regression mode for training using Sigmoid activation function
7. The trained ELM produces an output vector of size 1024 x 1 which is used for watermarking in the host image by using Cox's algorithm
8. Take the Inverse DCT (IDCT) of this matrix of coefficients to obtain signed image

3.4   Extracting the Watermark

In the extraction process, block wise DCT of host and signed images are computed and the coefficients of the original image which are used in the embedding process are subtracted from the respective coefficients of the signed image. This is according to the extraction scheme given by Cox et. al. [11]. In this manner, both the original and recovered watermark sequences X and X* are known. These watermarks are correlated by similarity correlation $SIM(X, X*)$ and normalized correlation $NC(X, X*)$ parameters given by Eqns. 8. Listing 2 depicts the watermark extraction algorithm.

$$SIM(X, X*) = \frac{\sum_{i=1}^{m}\sum_{j=1}^{n}[X(i,j).X*(i,j)]}{\sum_{i=1}^{m}\sum_{j=1}^{n}\sqrt{X*.X*}} \quad NC(X, X*) = \frac{\sum_{i=1}^{m}\sum_{j=1}^{n}[X(i,j).X*(i,j)]}{\sum_{i=1}^{m}\sum_{j=1}^{n}[X*(i,j)]^2} \quad (8)$$

Listing 2: Watermark Extraction Algorithm

1. Apply 8 x 8 block coding on original and watermarked images
2. Compute DCT coefficients of all blocks of both these images
3. Subtract the computed DCT coefficients of the original image from respective coefficients of watermarked image and thus recover the watermark
4. Compute $SIM(X, X*)$ and $NC(X, X*)$ parameters to check similarity between X and X*

In the second part of this experiment, the watermarked images are subject to five different image processing attacks to examine the robustness of the embedding algorithm. These are discussed in detail in section 3.6.

4   Experimental Results

4.1   Embedding and Extraction

Fig. 6(a-c) show original host images Lena, Baboon and Cameraman of size 256x256. Fig. 7(a-c) depict signed images - Lena, Baboon and Cameraman. Their MSE and PSNR values for $\hat{N}$ = 100 hidden neurons are mentioned on the top of these images. The PSNR values for these images indicate that the perceptible quality of the signed images is quite good. Table 1 compiles the original and extracted watermarks which are converted in the form of 32 x 32 size images for visual comparison. The $SIM(X, X*)$ and $NC(X, X*)$ values are also compiled in Table 1. A close observation of Table 1 indicates a high degree of similarity between embedded and recovered watermarks. This is also indicated by the fact that in case of all three images, we obtain high computed values of $SIM(X, X*)$ and $NC(X, X*)$ parameters. However, the best results are obtained for Baboon.

6(a)

6(b)

MSE= 0.00880, PSNR=49.2658 dB

6(c)

7(a)

MSE= 0.00790, PSNR=50.1656 dB   **MSE= 0.00106, PSNR=47.5689 dB**

7(b)

7(c)

Fig. 6. Original host images (a) Lena, (b) Baboon and (c) Cameraman

Fig. 7. Signed Images (a) Lena, (b) Baboon and (c) Cameraman

Table 2 - 4 compile PSNR, $SIM(X, X*)$ and $NC(X, X*)$ values for all three images as a function of number of hidden neurons $\hat{N}$. These values are found to be almost stable.

Table 1. Original and extracted watermarks for Lena, Baboon and Cameraman

Image	Original Watermark	Extracted Watermark	$SIM(X, X*)$	$NC(X, X*)$
Lena			21.5550	0.9302
Baboon			27.1711	0.9078
Cameraman			18.9264	0.9274

Table 2. PSNR, $SIM(X, X*)$ and $NC(X, X*)$ as a function of $\hat{N}$ for signed Lena

Hidden Neurons $\hat{N}$	ELM Train Time (sec)	Embed Time (sec)	Extract Time (sec)	Total Time (sec)	PSNR (dB)	SIM (X, X*)	NC (X, X*)
20	0.0156	13.3125	0.3594	13.6875	49.2641	21.5190	0.9287
40	0.0469	14.6094	0.3906	15.0469	49.2568	21.5399	0.9287
60	0.0781	13.7813	0.3906	14.25	49.2658	21.5550	0.9302
80	0.1094	15.0156	0.4375	15.5625	49.2803	21.5152	0.9364
100	0.2500	14.0313	0.4375	14.7188	49.2658	21.5587	0.9364

These tables also give the ELM training time, Embedding time and time for extracting the watermark from the signed images. Note that the ELM training time is of the order of millisecond which indicates that the machine is very fast. The total time which encapsulates training of ELM, embedding and extraction of watermark is of the order of seconds. This is unlike its other counterparts such as BPN which consumes more time to train and thus probably becomes unsuitable for this application on a real time scale. Note that this is as per our expectation mentioned in Section 3.2. It is therefore inferred that this machine may also be successfully used to embed and extract watermarks from videos. Fig. 8(a-b) respectively depict plots of PSNR and $SIM(X, X*)$ for all three images with respect to different values of scaling coefficient or embedding strength (α) considered in this work to optimize the formula given in Eqn. (7).

Table 3. PSNR, $SIM(X, X*)$ and $NC(X, X*)$ as a function of $\hat{N}$ for signed Baboon

Hidden Neurons $\hat{N}$	ELM Train Time (sec)	Embed Time (sec)	Extract Time (sec)	Total Time (sec)	PSNR (dB)	SIM (X, X*)	NC (X, X*)
20	0.0313	12.8906	0.3906	13.3125	50.1080	27.1854	0.9076
40	0.0469	12.9375	0.3906	13.375	50.1399	27.1678	0.9097
60	0.0781	12.7969	0.3908	13.2658	50.1475	27.1711	0.9078
80	0.0938	10.4375	0.3906	10.9219	50.1582	27.1728	0.9115
100	0.1563	10.8438	0.4219	11.422	50.1656	27.1354	0.9103

Table 4. PSNR, $SIM(X, X*)$ and $NC(X, X*)$ as a function of $\hat{N}$ for signed Cameraman

Hidden Neurons $\hat{N}$	ELM Train Time (sec)	Embed Time (sec)	Extract Time (sec)	Total Time (sec)	PSNR (dB)	SIM (X, X*)	NC (X, X*)
20	0.0469	10.9531	0.3906	11.3906	47.3442	19.0768	0.9332
40	0.0469	11.0938	0.4063	11.547	47.3469	18.9562	0.9296
60	0.0781	10.6563	0.3906	11.125	47.4232	18.9264	0.9274
80	0.1250	10.8438	0.4063	11.3751	47.5510	19.0030	0.9418
100	0.1406	10.7344	0.3906	11.2656	47.5689	18.8217	0.9368

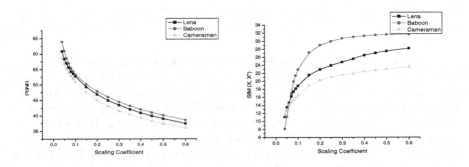

Fig. 8. Plot of (a) PSNR and (b) $SIM(X, X*)$ with respect to Scaling Coefficient (α)

From both these plots, it is evident that an optimized value of α is 0.15 as on either side from it, the PSNR and $SIM(X, X*)$ are found to be stabilized. In other words, if a tangent is drawn on these curves, it shall have a slope nearly equal to +1 or -1 at around α = 0.15. This is suggestive of optimized embedding and extraction to take place at $\alpha = 0.15$ and we, therefore, consider this value of α for all our practical computations executed in the course of this experiment.

Motwani et al. [3] have used three HVS characteristics namely brightness, texture and edge distance to be fed into a Mamdani type fuzzy inference system (FIS) driven by 27 inference rules. They have computed these fuzzy input variables for each wavelet coefficient in the image. The output of their fuzzy system is a single crisp value which gives a perceptual value for each corresponding wavelet coefficient. They argue that the fuzzy logic based watermarks are robust to attacks and at the same time achieve a high level of imperceptibility. However, they have not discussed their watermarking scheme in terms of time complexity of embedding and extraction modules. We, in this paper, present better visual quality of signed Lena image. As far as image processing attacks to examine robustness are concerned, we execute a number of attacks including Jpeg (QF = 50%). It is worth mentioning here that we obtain better results in case of JPEG attack as compared to those obtained by Motwani et al. [3]. A detailed description of these attacks is presented in the next section.

4.2   Executing Image Processing Attacks

As mentioned in section 3.3, in the second part of this experiment, the watermarked images have undergone five different image processing attacks to examine the robustness of the embedding algorithm. These are: (1) JPEG compression (QF = 90), (2) Gaussian Blur (Radius = 1.0 unit), (3) Brightness and Contrast (each 10%), (4) Median Filtering (Filtering aperture = 3 units) and (5) 10% Gaussian noise addition. The attacked images are subsequently examined by PSNR, $SIM(X, X*)$ and $NC(X, X*)$. Table 5 compiles these values for all three images. Note that the best results are achieved for attacked Baboon image while attacked Cameraman is the worst. Note that a similar behavior is obtained for signed images also.

Table 4. SIM(X,X*) and NC(X,X*) values for attacked images

Image / Attack	Lena			Baboon			Cameraman		
	PSNR	SIM (X, X*)	NC (X,X*)	PSNR (dB)	SIM (X, X*)	NC (X,X*)	PSNR (dB)	SIM (X, X*)	NC (X,X*)
JPEG Compression	39.0478	12.7153	0.8508	37.3231	15.9166	0.8611	38.1254	11.3129	0.8799
Gaussian Blur	29.2858	11.4209	0.8017	25.7309	14.5590	0.8032	26.1450	10.2500	0.8410
Brightness & Contrast	26.9411	11.7106	0.8011	26.8859	15.9128	0.8110	26.7155	10.7858	0.8510
Median Filter	30.9994	11.3914	0.8311	25.3321	14.3378	0.8222	27.3585	10.3687	0.8527
Gaussian Noi	30.5415	11.7827	0.8211	30.5167	16.0045	0.8557	30.5866	10.8160	0.8616

We report high computed values of PSNR, $SIM(X, X*)$ and $NC(X, X*)$ after executing attacks. This is indicative of the fact that the proposed watermark embedding and extraction scheme is very robust against selected attacks. As compared to JPEG (QF = 50%) attacked Lena image by Motwani et. al. [3], our PSNR = 32.4762 dB and $NC(X, X*)$ = 0.8311. The correlation value reported in [3] is 0.835 for the same JPEG attack which is almost equal to our value. The authors did not report PSNR values of the attacked images. Also that, they take the salience factor (α) = 0.001 which is very small as compared to our value of the scaling parameter (α = 0.15). On the other hand, they use a different formulation to embed the normalized sequence watermark into host images. It is therefore clear, that we obtain better results as compared to those reported in [3]. Hans-Arno Jacobsen [12] emphasized on a generic architecture for hybrid intelligent systems. He has argued in favor of integration of intelligent systems that aims at overcoming the limitations of individual techniques through hybridization or fusion of various techniques. He further argues that neural networks are suited for learning and adaptation while the fuzzy knowledge based systems show complementary characteristics. The integration of these intelligent systems helps out in overcoming limitations of individual techniques. As expected in Section 1, in the present case, the fusion of Fuzzy Inference System (FIS) and extreme learning machine (ELM) has given better results in terms of optimization of embedding and extraction processes. In view of this discussion, it can be concluded that the proposed optimized watermark embedding and extraction scheme for gray scale images offers less time complexity. This scheme is also found to be very robust against common image processing attacks.

5   Conclusions

A novel image watermarking algorithm is proposed in this paper in transform domain taking into account three different Human Visual System (HVS) parameters which are initially fed to the Mamdani type Fuzzy Inference System (FIS) for three gray scale images – Lena, Baboon and Cameraman. The three HVS parameters used in this work are luminance sensitivity, edge sensitivity and contrast sensitivity. The output of the FIS is a single crisp output known as weighing factor (WF). A dataset is prepared by using this WF as label. A newly developed fast neural network known as Extreme Learning Machine (ELM) is trained using this dataset. The ELM produces a row vector of size 1024 which is used as watermark embedded within the grayscale images.

The signed images are subject to quality assessment using PSNR before and after implementing image processing attacks to examine the issue of robustness. Besides this, the similarity correlation parameter $SIM(X, X*)$ and normalized correlation parameter $NC(X, X*)$ is computed for signed as well as attacked images. High values of SIM and NC parameters indicate that the extraction process is quite successful. It is concluded that the Fuzzy - ELM used to embed the watermark in this work produces very good results in terms of perceptibility and robustness assuming capacity to be constant. The watermark embedding and extraction procedures are found to be well optimized and as the proposed scheme has a time complexity of the order of milliseconds, it offers good real time practical applications.

References

[1] Huang, S., Zhang, W., Feng, W., Yang, H.: Blind watermarking scheme based on neural network. In: 7th World Congress on Intelligent Control and Automation (WCICA 2008), pp. 5985–5989 (2008)

[2] Piao, C.-R., Beack, S.-H., Woo, D.-M., Han, S.-S.: A Blind Watermarking Algorithm Based on HVS and RBF Neural Network for Digital Image. In: Jiao, L., Wang, L., Gao, X.-B., Liu, J., Wu, F. (eds.) ICNC 2006. LNCS, vol. 4221, pp. 493–496. Springer, Heidelberg (2006)

[3] Motwani, M.C., Harris Jr., F.C.: Fuzzy Perceptual Watermarking for Ownership Verification. In: Proceedings of the International Conference on Image Processing, Computer Vision, and Pattern Recognition (IPCV 2009), Las Vegas, Nevada, July 13-16 (2009)

[4] Mohanty, S.P., Ramakrishnan, K.R., Kankanhalli, M.: A Dual Watermarking Technique for Images. In: ACM Multimedia, Part 2, pp. 49–51 (1999)

[5] Mehta, R., Mishra, A., Singh, R., Rajpal, N.: Digital Image Watermarking in DCT Domain Using Finite Newton Support Vector Regression. In: Proceedings of 6th International Conference on Intelligent Information Hiding and Multimedia Signal Processing, pp. 123–126 (2010)

[6] Chaudhary, V., Mishra, A., Mehta, R., Verma, M., Singh, R., Rajpal, N.: Watemarking of Grayscale Images in DCT Domain Using Least-Squares Support Vector Regression. International Journal of Machine Learning and Computing 2(6), 725–728 (2012)

[7] Huang, G.-B., Zhu, Q.-Y., Siew, C.K.: Extreme Learning Machine: Theory and Applications. Neurocomputing (70), 489–501 (2006)

[8] Huang, G.-B., Zhu, Q.-Y., Siew, C.K.: Real-Time Learning Capability of Neural Networks. IEEE Transactions on Neural Networks 17(4), 863–878 (2006)

[9] Huang, G.-B.: The Matlab code for ELM (2004),
http://www.ntu.edu.sg/home/egbhuang

[10] Mishra, A., Goel, A., Singh, R., Singh, L., Chetty, G.: A Novel Image Watermarking Scheme Using Extreme Learning Machine. In: Conference Proceedings of International Joint Conference on Neural Networks (IJCNN), pp. 1–6 (June 2012)

[11] Cox, I.J., Kilian, J., Leighton, F.T., Shamoon, T.: Secure spread spectrum watermarking for multimedia. IEEE Transactions on Image Processing 6(12), 1673–1687 (1997)

[12] Jacobsen, H.A.: A Generic Architecture for Hybrid Intelligent Systems. In: Proceedings of The IEEE World Congress on Computational Intelligence (FUZZ IEEE), USA, vol. 1, pp. 709–714 (1998)

[13] Gonzalez, R.C., Woods, R.E., Eddins, S.L.: Digital Image Processing Using MATLAB, p. 406, 467. Pearson Education (2005)

[14] Lou, D.-C., Hu, M.-C., Liu, J.-L.: Healthcare Image Watermarking Scheme Based on Human Visual Model and Back-Propagation Network. Journal of C.C.I.T 37(1), 151–162 (2008)

[15] Agarwal, C., Mishra, A., Sharma, A.: Digital image watermarking in DCT domain using Fuzzy Inference System. In: Proceedings of 24th IEEE Conference - CCECE-2011, pp. 822–825 (2011)

Leveraging Two-Stage Weighted ELM for Multimodal Wearables Based Fall Detection

Zhenyu Chen[1,2,*], Yiqiang Chen[1], Lisha Hu[1,2],
Shuangquan Wang[1], and Xinlong Jiang[1,2]

[1] Beijing Key Laboratory of Mobile Computing and Pervasive Device,
Institute of Computing Technology, Chinese Academy of Sciences, Beijing, China
chenzhenyu@ict.ac.cn.
[2] University of Chinese Academy of Sciences, Beijing, China

Abstract. For the elderly people, timely detecting the fall accident is very critical to receive the first aid. In order to achieve high detection accuracy and low false-alarm rate at the same time, we propose a multimodal wearables based fall detecting and monitoring method leveraging two-stage weighted extreme learning machine. Experimental results show that our method is able to effectively implement on miniaturized wearable devices, and compared to state-of-the-art ELM classifier, we can also obtain higher detection accuracy and lower false-alarm rate simultaneously, which enables various kinds of mHealth applications in large-scale population, especially for the elderly people's healthcare in the field of fall detection.

Keywords: mHealth, Fall Detection, Wearable Device, Extreme Learning Machine, Weighted ELM.

1 Introduction

With the rapid development of sensor technology and the increasing availability of affordable sensor-embedded wearable devices, commonly massive sensor-based wearables enables various kinds of sensing and communication capabilities in large-scale applications. Especially for the elderly people who always live independently, the fall has become great threat to the public health and seriously diminish the quality of people's lives, accordingly leads to various psychological problems like psychological fear of movement, worry about independent living [1], etc. Therefore, timely detecting and monitoring the occurrence of fall accidents will great help injured elderly people get the valuable first-aid chance. Meanwhile, if the alarm for fall detection often happens wrongly, it probably heavily interrupts the people's regular livings and behaviors.

Noury [1] discusses the psychological consequences after the falling. The worst one is that the elderly cannot call for help after falling with unconscious. Therefore, the aim of their design for the fall detector is to guarantee that the elderly

---

* Corresponding author.

© Springer International Publishing Switzerland 2015
J. Cao et al. (eds.), *Proceedings of ELM-2014 Volume 2,*
Proceedings in Adaptation, Learning and Optimization 4, DOI: 10.1007/978-3-319-14066-7_16

can obtain the first aid after the falling, reducing their feared state of mind and encouraging them to more active activity. Narayanan et al. [2] invent a Waist-mounted rechargeable Triaxial Accelerometer-based device to detect and prevention the elderly's falling. The device in [2] is named "Prevent Fall Ambulatory Monitor", which denotes by PFAM for short. The system is not only a detector, but also supply the elderly with self-test for risk assessment of the falling. Grassi et al. [3] put forward the high-reliability fall detection framework, which uses 3D time-of-flight range camera, wearable Zigbee MEMS accelerometer and microphone, where the camera adopts person detection and tracking algorithm to detect the centroid height of the people. When this height is below a threshold, the people is regarded as the falling status. Shi et al. [4] implement an Android application called "uCare" to detect the elderly's falling and seek the first aid. They present the fall detection method by exploring a five-phase model which details the state alternation during the elderly's falling activity. The accelerometer data generated from mobile phones is used for the proposed five-phase model to improve the accuracy of fall detection. Zhou et al. [5] propose an activity transition based fall detection model, this model mainly extracts features from transition data between adjacent activities to recognize various kinds of normal activities and abnormal activities, which is able to reduce a number of normal activities but focuses on abnormal activity in the transition section.

Accordingly, how to build the classifier for achieving high accuracy and low false-alarm rate simultaneously, has become a key challenging issue for imbalance learning in large-scale wearable computing applications. In this paper, running on eyeglass and watch miniaturized wearable devices, our work is to accurately and timely detect and monitor the fall accident using weighted extreme learning machine. The empirical experimental results show that our proposed method is effective and does contribute good impact in the field of fall detecting and monitoring, especially for the elderly people's healthcare.

2   Our Method

The framework of our proposed method is shown in Figure 1, which mainly consists of the initial classification step, the result refining step in the following.

Step (1): the initial classification step. For the accelerometer sampling data from wearable eyeglass device, its related features are extracted accordingly, then the initial classification is trained on the labeled training dataset using weighted extreme learning machine [6], and the optimal parameters of this classifier such as C and number of hidden nodes are generated. Based on the trained classifier, the samples whose classification results are suspected fall will be used as the input of the next stage for result refining. Also, running on wearable watch device with another set of optimal parameters like C and number of hidden nodes, the initial weighted ELM classifier keeps detecting suspected fall activity and is utilized for result refining as the input of next stage as well.

Step (2): the result refining step. If wearable eyeglass device is detected as suspected fall activity, the feature data stored on eyeglass device will be triggered to transmit to the connected mobile phone via bluetooth interface. At the same time, in case when the suspected fall is detected on eyeglass, the feature the feature data stored on wearable watch device will be transmitted as well if suspected fall happened on watch. Accordingly, eyeglass generated features, or eyeglass and watch generated multimodal features will be further refined through another weighted extreme learning machine classifier again. Here, the weight value is different from the first stage, because samples of different classes of all activities are varying between the first stage and the second stage.

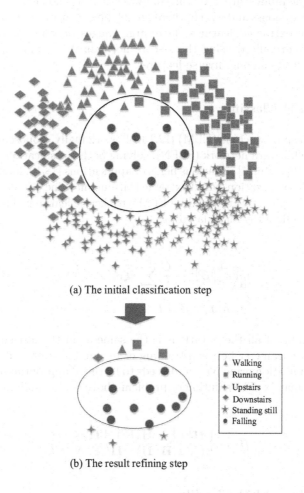

(a) The initial classification step

(b) The result refining step

▲	Walking
■	Running
✦	Upstairs
◆	Downstairs
✦	Standing still
●	Falling

Fig. 1. Our proposed method for fall detecting and monitoring

2.1 Feature Extraction

For the collected accelerometer magnitude series from eyeglass and watch devices, the sampling rate is 10Hz and the length of windows is 13, totally 14 acceleration features [7] are extracted without overlapping between consecutive windows. They are maximum, minimum, mean, standard deviation, energy, zero crossing rate, four amplitude statistics features and four shape statistics features of the power spectral density (PSD) [8].

Extracting these features from accelerometer data on eyeglass, there are 14 features as the input feature vector, and when features come from accelerometer data on both eyeglass and watch, there are 28 input features in total. Meanwhile, due to these extracted features have large ranges of value domain, all these features are normalized using the z-score normalization approach to eliminate the scaling effects among diverse features.

2.2 COELM Classifier

Extreme Learning Machine (ELM) [12,13,14] is originally proposed for single hidden layer feedforward network (SLFN), solving both classification and regression problems [9,10,11,16,17] in various fields of applications [18,19,20,21,22,23,24,25]. Constrained-optimization-based extreme learning machine (COELM) [15] is proposed with the purpose of extending ELM with kernel learning. The classification problem of COELM is formulated as:

$$\min_{\beta,\xi} \frac{1}{2}\|\beta\|^2 + \frac{C}{2}\sum_{i=1}^{N}\|\xi_i\|^2 \tag{1}$$
$$\text{s.t. } h(x_i) \cdot \beta = t_i - \xi_i, \ i = 1, \cdots, N$$

The meaning of all the notations is the same as in the last subsection. C is the trade-off parameter between training error minimization and generalization ability maximization principles, and needs to be tuned appropriately. $\mathbf{I}$ represents the identity matrix. The solution of problem above is also analytical determined as:

$$\beta^* = \begin{cases} \mathbf{H}^T(\frac{\mathbf{I}}{C} + \mathbf{H}\mathbf{H}^T)^{-1}\mathbf{T}, N < L \\ (\frac{\mathbf{I}}{C} + \mathbf{H}^T\mathbf{H})^{-1}\mathbf{H}^T\mathbf{T}, N > L \end{cases} \tag{2}$$

2.3 Weighted ELM Classifier

Weighted extreme learning machine (Weighted ELM) [6] is proposed based on constrained-optimization-based extreme learning machine (COELM) recently, with the purpose of extending ELM for imbalance learning. The classification problem of Weighted ELM is formulated as:

$$\text{minimize}\ \ L_{P_{ELM}} = \frac{1}{2}\|\beta\|^2 + C\mathbf{W}\frac{1}{2}\sum_{i=1}^{N}\|\xi_i\|^2 \tag{3}$$
$$\text{s.t.}\ \ h(x_i)\cdot\beta = t_i^T - \xi_i^T,\ i = 1,\cdots,N$$

Similarly, the solution of problem above is also determined as:

$$\beta = \begin{cases} \mathbf{H}^T(\frac{\mathbf{I}}{C} + \mathbf{WHH}^T)^{-1}\mathbf{WT},\ \text{when N is small} \\ (\frac{\mathbf{I}}{C} + \mathbf{H}^T\mathbf{WH})^{-1}\mathbf{H}^T\mathbf{WT},\ \text{when N is large} \end{cases} \tag{4}$$

They [6] firstly use a weighting scheme $\mathbf{W}1$ automatically generated from the class information, which is in fact a special case of the cost sensitive learning:

$$\mathbf{W}1:\ \ W_{ii} = 1/\#(t_i) \tag{5}$$

Accordingly, for our fall detection problem, we adopt different weights for the initial classification step and the result refining step as the weighting scheme $\mathbf{W}2$ [6]:

$$\mathbf{W}2:\ \ W_{ii} = \begin{cases} 0.618/\#(t_i),\ t_i > AVG(t_i) \\ 1/\#(t_i),\ \ \ \ \text{otherwise} \end{cases} \tag{6}$$

3   Experiment

In the data collection, our study population contains 1 female and 6 males ranging from 23 to 29 years old, each participant carries a smartphone and wears an eyeglass device during the data collection and collects six kinds of activities, including the fall, standing still, walking, running, downstairs and upstairs, where the fall consists of front, back, left and right-toward falls. The duration of standing still is about 15 seconds each time, and other activities except the fall are collected for 2-3 minutes. The duration of fall activity is about 2.6 seconds on average, and each participant is required to lie immediately before and after the fall activity, also keep still for 5 seconds.

In the experimental evaluation, the precision and recall are used to measure the accuracy, which are defined in the following:

$$Precision = \frac{\sharp\ of\ true\ positive}{\sharp\ of\ true\ positive + \sharp\ of\ false\ positive} \tag{7}$$

$$Recall = \frac{\sharp\ of\ true\ positive}{\sharp\ of\ true\ positive + \sharp\ of\ false\ negative} \tag{8}$$

In the classifier training on wearable eyeglass and watch devices, there are two parameters, i.e. the regularization coefficient C and the number of hidden nodes, need to be determined. We utilize grid-search method to select the optimal parameters from 100 pairs of parameters and employ 10-fold cross-validation

Table 1. The optimal parameters of Weighted ELM classifier for eyeglass and watch features in the first stage

	Optimal parameter	Value	Index
Eyeglass	C	16	4
	Hidden nodes	256	8
Watch	C	16	4
	Hidden nodes	1024	10

Table 2. The optimal parameters of Weighted ELM classifier for eyeglass and eyeglass-watch features in the second stage

	Optimal parameter	Value	Index
Eyeglass	C	2	1
	Hidden nodes	512	9
Eyeglass-Watch	C	4	2
	Hidden nodes	64	6

Table 3. The performance comparison of precision and recall between the original method of ELM and two-stage method of weighted ELM

	Original ELM	Two-stage weighted ELM
Precision	73.68%	95.74%
Recall	90.74%	93.67%

method, then we obtain optimal parameters of the regularization coefficient C and the number of hidden nodes in Table 1 and Table 2 for different two stages, respectively.

Accordingly, in the first classification step, the initial fall detection indicates the suspected fall detection through the weighted ELM classifier over wearable eyeglass device. Then, the initial detection result is further recognized for determining the final fall activity, this detecting result by weighted ELM classifier is utilized as the input to the next step. Next, in the second result fining step, based on the feature data generated from wearable eyeglass device or eyeglass-watch devices, the suspected fall is further refined through another weighted ELM classifier again, through adopting different weights from the first step. At last, we can see from Table 3, our proposed two-stage weighted ELM classifier can obtain higher recall rate (the recall rate increases from 90.74% to 93.67%) than original ELM classifier, and significantly improve the precision rate from 73.68%

to 95.74%, which shows that the false-alarm rate is lower than 5%. Accordingly, compared to the result of original ELM classifier, our method can achieve the high detection accuracy and low false-alarm rate simultaneously, also verify the effectiveness of imbalance learning for the weighted ELM classifier.

4 Conclusion

In this paper, we propose a multimodal wearable devices (i.e. eyeglass and watch) based fall detection method using two-stage weighted ELM. In particular, we explore the initial classification to detect the suspected fall firstly, then another weighted ELM classifier is employed again to refine the detection result. Compared to the original ELM classifier, our proposed method is able to detect the fall activity with high detection accuracy and low false-alarm rate simultaneously. Our encouraging results contribute towards the on-going development of mHealth technology which is effective to track the key fall activity in the field of healthcare, especially for the elderly people in the real-world.

Acknowledgments. This work is supported in part by National Natural Science Foundation of China (61173066, 61472399), National Science and Technology Major Project (2012ZX07205-005), Beijing Natural Science Foundation (4144085), Open Project of Beijing Key Laboratory of Mobile Computing and Pervasive Device.

References

1. Noury, N.: A smart sensor for the remote follow up of activity and fall detection of the elderly. In: 2nd Annual International IEEE-EMB Special Topic Conference on Microtechnologies in Medicine & Biology, pp. 314–317. IEEE (2002)
2. Narayanan, M.R., Lord, S.R., Budge, M.M., Celler, B.G., Lovell, N.H.: Falls management: detection and prevention, using a waist-mounted triaxial accelerometer. In: 29th Annual International Conference of the IEEE Engineering in Medicine and Biology Society, EMBS 2007, pp. 4037–4040. IEEE (2007)
3. Grassi, M., Lombardi, A., Rescio, G., Malcovati, P., Malfatti, M., Gonzo, L., Leone, A., Diraco, G., Distante, C., Siciliano, P., et al.: A hardware-software framework for high-reliability people fall detection. In: 2008 IEEE Sensors, pp. 1328–1331. IEEE (2008)
4. Shi, Y., Shi, Y., Wang, X.: Fall detection on mobile phones using features from a five-phase model. In: 2012 9th International Conference on Ubiquitous Intelligence & Computing and 9th International Conference on Autonomic & Trusted Computing (UIC/ATC), pp. 951–956. IEEE (2012)
5. Zhou, M., Wang, S., Chen, Y., Chen, Z., Zhao, Z.: An activity transition based fall detection model on mobile devices. In: Park, J.H.(J.), Jin, Q., Yeo, M.S.-S., Hu, B. (eds.) Human Centric Technology and Service in Smart Space. LNEE, vol. 182, pp. 1–8. Springer, Heidelberg (2012)
6. Zong, W.W., Huang, G.B., Chen, Y.: Weighted extreme learning machine for imbalance learning. Neurocomputing 101, 229–242 (2013)

7. Figo, D., Diniz, P.C., Ferreira, D.R., Cardoso, J.M.P.: Preprocessing techniques for context recognition from accelerometer data. Personal Ubiquitous Comput. 14 (October 2010)
8. Wang, X.L.: High accuracy distributed target detection and classification in sensor networks based on mobile agent framework (2004)
9. Gao, X., Hoi, S.C.H., Zhang, Y., Wan, J., Li, J.: SOML: Sparse online metric learning with application to image retrieval. In: AAAI, pp. 1206–1212 (2014)
10. Chen, Z., Chen, Y., Wang, S., Liu, J., Gao, X., Campbell, A.T.: Inferring social contextual behavior from bluetooth traces. In: UbiComp, pp. 267–270 (2013)
11. Chen, Z., Lin, M., Chen, F., Lane, N.D., Cardone, G., Wang, R., Li, T., Chen, Y., Choudhury, T., Campbell, A.T.: Unobtrusive sleep monitoring using smartphones. In: PervasiveHealth, pp. 145–152 (2013)
12. Huang, G.B., Zhu, Q.Y., Siew, C.K.: Extreme learning machine: a new learning scheme of feedforward neural networks. In: Proceedings of the 2004 IEEE International Joint Conference on Neural Networks, vol. 2, pp. 985–990. IEEE (2004)
13. Huang, G.B., Zhu, Q.Y., Siew, C.K.: Extreme learning machine: theory and applications. Neurocomputing 70(1), 489–501 (2006)
14. Huang, G.B., Chen, L., Siew, C.K.: Universal approximation using incremental constructive feedforward networks with random hidden nodes. IEEE Transactions on Neural Networks 17(4), 879–892 (2006)
15. Huang, G.B., Zhou, H., Ding, X., Zhang, R.: Extreme learning machine for regression and multiclass classification. IEEE Transactions on Systems, Man, and Cybernetics, Part B: Cybernetics 42(2), 513–529 (2012)
16. Chen, Y., Chen, Z., Liu, J., Hu, D.H., Yang, Q.: Surrounding context and episode awareness using dynamic bluetooth data. In: UbiComp, pp. 629–630 (2012)
17. Zhao, Z., Chen, Y., Liu, J., Shen, Z., Liu, M.: Cross-people mobile-phone based activity recognition. In: IJCAI, pp. 2545–2550 (2011)
18. Chen, Y., Zhao, Z., Wang, S., Chen, Z.: Extreme learning machine-based device displacement free activity recognition model. Soft Computing 16(9), 1617–1625 (2012)
19. Zhao, Z., Chen, Y., Wang, S., Chen, Z.: Fallalarm: Smart phone based fall detecting and positioning system. In: ANT/MobiWIS, pp. 617–624 (2012)
20. Chen, Z., Wang, S., Shen, Z., Chen, Y., Zhao, Z.: Online sequential elm based transfer learning for transportation mode recognition. In: IEEE Conference on Cybernetics and Intelligent Systems (CIS), pp. 78–83. IEEE (2013)
21. Wang, S., Chen, Y., Chen, Z.: Recognizing transportation mode on mobile phone using probability fusion of extreme learning machines. International Journal of Uncertainty, Fuzziness and Knowledge-Based Systems 21(suppl. 02), 13–22 (2013)
22. Zhao, Z., Chen, Z., Chen, Y., Wang, S., Wang, H.: A class incremental extreme learning machine for activity recognition. Cognitive Computation 6(3), 423–431 (2014)
23. Hu, L., Chen, Y., Wang, S., Chen, Z.: b-COELM: A fast, lightweight and accurate activity recognition model for mini-wearable devices. Pervasive and Mobile Computing (2014)
24. Chen, Z., Chen, Y., Hu, L., Wang, S., Jiang, X., Ma, X., Lane, N.D., Campbell, A.T.: Contextsense: Unobtrusive discovery of incremental social context using dynamic bluetooth data. In: UbiComp 2014, pp. 23–26. ACM (2014)
25. Jiang, X., Liu, J., Chen, Y., Liu, D., Gu, Y., Chen, Z.: Feature adaptive online sequential extreme learning machine for lifelong indoor localization. Neural Computing and Applications (2014)

A Novel Web Service Quality Prediction Framework Based on F-ELM

Ying Yin, Yuhai Zhao, Gang Sheng, Bin Zhang, and Guoren Wang

College of Information Science and Engineer
Northeastern University, Shenyang, China 110819
yinying@ise.neu.edu.cn

Abstract. With the prevalence of service computing and cloud computing, more and more services are emerging and running on highly dynamic and changing environments (The Web). Under these uncontrollable circumstances, these services will generate huge volumes of data, such as trace log, QoS (Quality of Service) information and WSDL files. It is impractical to monitor the changes in QoS parameters for each and every service in order to timely trigger precaution, due to high computational costs associated with the process. In order to overcome the above problem, this paper proposes a web service quality prediction method based on improved Extreme Learning Machine with feature optimization. First, we extract web service trace logs and QoS information from the service log and convert them into feature vectors. Furthermore, in order to actively respond ELM more quickly, we mine early feature subsets in advance by developing a feature mining algorithm, named FS, and apply such feature subsets to ELM for training (F-ELM). Experimental results prove that F-ELM (trained by the selected feature subsets) can efficiently lift the reliability of service quality and improve the earliness of prediction.

Keywords: Extreme Learning Machine, Prediction, Feature Selection.

1 Introduction

The advantage of composite Web services is that it realizes a complex application by connecting multiple component services seamlessly. However, in real applications, Web service lives in a highly dynamic environment, both the network condition and the operational status of each of the constituent Web Services(WSs) may change during the life-time of a business process itself. The instability brought on by various uncertain factors often makes the composite services to be interrupted or interrupted temporally. Therefore, it is very important to ensure the normal execution of the composite service applications and to provide a reliable software system[1].

Web service Prediction is an important technology which aims to optimally select and integrate reliable Web services and to ensure reliable operation of these services. For a series of Web services with similar functionality, QoS (Quality of

© Springer International Publishing Switzerland 2015
J. Cao et al. (eds.), *Proceedings of ELM-2014 Volume 2*,
Proceedings in Adaptation, Learning and Optimization 4, DOI: 10.1007/978-3-319-14066-7_17

Service) plays a key factor to differ the different Web services quality from their providers. In predicting the quality of a Web service for user, it is important to consider non-functional properties of the Web service so as to satisfy the constraints or requirements of the users. On the other hand, prediction can also ensure the reliability of service-based application system. For example, once an executing composite service meets an error or fault, a series of events including reselecting and pre-computing the QoS of the replacement services. These events however will increase the waiting time and expend system resources. Triggering precaution by prediction has gradually become a valid exceptional diagnosis method to efficiently decrease the waiting time and system overhead. Therefore, the ability to predict reliability of invoked Web service early not only ensures more reliable quality of service-oriented application systems but also helps to reduce re-engineering cost.

In service computing, Web service quality prediction has become an important research problem and has attracted a lot of attention in recent years. A number of Web service prediction models have been proposed, such as ML-based methods[2,3], QoS-aware based methods[4,6] and collaborative filtering-based methods[9,7,8] and so on. The key of these models is realized by monitoring and evaluating the quality of composite services. Despite the fact that these models improve the quality of composite services to some extent, current methods still three major challenges remain, as defined below. First of all, most of the traditional ML-based prediction models[3], such as Support Vector Machines (SVM) and Artificial Neural Networks (ANN) are more sensitive to user specified parameters. Secondly, prediction models based on QoS monitoring, such as Naive Bayes and Markov model[10,5], which assume that sequences in a class are generated by an underlying model M and the probability distributions are described by a set of parameters. These parameters however, are obtained by predicting the QoS during the whole lifecycle of the services and will therefore lead to high overhead costs. Another method is prediction based on sequence distance, such as collaborative filtering[9,7,8], this type of methods require the definition of to define a distance function to measure the similarity between a pair of sequences. However, how to select an optimal similarity function is far from trivial, as it will introduce numerous parameters and measures for distances which can be rather subjective.

Extreme Learning Machine (ELM) is becoming popular due to it generally requires far less training time than the conventional learning machines [13,15,14,16,17,18]. ELM has originally been developed based on Single-hidden Layer Feedforward Neural Networks (SLFNS)in[11,12]. In ELM, hidden nodes parameters are chosen randomly. A Survey has been done on ELM and its variants in[19]. Prediction base on ELM has been applied to many domains successfully, such as bioinformatics, XML document classification, mobile object prediction, and so on[19].

Web service reliability prediction is similar to protein secondary structure prediction for the view that the executed service sequences are going to be classified automatically to respective categories (such as Bronze, Sliver, Gold, Platinum and Exception) according to their corresponding execution behaviors. Unlike protein data however, the character of the data describing of Web service is more complex, since the data generated from service-based system are typically heterogeneous, of multiple data types and high dynamic. Example of service-generated data includes service trace log, Quality of Service (QoS) information, service invocation relationship, etc. Beyond that each of the QoS attributes defines different dimensions of the quality of Web service, some of them depend environment absolutely. Traditional prediction methods, most of which only focus on single-type data, can not be implemented directly for the complex service data generated in this context. Most important, previous prediction approaches are weak on active and quick response ability, especially in online and real time process environment. In this paper, we extract Web service execution information and represent them as feature vectors. Furthermore, Proposal of a mining method to select early and concise but high discriminating features to predict the service quality efficiently.

The main contributions in this paper are as follows: (1) Methods for extracting various execution information from the service log data and turning them into service sequences to be mined. (2) Proposal of a method to mine early and concise but highly discriminate feature subsets for different categories. (3) Obtain F-ELM classifier by training ELM using the extracted feature subsets.

The remainder of this paper is organized as follows: Section 2 gives a brief overview of ELM. Section 3 presents the prediction architecture based on F-ELM. Section 4 studies the feature vectors representation of Web services. Section 5 proposes a method to determine the Feature the mining algorithm, and finally, we give our conclusions in Section 6.

2   Brief Introduction of Extreme Learning Machine

Three common approaches of feedforward networks training were summarized in[21]: 1) gradient-descent based (e.g., back propagation method for multi-layer feedforward neural networks); 2) least square based (e.g., ELM for the generalized single-hidden layer feedforward networks); 3) standard optimization method based (e.g., SVM for a specific type of Single-hidden Layer Feedforward Networks). ELM and its variants[15,22] based on SLFNs for classification and can achieve better generalization performance than conventional learning algorithms. Moreover, ELM is less sensitive to user specified parameters, and can be deployed faster and more conveniently[21].

As mentioned, ELM is based on SLFN type classifiers. Standard SLFNs with N arbitrary samples $(\mathbf{x_i}, \mathbf{t_i}) \in \mathbf{R^{n \times m}}$ and activation function $g(x)$ are modeled in[12] as

$$\sum_{i=1}^{L} \beta_i g_i(\mathbf{x_j}) = \sum_{i=1}^{L} \beta_i g(\mathbf{w_i} \cdot \mathbf{x_j} + b_i) = \mathbf{o_i}, (j = 1, ..., N) \quad (1)$$

where L is the number of hidden layer nodes, $\mathbf{w_i} = [w_{i1}, w_{i2}, ..., w_{in}]^T$ is the weight vector between the ith hidden node and the input nodes, $\beta_i = [\beta_{i1}, \beta_{i2}, ...,$

$\beta_{im}]^T$ is the weight vector between the ith hidden node and the output nodes, and b_i is the threshold of the ith hidden node. Then we have the output of ELM

$$f(x) = \sum_{i=1}^{L} \beta_i g(\mathbf{a_i}, b_i, \mathbf{x}) \tag{2}$$

where

$$H(\mathbf{w_1}, \ldots, \mathbf{w_L}, b_1, \ldots, b_L, \mathbf{x_1}, \ldots, \mathbf{x_L}) = \begin{bmatrix} g(\mathbf{w_1} \cdot \mathbf{x_1} + b_1) & \ldots & g(\mathbf{w_L} \cdot \mathbf{x_1} + b_L) \\ \vdots & \ldots & \vdots \\ g(\mathbf{w_1} \cdot \mathbf{x_N} + b_1) & \ldots & g(\mathbf{w_L} \cdot \mathbf{x_N} + b_L) \end{bmatrix}_{N \times L},$$

$$\beta = \left[\beta_1^T, \ldots, \beta_L^T \right]^T_{m \times L}$$

The decision function for binary classification [8] is

$$d(x) = sign(\sum_{i=1}^{L} \beta_i g(\mathbf{a_i}, b_i, \mathbf{x})) = sign(\beta \cdot \mathbf{H}) \tag{3}$$

When g(x) approximates the N samples with zero error that $\Sigma_{j=1}^{L} \|o_j - t_j\| = 0$, their outputs β_i, w_i and b_i such that

$$\sum_{i=1}^{L} \beta_i g(\mathbf{w_i} \cdot \mathbf{x_j} + b_i) = \mathbf{t_j}, j = 1, \ldots, N \tag{4}$$

The equation above can be expressed compactly as following

$$\mathbf{H}\beta = \mathbf{T} \tag{5}$$

where $\mathbf{T} = [\mathbf{t}_1^T, \ldots, \mathbf{t}_L^T]^T_{m \times L}$.

The ELM algorithm is a relatively fast method as compared to conventional learning algorithms. ELM not only tends to reach the smallest training error but also the smallest norm of weights [1]. Given a training set $\aleph = \{(\mathbf{x_i}, \mathbf{t_i}) | \mathbf{x_i} \in \mathbf{R^n}, \mathbf{t_i} \in \mathbf{R^m}, i = 1, \ldots, N\}$, activation function $g(x)$ and hidden node number L, algorithm ELM is described as following[12].

Algorithm 1. ELM
---
1. **for**(i=1 to L)
2.   randomly assign input weight $\mathbf{w_i}$
3.   randomly assign bias b_i
4. **endfor**
5. calculate $\mathbf{H}$
6. calculate $\beta = \mathbf{H^\dagger T}$

---

3   System Architecture

Figure 1 shows the whole system architecture based on F-ELM. There are four phases in the prediction process: (1) recording of the composite service execution log information, extracting multi-dimensional QoS features and converting them into service sequences to be mined. (2) mining early feature subsets by means of prefix trees. (3) Inputting training data with selected features and obtaining the F-ELM classifier. (4) Periodically updating early features with trend values change periodically. Below each of the steps will be briefly discussed.

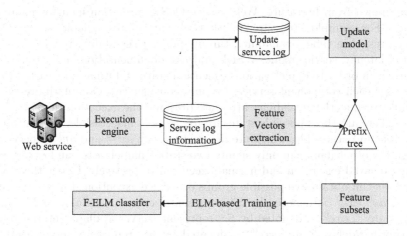

Fig. 1. The framework

At first, the system needs to collect large amounts of composite service execution information, aiming to mine useful knowledge for reliable prediction. The original service log includes a variety of structural and un-structural data information, such as trace logs, Quality of Service (QoS) information, service invocation relationships, Web service Description Language (WSDL) and so on. These information is typically heterogeneous, of multiple data types, and high dynamic. Then, the system converts these infomation into service services to be mined. The content of this part will be discussed in Section 2.

As we know, various behavioral features are hidden in service execution sequences. Different behavioral features can lead to different service quality under different environment. However, some features may appear in several different categories (such as successful or failed) simultaneously, then we say these features have no or less contribution to discriminate different categories. Therefore, not all features are useful for discriminating different class. Besides, both the time when these features occur and the length of the behavioral features influences the prediction result. Under various environmental conditions, mining early feature subsets with minimal cost and triggering pre-caution as early as possible are important issues in our work. We will mine early feature subsets by means of prefix tree. The content of this part will be described in Section 5.

Further, when a new service sequence is inputted, the update module judges QoS status of the service sequences. If the status of the service properties change, the update module sends update request to the prefix tree according to the update strategy. These procedures and strategies will be described in the future work section. Besides judging the status, the feature values of each node in the prefix tree are recalculated periodically.

4   Sequence Conversion from Service Execution Log

As we known from literature, Web service QoS information includes many attributes. For example, literature[20] lists twelve attributes to depict service QoS, such as response time, availability, throughput, successability, reliability, compliance, latency, service name, WSDL address, documentation and service classification. In order to depict various execution status information clearly in this paper, for each component service, we only consider two QoS attributes as an example, such as the availability attribute (av) and the execution time attribute (exe). There are three possible states for av, namely inaccessible, intermittently accessible and accessible, which are denoted by av^0, $av^{0.5}$ and av^1 respectively. exe on the other hand can only signify two states, namely exe^0 and exe^1 representing delayed execution and normal execution respectively. Under these situations, we can obtain five possible groups of service execution statuses information: $< av^0, exe^0 >$, denoted by S^0, represents server unavailable and runtime delay, $< av^{0.5}, exe^0 >$, denoted by S^1, represents server available intermittently and runtime delay, $< av^1, exe^0 >$, denoted by S^2, represents server available and runtime delay, $< av^{0.5}, exe^1 >$, denoted by S^3, represents server available intermittently but normal execution and $< av^1, exe^1 >$, denoted by S^4, represents server available but normal execution. However, for the status information $< av^0, exe^1 >$, denoted by S^0, represents server unavailable but normal execution. We do not consider this status, as it cannot occur in practice.

Table 1. An example of service execution instances

ID	ExecuteLogEntry	Count	Class
1	S_1^0	3	failed with type A
2	S_1^1, S_3^1	3	failed with type B
3	S_1^2, S_2^2, S_3^2	2	failed with type C
4	S_1^1, S_2^1, S_3^1	3	failed with type B
5	$S_1^4, S_2^2, S_3^2, S_4^2$	2	failed with type C
6	S_1^4, S_2^2, S_3^2	2	failed with type C
7	$S_1^4, S_2^3, S_3^4, S_4^2$	2	successful
8	S_1^4, S_2^3, S_3^4	3	successful

Based on the description assumption above, large collections of composite service execution records D are collected, which are stored as sequential data as shown in Table 2. Let $S=\{S_1, S_2,...,S_n\}$, then each record consists of one or more component services from S. Note that every component service will have different QoS expression values, which can evaluate describe the execution status of that component service. For the sake of simplification, we let the capital letters with subscript denote the different abstract service classes and the superscript denotes the different QoS statuses of the component service. Table 1 only shows a service execution data with 15 failed user types, 5 successful user types and 4 abstract service classes.

5   Feature Subsets Selection

Previous existing feature mining methods need to set some thresholds as quality measures such as support, confidence, so that only those relatively frequent rules are selected for classifier construction. However, the mining method is ineffective in practice and the mined rules can still be very large. It is very hard for a user to seek out appropriate rules. In our case, we consider early, concise and discriminative properties to set a utility measure and only preserve those early, concise and discriminative feature subsets passing the measure from the training database.

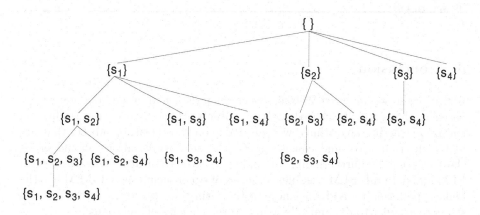

Fig. 2. The FS enumeration tree

We discuss the detail of Feature Subsets(FS) selection algorithm below, taking Table 1 as the example, where the minimum weight intra-class support with early factor threshold wis^e.

The mining process is conducted on a prefix tree as shown in Figure 2, which is build on Table 1. Limited by space, we omit the description of constructing such a prefix tree structure. From Figure 2, we can see there are four abstract service s_1, s_2, s_3 and s_4 at level 1. We obtain different status information for each abstract service in descending order at level 2. A support computing method is different from traditional one, we consider both concise and early characters. Therefore, the obtained order using wis^e for each ith item is also different from traditional approaches. For example, at first, the algorithm scans Table2 once and computes the wis^e of each one item. After computing, we generate the candidate early feature for the second level.

We can use the wis^e-based pruning to prune some rules. We can pruning some redundancy rules by applying pruning rule. Further, if all features under threshold are pruned, then rules containing these features will be pruned.

Next, we propose an ELM based feature selection method in this section. Due to there may be several sequences of the largest confidence value and there may be more than one nodes of the same number of neighbors. In such case, we use ELM to evaluate every possible candidate. The one of the largest prediction accuracy is selected. The process is formalized in Algorithm 1.

Algorithm 1. The Fss Algorithm
Input: feature sets
Output: The selected feature subset(Fss)

1: $Fss=X$
2: Let X be the candidate feature sets
3: For each feature X belongs to class C, $X \in Fss$
4: using ELM evaluates every possible X
5: the X of the highest accuracy on ELM;
6: return Fss

---

6   Conclusion

In this paper, we analyzed Web service characteristic and proposed a prefix tree-based methods to extract early feature subsets. Different from the traditional feature extracting algorithms, we only focus on those feature subsets that are highly discriminative and concise, that is these feature subsets should be as short as possible. Furthermore, we apply the extracted feature subsets to train ELM and obtain F-ELM classifier. A series of experiments show F-ELM has the highest performance and efficiency. In the future, we will focus on the update strategy in order to dynamically adapt to the changes of environment.

Acknowledgment. National Natural Science Foundation of China (61272182, 61100028, 61100027), New Century Excellent Talents (NCET-11-0085), the Ph.D. Programs Foundation of Ministry of Education of China(young teacher) (No.20110042120034) and the Fundamental Research Funds for the Central Universities under grants (No.130504001).

References

1. Zhang, L., Zhang, J., Hong, C.: Services Computing. TSingHua University Press, Beijing (2007)
2. Wang, F., Liu, L., Dou, C.-X.: Stock Market Volatility Prediction: A Service-Oriented Multi-kernel Learning Approach. In: IEEE SCC, pp. 49–56 (2012)

3. Han, J.W., Kamber, M.: Data Mining: Concepts and Techniques, 3rd edn. Machine Learning Press (2012)

4. Goldman, A., Ngoko, Y.: On graph reduction for QoS prediction of very large Web service compositions. In: IEEE SCC, pp. 258–265 (2012)

5. Wu, J., Chen, L., Jian, H.-Y., Wu, Z.-H.: Composite service recommendation based on bayes theorem. International Journal of Web Services Research (IJWSR) 9(2), 69–93 (2012)

6. Tao, Q., Chang, H.-Y., Gu, C.-Q., Yi, Y.: A novel prediction approach for trust-worthy QoS of Web services. Expert Syst. Appl. (ESWA) 39(3), 3676–3681 (2012)

7. Zheng, Z.-B., Chen, J.-L., Lyu, M.-R.: Personalized Web service Recommendation via Normal Recovery Collaborative Filtering. IEEE Transactions on Services Computing (TKDE) (acccepted)

8. Lo, W., Yin, J.-W., Deng, S.-G., Li, Y., Wu, Z.-H.: Collaborative Web service QoS Prediction with Location-Based Regularization. In: ICWS, pp. 464–471 (2012)

9. Zheng, Z.-B., Ma, H., Lyu, M.-R., King, I.: QoS-Aware Web service Recommendation by Collaborative Filtering. IEEE Transactions on Service Computing (TSC) 4(2), 140–152 (2011)

10. Park, J.-S., Yu, H.-C., Chung, K.-S., Lee, E.-Y.: Markov Chain Based Monitoring Service for Fault Tolerance in Mobile Cloud Computing. In: AINA Workshops, pp. 520–525 (2011)

11. Huang, G.-B., Siew, C.K.: Extreme learning machine: Rbf network case. In: ICARCV, pp. 1029–1036 (2004)

12. Huang, G.B., Zhu, Q.Y., Siew, C.K.: Extreme learning machine: Theory and applications. Neurocomputing 70, 489–501 (2006)

13. Huang, G.-B., Ding, X., Zhou, H.: Optimization method based extreme learning machine for classification. Neurocomputing 74(1-3), 155–163 (2010)

14. Wei, X.-K., Li, Y.-H., Feng, Y.: Comparative study of extreme learning machine and support vector machine. In: Wang, J., Yi, Z., Žurada, J.M., Lu, B.-L., Yin, H. (eds.) ISNN 2006. LNCS, vol. 3971, pp. 1089–1095. Springer, Heidelberg (2006)

15. Li, M.-B., Huang, G.-B., Saratchandran, P., Sundararajan, N.: Fully complex extreme learning machine. Neurocomputing 68, 306–314 (2005)

16. Jun, W., Shitong, W., Chung, F.-L.: Positive and negative fuzzy rule system, extreme learning machine and image classification. International Journal of Machine Learning and Cybernetics 2, 261–271 (2011)

17. Wang, X., Chen, A., Feng, H.: Upper integral network with extreme learning mechanism. Neurocomputing 74(16), 2520–2525 (2011)

18. Chacko, B., Krishnan, V.V., Raju, G., Anto, P.B.: Handwritten character recognition using wavelet energy and extreme learning machine. International Journal of Machine Learning and Cybernetics, 1–13, doi:10.1007/s13042-011-0049-5

19. Huang, G.-B., Wang, D., Lan, Y.: Extreme learning machines: a survey. International Journal of Machine Learning and Cybernetics 2, 107–122 (2011), doi:10.1007/s13042-011-0019-y

20. Mohanty, R., Ravi, V., Patra, M.-R.: Web-services classification using intelligent techniques. Expert Systems with Applications 37, 5484–5490 (2010)

21. Cortes, C., Vapnik, V.: Support vector networks. Machine Learning 20, 273–297 (1995)

22. Feng, G., Huang, G.-B., Lin, Q., Gay, R.: Error minimized extreme learning machine with growth of hidden nodes and incremental learning. IEEE Transactions on Neural Networks 20(8), 1352–1357 (2009)

23. Zhao, Y.-H., Wang, G.-R., Zhang, X., Xu Yu, J., Wang, Z.-H.: Learning Pheno-type Structure Using Sequence Model. IEEE Transactions on Knowledge and Data Engineering (2013) (accepted)
24. Zhao, Y.-H., Wang, G.-R., Yu, J.X., Chen, L., Yu, G.: Maximal Subspace Coregulated Gene Clustering 20(1), 83–98 (2008)
25. Zhao, Y.-H., Wang, G.-R., Li, Y., Wang, Z.-H.: Finding Novel Diagnostic Gene Patterns Based on Interesting Non-redundant Contrast Sequence Rules. In: IEEE ICDM, pp. 972–981 (2011)

Multi-class AdaBoost ELM

Yefeng Shen[1], Yunliang Jiang[3], Weicong Liu[2], and Yong Liu[2,*]

[1] Hangzhou Dianzi University, Zhejiang, Hangzhou, China, 310018
[2] Institute of Cyber-Systems and Control, Zhejiang University, Zhejiang, Hangzhou, China, 310027
[3] Huzhou Teachers College, Zhejiang, Huzhou, China, 313000
yongliu@iipc.zju.edu.cn, cckaffe@yahoo.com.cn

Abstract. Extreme learning machine (ELM) is a competitive machine learning technique, which is simple in theory and fast in implementation, it can identify faults quickly and precisely as compared with traditional identification techniques. As verified by the simulation results, ELM tends to have better scalability and can achieve much better generalization performance and much faster learning speed comparing with traditional SVM. In this paper, we introduce a Multi-class AdaBoost based ELM ensemble method. In our approach, the ELM algorithm is selected as the basic ensemble predictor due to its rapid speed and good performance. Compared with the existed boosting ELM algorithm, our algorithm can be directly used in multi-class classification problem. We also carried out comparable experiments with face recognition datasets, the experimental results show that the proposed algorithm can not only make the predicting result more stable, but also achieve better generalization performance.

Keywords: Extreme Learning Machine, Multi-class AdaBoost, classification, Face Recognition.

1 Introduction

Many research works have been done in feedforward neural networks, which pointed out that the feedforward neural networks are able to not only approximate complex nonlinear mapping, but also provide models for some natural and artificial problems which classic parametric technics are unable to handle.

Recently, Huang et al. [1] proposed a new simple algorithm based on single layer feedforward networks (SLFNs) called Extreme Learning Machine (ELM). For ELM randomly generates parameters of the networks, its learning speed can be thousands of times faster than traditional feedforward network learning algorithms like backpropagation (BP) algorithm, which needs to iterate many times to get optimal parameters.

AdaBoost [2] is one of the most popular algorithms of classifier ensemble to improve the generalization performance. Wang and Li in [3] proposed an algorithm named Dynamic AdaBoost Ensemble ELM (named DAEELM in this paper). The proposed algorithm takes the ELM as the basic classifier and applies AdaBoost to

---

* Corresponding author.

© Springer International Publishing Switzerland 2015
J. Cao et al. (eds.), *Proceedings of ELM-2014 Volume 2,*
Proceedings in Adaptation, Learning and Optimization 4, DOI: 10.1007/978-3-319-14066-7_18

solve binary classification problem. Similarly, Tian and Mao in [4] combined the modified AdaBoost.RT [5] with ELM to propose a new hybrid artificial intelligent technique called ensemble ELM. Ensemble ELM aims to improve ELM's performance in regression problem.

However, until now, not so much works have been done to apply AdaBoost to ELM for multi-class classification problem directly. In Freund's work[6], they give two extensions of their boosting algorithm to multi-class prediction problems in which each example belongs to one of several possible classes (rather than just two).Since ELM can directly work for multi-class classification problem, this paper proposes an algorithm named Multi-class AdaBoost ELM (MAELM). This new algorithm applies Multi-class AdaBoost as an ensemble method to a number of ELMs. In addition, this paper proposes a structure to apply ELM and MAELM to Local Binary Patterns (LBP) [7] based face recognition problem. Experiments in LBP based face recognition will show that the proposed algorithm outperforms the original ELM.

The rest of the paper is organized as follows: Section 2 gives a brief review of the ELM, original and Multi-class AdaBoost and LBP. The proposed MAELM is presented in Section 3. The experimental result will be shown in Section 4 and a short discussion about the proposed algorithm will be presented in Section 5. Finally, in section 6, we conclude the paper.

2   A Review of Related Work

In this section, a review of the original ELM algorithm, Multi-class AdaBoost and the LBP based face recognition is presented.

2.1   ELM

For N arbitrary distinct samples (x_i, t_i), where $x_i = [x_{i1}, x_{i2}, ..., x_{id}]^T \in R^d$ and $t_i = [t_{i1}, t_{i2}, ..., t_{iK}]^T \in R^K$, standard SLFNs with L hidden nodes and activation function $h(x)$ are mathematically modeled as:

$$\sum_{i=1}^{L} \beta_i h_i(x_j) = \sum_{i=1}^{L} \beta_i h_i(w_i \cdot x_j + b_i) = o_j \tag{1}$$

where $j = 1, 2, ..., N$.

Here $w_i = [w_{i1}, w_{i2}, ..., w_{id}]^T$ is the weight vector connecting the ith hidden node and the input nodes, $\beta_i = [\beta_{i1}, ..., \beta_{iK}]^T$ is the weight vector connecting the ith hidden node and the output nodes, and b_i is the threshold of the ith hidden node.

The standard SLFNs with L hidden nodes with activation function $h(x)$ can be compactly written as:

$$H\beta = T \tag{2}$$

where

$$H = \begin{bmatrix} h_1(w_1 \cdot x_1 + b_1) & ... & h_L(w_L \cdot x_1 + b_L) \\ \vdots & \vdots & \vdots \\ h_1(w_1 \cdot x_N + b_1) & ... & h_L(w_L \cdot x_N + b_L) \end{bmatrix} \tag{3}$$

$$\beta = \begin{bmatrix} \beta_1^T \\ \vdots \\ \beta_L^T \end{bmatrix}, \text{ and } \quad T = \begin{bmatrix} t_1^T \\ \vdots \\ t_N^T \end{bmatrix} \tag{4}$$

Different from the conventional gradient-based solution of SLFNs, ELM simply solves the function by:

$$\beta = H^+ T \tag{5}$$

H^+ is the Moore-Penrose generalized inverse of matrix H. As Huang has pointed out in [8], H^+ can be represented by:

$$H^+ = H^T (\frac{I}{C} + HH^T)^{-1} \tag{6}$$

where I is an identity matrix, which has the same dimension with HH^T. C is a constant number which can be set by the user. Adding $\frac{1}{C}$ can avoid the situation that HH^T is singular. Huang et al. [1] successfully applied ELM to solve binary classification problem and Huang et al. [8] extended the ELM to directly solve the multi-class classification problem.

2.2   Original AdaBoost and Multi-class AdaBoost

AdaBoost has been very successfully applied in binary classification problem. Original AdaBoost is proposed in [2]. Before proposing the AdaBoost algorithm, the function I(x) is pre-defined as:

$$I(x) = \begin{cases} 1, if \ x = true \\ 0, if \ x = false \end{cases} \tag{7}$$

AdaBoost algorithm is summarized as follows:

Given the training data $\{(x_1, y_1), (x_2, y_2), \dots, (x_N, y_N)\}$, where $x_i \in R^d$ denotes the ith input feature vector with d dimensions, y_i denotes the label of the ith input feature vector, where $y_i \in \{-1, +1\}$. Use $T_j(x)$ to denote the jth weak classifier and suppose M weak classifiers will be combined.

1. Initialize the observation weights $\omega_i = \frac{1}{N}, i = 1, 2, \dots, N$
2. For $m = 1:M$
(a) Fit a classifier $T_m(x)$ to the training data using weights ω_i
(b) Compute the weighted error

$$err_m = \frac{\sum_{i=1}^N \omega_i \, I(y_i \neq T_m(x_i))}{\sum_{i=1}^N w_i} \tag{8}$$

(c) Compute the weight of the mth classifier

$$\alpha_m = log \frac{1 - err_m}{err_m} \tag{9}$$

(d) Update the weights of sample data, for all i=1, 2,...,N

$$\omega_i = \omega_i \cdot \exp\left(\alpha_m \cdot I\left(y_i \neq T_m(x_i)\right)\right) \tag{10}$$

(e) Re-normalize ω_i, for all i=1, 2,...,N

1. Output

$$C(x) = \arg max_k \sum_{m=1}^{M} \alpha_m \cdot I(T_m = k) \tag{11}$$

Here k is +1 or−1. In binary classification, any classifier whose generalization performance is better than $\frac{1}{2}$ is a weak classifier. For the original AdaBoost, we have:

1) For the ith and the jth classifier, if $err_i < err_j < \frac{1}{2}$, we have $\alpha_i > \alpha_j > 0$, which means the final ensemble classifier values more of the ith classifier's result. Especially, if $err_j = \frac{1}{2}$, $\alpha_j = 0$, which means the final ensemble classifier just ignores the classifier since its effect is the same as random guess.
2) If the pth classifier misclassifies the qth sample, the qth sample will have a big weight in the next iteration. As a result, the $(p + 1)^{th}$ classifier will pay more attention to it. On the contrary, if the pth classifier classifies the qth sample correctly, the qth sample will have a small weight in the next iteration, which means the $(p + 1)^{th}$ classifier will pay less attention to it.

However, for a K-class classification problem, we have $y_i \in \{1,2, ..., K\}$ and $K > 2$. If a classifier's generalization performance is better than $\frac{1}{K}$ (maybe much smaller than $\frac{1}{2}$), it can be called a weak classifier. Since original AdaBoost only takes a classifier whose generalization performance is better than $\frac{1}{2}$ as a weak classifier, obviously, it cannot be directly implemented to multi-class conditions that K is bigger than 2. Freund et al. [6] extends the original AdaBoost to multi-class condition. The weight of the mth classifier is modified as:

$$\alpha_m = log\frac{1 - err_m}{err_m} + log(K - 1) \tag{12}$$

Similarly to the binary condition, for the ith and the jth classifier, if $err_i < err_j < 1 - \frac{1}{K}$, we have $\alpha_i > \alpha_j > 0$, which means the final ensemble classifier values more of the ith classifier's result. Especially if $err_j = 1 - \frac{1}{K}$, $\alpha_j = 0$.

2.3   LBP Based Face Recognition

The original LBP operator goes through each 3×3 neighborhood in a picture. It takes the center pixel as the threshold value of the neighborhood and considers the result as a decimal number. Then the texture of the picture can be represented by the histogram of all the decimal numbers.

To apply LBP operator in face recognition problem, Ahonen et al. [7] divided the face image into several windows and calculated the histogram of each windows by

LBP operator. The final feature vector is got by combining the histograms into a spatially enhanced histogram. The spatial enhanced histogram provides with three levels of information: the patterns of pixel level; the patterns of regional level; the global patterns of the face image. Experiments in [7] have shown that the LBP description is more robust against variants in pose or illumination than holistic methods. All our experiments in Section 4 are done with most original LBP operator.

3   MAELM and Face Recognition Structure

In this part, the Multi-class AdaBoost ELM (MAELM) algorithm is proposed and a structure of face recognition based on LBP and ELM is also included.

3.1   Proposed MAELM Algorithm

By applying the Multi-class AdaBoost to ELM, this paper proposes the Multi-class AdaBoost ELM (MAELM) algorithm. The algorithm takes a number of ELM classifiers as the weak classifiers. $ELM_i(x)$ denotes the ith ELM classifier. The proposed algorithm puts as follows:

1 Initialize the observation weights $\omega_i = \frac{1}{N}, i = 1,2 \dots, N$

2 For m=1: M

(a) Fit a classifier $ELM_m(x)$ to the training data using weights ω_i

(b) Compute the weighted error

$$err_m = \frac{\sum_{i=1}^{N} \omega_i\, I\big(c_i \neq ELM_m(x_i)\big)}{\sum_{i=1}^{N} w_i} \tag{13}$$

(c) Compute the weight of the mth classifier

$$\alpha_m = log\frac{1 - err_m}{err_m} + log(K - 1) \tag{14}$$

(d) Update the weight of sample data, for all i=1,2,...N

$$\omega_i = \omega_i \cdot \exp\big(\alpha_m \cdot I\big(c_i \neq ELM_m(x_i)\big)\big) \tag{15}$$

(e) Re-normalize ω_i

3 Output

$$C(x) = \arg max_k \sum_{m=1}^{M} \alpha_m \cdot I(ELM_m(x) = k) \tag{16}$$

Part 2.(a) of the proposed algorithm should be paid more attention. Both references [3] and [4] did not give any detail of how to fit the basic classifier $ELM_m(x)$ with weighted samples, but it is a very important part of AdaBoost. Zong et al. [9] proposed an algorithm named weighted ELM by introducing a diagonal matrix

$W \in R^{N \times N}$, whose element $W_{i,i}$ denotes the weight of the ith training sample. In view of some special situations, we introduce the weighted ELM algorithm. Obvious, it boils down to the original one when the weighted matrix is the identity matrix.

Under the weighted circumstance, the solution of β becomes:

$$\beta = H^T(\frac{I}{C} + WHH^T)^{-1}WT \tag{17}$$

3.2   Application in LBP Based Face Recognition

This paper combines LBP based feature vectors with ELM to build a face recognition structure. There have been some papers ([10, 11]) about applying ELM in face recognition problem.

In order to get better generalization performance, the proposed face recognition structure implements the LBP based method to get the feature vector and ELM as the classifier. At the same time, ELM is very fast in classification and has very good generalization performance. So, it is reasonable to combine LBP method and ELM to build the face recognition structure.

There are two steps of the proposed face recognition structure. The first step is to train the training samples by ELM or MAELM. In this step, the training samples are represented by LBP based feature vectors. Then the feature vectors are used to train the classifier model by ELM or MAELM. The second step is to predict the labels of the test samples. The test samples are also represented by the LBP based feature vectors. Then the classifier model trained in the first step is implemented to predict the labels of the test samples.

4   Experiments

In this paper, two of the mostly used face recognition datasets Yale and ORL are used to prove the efficiency of the proposed algorithm. To make the results valid, except the Section 4.2, the average testing accuracy is obtained on 20 trials randomly generated training set and test set. This paper chooses the sigmoid function as the activation function for it is most commonly used.

M is the number of the basic classifier, C is the constant value in generalized inverse of H, L is the number of hidden nodes in ELM, t is the number of training image of each person and we divide each face image into w*w windows.

4.1   Performance Changes with C and L

Although ELM is comparatively not that sensitive to the arguments as SVM, its performance still changes with the hidden layer number L and the constant value C.

Suppose we have N training samples, Huang et al. [1] rigorously proves that SLFNs (with N hidden nodes) with random bias and input weights can exactly learn the N distinct observations. If the training error is allowed, the number of hidden nodes can be much smaller than N. At the same time, the constant value C also has some impacts of the solution of H's Moore-Penrose generalized inverse.

In this part, the experiment is conducted in Yale dataset. We set $M = 20, t = 5, w = 3$. In addition, the L is set as $100, 400, 700, \ldots, 1900$ and the C is set as $10^{-5}, 10^{-4}, \ldots, 10^{5}$. The performance of ELM and MAELM is shown in Figure 1.

It is obvious that both ELM and MAELM are not sensitive to the change of arguments. The difference between ELM and MAELM is mainly in the region where L is very small and C is very large. From Figure 1, one can conclude that ELM performs badly in this region. On the contrary, MAELM is still very stable in this region.

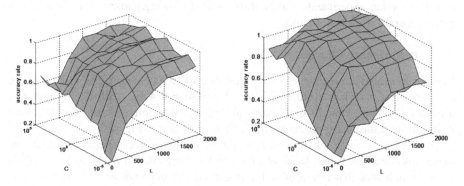

Fig. 1. The performance of ELM (left)   The performance of MAELM (right)

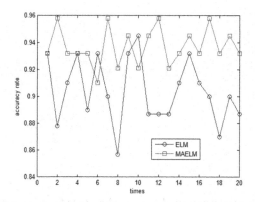

Fig. 2. Performance of ELM and MAELM under the same training set and test set

4.2 Prediction Stability Analysis

Since the original ELM randomly generates the weights between the input layer and the hidden layer, as well as the bias of the activation function, its performance even for the same training and test set changes each time. The proposed algorithm successfully reduces the instability.

From Figure 1, one is able to conclude that ELM tends to get better performance when $C = 1$, while $C = 10^{3}$ is better for MAELM. Let $M = 10, C = 10^{3}, L = 1000, t = 5, w = 3$ for MAELM, and $C = 1, L = 1000, t = 5, w = 3$ for ELM.

Experiments are done in Yale data sets. In order to prove the proposed algorithm is more stable than the original ELM, experiments are done in the same training set and test set (randomly generated) for 20 times. The result is shown in Figure 2.

In Figure 2, it is obvious that the performance of MAELM is much more stable than the original ELM. Please notice that although the generalization performance of MAELM seems to be much better than ELM, it is improper to conclude MAELM performs better. The reason is that the training set and test set are fixed. One cannot exclude the possibility that MAELM performs better than ELM only under this data sets. Some other experiments will be done in the following parts to show MAELM's better generalization performance.

4.3   Performance Changes with M

In order to evaluate the changes of performance when M changes, the experiment in this part lets $C = 1, L = 1000, t = 5, w = 4,$ and $M = 2, 4, 6, ..., 50$. The average test accuracy is obtained on 20 trials randomly generated training set and test set. Yale dataset is used for the experiment. The result is presented in Figure 3.

From Figure 3, it is obvious that as the M increases, the generalization performance also becomes better. However, the trend becomes slower as M increases. This situation indicates that in real-world applications, M does not need to be very big. Good generalization performance can be obtained by setting M less than 30.

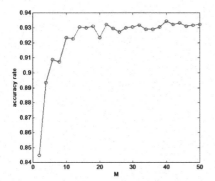 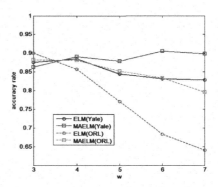

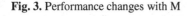

Fig. 3. Performance changes with M   **Fig. 4.** performances in Yale and ORL

4.4   Better Generalization Performance than ELM

In this part, experiments are done both in Yale and ORL datasets. The experiments set the parameters of ELM and MAELM as follows: $C = 1, L = 1000, t = 5, M = 20$ (MAELM only), $w = 3, ..., 7$. The average testing accuracy is obtained on 20 trials randomly generated training set and test set.

The experiment indicates that MAELM has better generalization performance both in Yale and ORL datasets under different window sizes. See Figure 4 for details.

5   Discussion

5.1   Complexity Comparison

Very similar to MAELM, the DAEELM [3] also considers taking the ELM as the weak classifier and implement AdaBoost as the ensemble method. The difference is that MAELM implement multi-class AdaBoost which can be directly used in multi-class classification problem, while DAEELM implements dynamic ensemble Ada-Boost [12], which aims to solve the binary classification problem.

Many methods have been developed to apply binary classifier to multi-label problem. One-against-all (OAA) [13] and one-against-one (OAO) [14] are mostly used. For a K-class classification problem, under OAA condition, K classifiers have to be trained. Each of them separates a single class from all the remaining. Under the OAO condition, $\frac{K(K-1)}{2}$ classifiers have to be trained. Each one separates a pair of classes.

Suppose both MAELM and DAEELM have M iterations. For a K-class classification problem, MAELM only needs to train M ELMs, while DAEELM needs to train $M \times K$ and $\frac{M \times K \times (K-1)}{2}$ classifiers for OAA and OAO condition respectively. Although DAEELM may stop the iteration earlier, it is obvious that, in theory, MAELM's computation complexity is much lower than DAEELM for K-class classification problem, especially when K is a very big number.

The authors of DAEELM have not published its codes and DAEELM has its own arguments which MAELM does not have. DAEELM also doesn't provide details of how it trains weighted data with ELM, so it will be unfair to compare the performance of MAELM and DAEELM. However, the conclusion that MAELM is much faster in multi-class classification problem can be drawn from the complexity analysis above.

5.2   Train ELM with Weighted Data

Section 3.1 has mentioned that training ELM with weighted data is a key problem when applying AdaBoost. However, references [3] and [4] didn't mention the key point at all. Toh in [15] first applied ELM to classify imbalanced data with two classes. ELM tries to minimize the training error of the data while the proposed algorithm tends to minimize the total-error-rate (TER), which takes the weights of the positive and negative data into consideration.

In Section 3.1, the weighted ELM is applied in MAELM. Actually, the weighted ELM is inspired and very similar to Regularized ELM proposed by Deng in [16]. The Regularized ELM aims to minimize the weighted training error of the weighted data.

6   Conclusion

This paper proposes a new boosting ELM named MAELM, which applies the Multi-class AdaBoost in ELM ensemble to directly solve multi-class classification problem. A face recognition structure combined LBP based method and ELM is also presented in the paper. Experiments show that in LBP based face recognition problem, the recognition result of MAELM is more stable and better than the original ELM.

Also, MAELM is compared with DAEELM in multi-class classification problem in theory, which indicates MAELM has much lower computation complexity. Moreover, this paper makes the problem how to train weighted data by ELM clear.

Acknowledgements. This research is based upon work supported in part by National Natural Science Foundation of China (61370173, 61173123) and the Natural Science Foundation Project of Zhejiang Province under Project LR13F030003.

References

1. Huang, G.B., Zhu, O.Y., Siew, C.K.: Extreme learning machine: theory and applications. Neurocomputing 70(1), 489–501 (2006)
2. Friedman, J., Hastie, T., Tibshirani, R.: Additive logistic regression: a statistical view of boosting 28(2), 337–407 (2000)
3. Wang, G., Li, P.: Dynamic Adaboost ensemble extreme learning machine. In: Proceedings of the 2010 IEEE International Conference on Advanced Computer Theory and Engineering, ICACTE, pp. V3-54–V3-58 (2010)
4. Tian, H.X., Mao, Z.Z.: An ensemble ELM based on modified AdaBoost. RT algorithm for predicting the temperature of molten steel in ladle furnace. IEEE Transactions on Automation Science and Engineering 7(1), 73–80 (2010)
5. Solomatine, D.P., Shrestha, D.L., AdaBoost, R.T.: A boosting algorithm for regression problems. In: Proceedings of the 2004 IEEE International Joint Conference, pp. 1163–1168 (2004)
6. Zhu, J., Rosset, S., Zou, H.: Multi-class adaboost. Ann Arbor, 1612 (2006)
7. Ahonen, T., Hadid, A., Pietikäinen, M.: Face recognition with local binary patterns. In: Pajdla, T., Matas, J(G.) (eds.) ECCV 2004. LNCS, vol. 3021, pp. 469–481. Springer, Heidelberg (2004)
8. Huang, G.B., Zhou, H., Ding, X.: Extreme learning machine for regression and multiclass classification. IEEE Transactions on Systems, Man, and Cybernetics, Part B: Cybernetics 42(2), 513–529 (2012)
9. Zong, W., Huang, G.B., Chen, Y.: Weighted extreme learning machine for imbalance learning. Neurocomputing (2012)
10. Zong, W., Huang, G.B.: Face recognition based on extreme learning machine. Neurocomputing 74(16), 2541–2551 (2011)
11. Mohammed, A.A., Minhas, R., Jonathan Wu, Q.M.: Human face recognition based on multidimensional PCA and extreme learning machine. Pattern Recognition 44(10), 2588–2597 (2011)
12. Li, R., Lu, J., Zhang, Y.: Dynamic Adaboost learning with feature selection based on parallel genetic algorithm for image annotation. Knowledge-Based Systems 23(3), 195–201 (2010)
13. Heisele, B., Ho, P., Wu, J.: Face recognition: component-based versus global approaches. Computer Vision and Image Understanding 91(1), 6–21 (2003)
14. Allwein, E.L., Schapire, R.E., Singer, Y.: Reducing multiclass to binary: A unifying approach for margin classifiers. The Journal of Machine Learning Research 1, 113–141 (2001)
15. Toh, K.A.: Deterministic neural classification. Neural Computation 20(6), 1565–1595 (2008)
16. Deng, W., Zheng, Q., Chen, L.: Regularized extreme learning machine. In: IEEE Symposium on Computational Intelligence and Data Mining, pp. 389–395 (2009)

Copy Detection among Programs Using Extreme Learning Machines*

Bin Wang, Xiaochun Yang, and Guoren Wang

College of Information Science and Engineering,
Northeastern University, Liaoning 110819, China
{binwang,yangxc,wanggr}@mail.neu.edu.cn

Abstract. Because of the complexity of software development, some software developers may plagiarize source code from other projects or open source software in order to shorten development cycle. Many methods have been proposed to detect plagiarism among programs based on the program dependence graph, a graph representation of a program. However, the accuracy and efficiency of the detection approaches need to be improved. By employing extreme learning machine (ELM), we construct feature space for describing features of every two programs with possible plagiarism relationship. Such feature space could be large and time consuming, so we propose approaches to construct a small feature space by pruning isolated control statements and removable statements from each program to accelerate both training and classification time. We conducted a thorough experimental study of this technique on real C programs collected from Internet. The experimental results show the high accuracy and efficiency of our ELM-based approach.

1 Introduction

The Internet and open source software are developing rapidly nowadays, providing developers easier accesses to get various open source software code. Meanwhile with the increase of users' need, software developers have to develop more complicated software product, leading to longer development cycle which directly determines development cost as well as enterprise profit. To save development cost, some illegal developers inevitably utilize all possible methods to shorten development cycle. One of the fastest method is to develop their project on existing projects, open source softwares, or prototype products. Such behavior can be infringement to original authors. In order to prevent such infringement, a copy detection tool is needed to detect software source code plagiarism.

Many methods have been proposed to detect plagiarism among programs based on the program dependence graph (PDG for short) [1], a graph representation of a program. However, the accuracy and efficiency of the detection approaches need to be improved.

---

* The work is partially supported by the National Basic Research Program of China (973 Program) (No. 2012CB316201), the National Natural Science Foundation of China (Nos. 61272178, 61322208, 61129002), the Doctoral Fund of Ministry of Education of China (No. 20110042110028), and the Fundamental Research Funds for the Central Universities (Nos. N120504001, N110804002).

J. Cao et al. (eds.), *Proceedings of ELM-2014 Volume 2,*
Proceedings in Adaptation, Learning and Optimization 4, DOI: 10.1007/978-3-319-14066-7_19

In this paper, we propose a detection approach based on ELM. The challenges and contributions are as follows.

(i) A source program and its plagiaristic program could be partially similar since the plagiaristic program could contain some useless statements to make it different from the source program. As we know, finding common statements between two programs or common subgraphs between their corresponding PDGs is very time consuming. Therefore, we adopt extreme learning machine (ELM) to learn this potential similarity and classify PDGs accordingly since it is well-known very efficient for classification with high accuracy.

(ii) In order to enhance the accuracy of the classification, we further utilize control dependencies in PDGs to remove isolated control dependence subgraph and removable nodes to shrink the size of PDGs in Section 4. We show using ELM to classify PDGs with small sizes not only increases the classification accuracy but also improves classification time, since using smaller PDGs as training set means a smaller feature space to accelerate both training and classification time. This ELM-based approach greatly exceeds the detection ability of existing algorithms.

In Section 5 we present experimental results on real data sets to demonstrate the accuracy and time efficiency of the proposed technique.

2 Preliminary and Background

2.1 Computer Program

A computer program is a sequence of statements, written to perform a specific task with a computer. In a program, we mainly focus on two kinds of statements, control statement and other statement. A control statement always generates a boolean value, whereas other statements have no such requirement. Two kinds of dependencies are defined as follows:

- *Control dependency.* A statement S is control-dependent on a control statement C, if execution of S depends on the boolean value of C.
- *Data dependency.* A statement S_1 is data-dependent on another statement S_2, if S_1 reads the value of a variable v that has been changed by its preceding statement S_2.

Fig. 1 shows that statement `inputVal = paserInt(inputStr)` (line 13) is control-dependent on statement `inputStr = getNext(input)` (line 11), because the execution of the former depends on whether the boolean value of the latter is 1. Statement `findVal=temp/2` (line 8) is data-dependent on statement `temp = c*(valLeft-revise)+valRight` (line 7), since the former reads the value of variable `temp`, and the later does an assignment on `temp`.

2.2 Program Dependence Graph

A program dependence graph (PDG) [1] is a graph structure consisting of a program unit, such as statements and relations among variables in a C language function. Given a

```
1. int count(Istream*input, input c, int f1, int f2)
2. {
3.        int count, revise, valLeft, valRight, temp, findVal, count;
4.        valLeft = getCrest(f1);
5.        valRight = getCrest(f2);
6.        revise = GlobalVal;
7.        tmp = c*(valLeft-revise) +  valRight;
8.        findVal = temp / 2;
9.        count = 0;
10.       char * inputStr;
11.       while(inputStr = getNext(input))
12.       {
13.             int inputVal = paserInt(inputStr);
14.             if(inputVal == 0)
15.             {
16.                   int any = findVal + inputVal;
17.                   int any2 = any + findVal;
18.             }
19.             if(inputVal == findVal)
20.                   count++;
21.       }
22.       return count;
23. }
```

Fig. 1. An example of a partial program

program P, let $G(P)$ be the program dependence graph of P. Each node in a PDG represents a program statement, and edges between nodes represent dependencies among statements.

Definition 1. *A PDG is a graph $G(V, E)$ and it satisfies:*

- *The node set V in $G(P)$ represents the set of statements in program P.*
- *The edge set E in $G(P)$ represents the set of data dependencies and control dependencies between statements in program P.*

Fig. 2 shows the PDG for the program in Fig. 1. The PDG of every function starts from conditional statement "function entry." Each rhombic node in Fig. 2 represents a conditional control statement, and the dotted line with a circle at the node represents a control dependency. Each edge represents a data dependence.

Problem Description. Given a set of PDGs $\{G(P_1), \ldots, G(P_n)\}$ corresponding to the set of programs $\{P_1, \ldots, P_n\}$, find all pairs of PDGs $\langle G(P_i), G(P_j) \rangle$ such that the program P_i and program P_j have copy relationship ($1 \leq i, j \leq n, i \neq j$).

2.3 Extreme Learning Machine – ELM

ELM [2] has been widely applied in many classification applications, which originally developed for single hidden-layer feedforward neural networks (SLFNs) and then extended to the "generalized" SLFNs where the hidden layer need not be neuron alike. ELM first randomly assigns the input weights and the hidden layer biases, and then analytically determines the output weights of SLFNs. It can achieve better generalization performance than other conventional learning algorithms at a extremely fast learning speed. Besides, ELM is less sensitive to user-specified parameters and can be deployed faster and more conveniently [3].

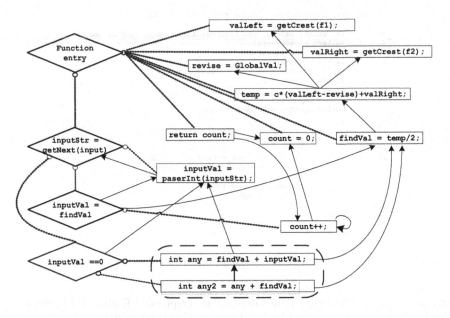

Fig. 2. The PDG for the program shown in Fig. 1

For N arbitrary distinct samples (x_j, t_j), where $\boldsymbol{x}_j = [x_{j1}, x_{j2}, \ldots, x_{jn}]^T \in \mathbb{R}^n$ and $\boldsymbol{t}_j = [t_{j1}, t_{j2}, \ldots, t_{jm}]^T \in \mathbb{R}^m$, standard SLFNs with L hidden nodes and activation function $g(x)$ are mathematically modeled as:

$$\sum_{i=1}^{L} \beta_i g_i(\boldsymbol{x}_j) = \sum_{i=1}^{L} \beta_i g(\boldsymbol{w}_i \cdot \boldsymbol{x}_j + b_i) = \boldsymbol{o}_j \quad (1 \leq j \leq N), \qquad (1)$$

where L is the number of hidden layer nodes, $\boldsymbol{w}_i = [w_{i1}, w_{i2}, \ldots, w_{in}]^T$ is the weight vector between the i-th hidden node and the input nodes, $\beta_i = [\beta_{i1}, w_{i2}, \ldots, w_{im}]^T$ is the weight vector between the i-th hidden node and the output nodes, b_i is the threshold of the i-th hidden node, and $\boldsymbol{o}_j = [o_{j1}, o_{j2}, \ldots, o_{jm}]^T$ is the j-th output vector of the SLFNs.

The standard SLFNs with L hidden nodes and activation function $g(x)$ can approximate these N samples with zero error. It means $\sum_{j=1}^{L} \|\boldsymbol{o}_j - \boldsymbol{t}_j\| = 0$ and there exist β_i, $\boldsymbol{w}_i$ and b_i such that

$$\sum_{i=1}^{L} \beta_i g(\boldsymbol{w}_i \cdot \boldsymbol{x}_j + b_i) = \boldsymbol{t}_j \quad (1 \leq j \leq N). \qquad (2)$$

Equation 2 can be expressed compactly as follows:

$$H\beta = T, \qquad (3)$$

where H is the hidden layer output matrix of the neural network, $H(w_1, w_2, \ldots, w_L, b_1, b_2, \ldots, b_L, x_1, x_2, \ldots, x_L) =$

$$[h_{ij}] = \begin{bmatrix} g(w_1 \cdot x_1 + b_1) & g(w_2 \cdot x_1 + b_2) & \ldots & g(w_L \cdot x_1 + b_L) \\ g(w_1 \cdot x_2 + b_1) & g(w_2 \cdot x_2 + b_2) & \ldots & g(w_L \cdot x_2 + b_L) \\ \ldots & \ldots & \ldots\ldots \\ g(w_1 \cdot x_N + b_1) & g(w_2 \cdot x_N + b_2) & \ldots & g(w_L \cdot x_N + b_L) \end{bmatrix}_{N \times L}, \quad (4)$$

$\beta = [\beta_1^T, \ldots, \beta_L^T]_{m \times L}^T$, and $T = [t_1^T, ldots, t_L^T]_{m \times N}^T$. Algorithm 1 shows the pseudocode of ELM.

Algorithm 1. ELM training

Input: Training set $\mathcal{N} = \{(x_j, t_j) | x_j \in \mathbb{R}^n (1 \le j \le N)\}$;
      Hidden node output function $g(w_i, b_i, x_j)$;
      Number of hidden nodes L;
Output: Weight vector β;
1 **for** $i = 1$ *to* L **do**
2   Randomly generate hidden node parameters (w_i, b_i);
3 Calculate the hidden layer output matrix H;
4 return the calculated output weight vector $\beta = H^\dagger T$, where $H^\dagger$ is the Moore-Penrose generalized by the inverse of matrix H;

---

3 ELM-Based Copy Detection

In this section, we propose an ELM-based framework for determining plagiarism between any two PDGs. Generally, a source PDG and its plagiarist PDG should be similar, so given a set of PDG, we first use an ELM classifier to find all similar PDGs.

In this section, we first show how to generate feature vectors for the ELM.

For a set of PDGs $\{G(P_1), \ldots, G(P_n)\}$, using ELM, the task is to learn a prediction rule from the training examples $\{D_j, y_j\}_{j=1}^n$, where D_j is the PDG in the training PDGs and $y_j \in \{+1, -1\}$ is the associated class label. Any similar PDGs have the same class label. Let $\mathcal{T}$ be the set of all patterns, i.e. the set of all subgraphs included in at least one training PDG. Then each training PDG $G(P_i)$ is encoded as a $|\mathcal{T}|$-dimensional vector $x_j = (x_{j,1}, \ldots, x_{j,|\mathcal{T}|})$,

$$x_{j,t} = I(t \subseteq G(P_i)),$$

where $I(\cdot)$ is 1 if the condition inside is true and 0 otherwise. For example, Fig. 3(e) shows an example of this feature space. Figs. 3(a) and 3(b) show an original code and its plagiaristic code, respectively. $G(P_1)$ and $G(P_2)$ are their corresponding PDGs shown in Figs. 3(c) and 3(d). For simplicity, we only use a line number in each node to express its corresponding statement. A $|\mathcal{T}|$-dimensional vector x_j is shown in Fig. 3(e), where every value is corresponding to a subgraph pattern in the training set.

We choose ELM to train these training examples $\{D_j, y_j\}_{j=1}^n$ since ELM is very efficient for large vector space.

```
4. valLeft = getCrest(f1);
5. valRight = getCrest(f2);
6. revise = GlobalVal;
7. tmp = c*(valLeft-revise) + valRight;
8. findVal = temp / 2;
```

(a) Original program P_1.

```
4. valLeft = getCrest(f1);
5. valRight = getCrest(f2);
6. revise = GlobalVal;
7. tmp = c*(valLeft-revise) + valRight;
8. mixVal = tmp;
9. findVal = mixVal / 2;
```

(b) Plagiaristic program P_2.

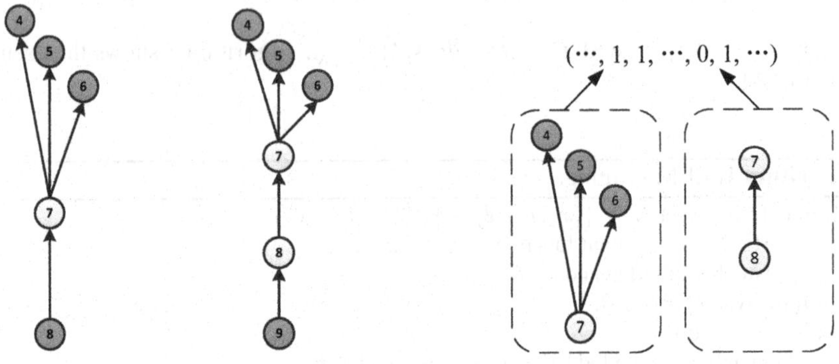

(c) $G(P_1)$.   (d) $G(P_2)$.   (e) Feature space and feature vector.

Fig. 3. An example of vector space

4   Achieving High Accuracy with Small Feature Space

For any two source and plagiarist programs pair (P_i, P_j), the plagiarist program P_j might contain useless control statements like loops, selective structures. If P_j contains a large number of such useless subgraphs, $G(P_i)$ might be dissimilar with $G(P_j)$ using ELM classifier.

In this section, we show removing subgraphs' corresponding useless control statements could increase the classification accuracy as well as accelerate the training efficiency since the number of unique patterns in $\mathcal{T}$ could be decreased.

4.1   Isolated Control Dependence Subgraphs

Definition 2. *Isolated control dependence subgraph. Given a PDG G, a subgraph G' of G is an isolated control dependence subgraph, if the following two conditions hold:*

(1) A variable v in a statement in G' is not used by its successive statements, and
(2) for any subgraph G'' of G', v does not appear in G''.

For example, in Fig. 2, the subgraph marked by dotted line is an isolated control dependence subgraph, whose corresponding statements are useless that could not affect the program's semantics, but would change the PDG structure.

For the source-plagiarist PDG pair $\langle G(P_i), G(P_j) \rangle$, where $P_i \rightsquigarrow P_j$, in order to keep the correctness of execution semantics of P_i in P_j, copyists need to insert or disturb some nodes in $G(P_i)$ to generate $G(P_j)$ as follows.

Feature 1. *A plagiarist PDG $G(P_j)$ and its source PDG $G(P_i)$ could be similar locally but dissimilar globally.*

For a source PDG $G(P_i)$ and its copy $G(P_j)$, $G(P_i)$ and $G(P_j)$ could share a common subgraph, but globally dissimilar since a copyist may insert some useless control dependence edges in $G(P_j)$.

4.2 Pruning Isolated Control Dependence Subgraphs

In order to decrease the size of feature space, we remove isolated control dependence subgraphs.

Notice that pruning isolated control dependence subgraphs might affect the execution of original program, because it can affect the function callings. As a consequence, such pruning might also remove subgraphs in an original PDG. Fortunately, according to our observation, this situation has no significant influence on detection result since if a control subgraph in the original program is removed, its corresponding control subgraph in the plagiaristic program would also be removed, and these two programs are still considered to contain plagiarism according to graph isomorphism algorithm.

Generally, we are not allowed to remove too many isolated control dependence subgraphs so that most of nodes can be kept in a PDG. Therefore, the number of total removed subgraph nodes $|V(G)|$ should not surpass a certain threshold value λ.

Algorithm 2 describes the process of pruning isolated control dependence subgraphs. It first generates a set of control dependence subgraphs. For each subgraph g, it records the number of dependencies with the statements outside g (lines $2-6$). It then only keeps isolated control dependence subgraphs in CG (line 7). The algorithm keeps removing remaining subgraphs in CG until the number of removed nodes is greater than a given threshold $|V(G)| \times \lambda$ (lines $8-14$). It finally returns the new graph as well as the number of removed nodes.

5 Experiments

All the algorithms were implemented using GNU C++. The experiments were run on a PC with an Intel(R) Core(TM)2 Duo CPU T6600@2.20GHz, 2GB RAM, running a Ubuntu (Linux) 32-bit operating system. We adopted Frama-c as a source code analysis tool to generate PDGs.

We ran our experiments on C program sets collected from the Internet. To ensure the validity of algorithm, we copied part of the functions in this programs, adopted common plagiarism methods to disguise them and put them together with original program sets to get test data set, including 250 functions and nearly 4000 lines of program.

In order to correctly reflect recall and precision of our algorithm ELM-PCD, we compare our algorithm with the two state-of-the-art tools: GPLAG – a classic source

---

Algorithm 2. Pruning isolated control dependence subgraphs

Input: PDG G, pruning threshold λ;
Output: PDG after pruning isolated control dependence subgraphs;

1   Generate control dependence subgraph set $CG \leftarrow \{g_1, \ldots, g_k\}$ in G;
2   **foreach** *control dependence subgraph g in CG* **do**
3     $g.count \leftarrow 0$;
4   **foreach** *statement S in g* **do**
5   **if** *S has dependence relations with statements outside g* **then**
6         $g.count \leftarrow g.count + 1$;

7   Remove subgraphs in CG whose number of dependencies are greater than 0;
8   $maxNumDelNodes \leftarrow |V(G)| \times \lambda$;
9   $deletCount \leftarrow 0$;
10   $G.removeNode \leftarrow 0$;
11   **foreach** *subgraph g in CG* **do**
12   **if** $deletCount + |V(g)| \leq maxNumDelNodes$ **then**
13   delete g from G;
14       $deletCount \leftarrow deletCount + |V(g)|$;

15   $G.removeNode \leftarrow deleteCount$;
16   **return** G;

---

code plagiarism detection software [4], and JPlag – a tool of finding plagiarisms among a set of programs [5].

Accuracy of Finding Plagiarisms. Fig. 4(a) shows precision of different detection approaches. ELM-Full represents the algorithm using full feature space and ELM-Small represents the algorithm using pruned feature space. Both ELM-based approaches had a higher accuracy than GPLAG. Since GPLAG is also based on PDG, the results show that PDG-based approaches can reflect programs' semantical features well. JPlag, which is based on sequence, was influenced by some small functions, leading to the decrease of accuracy.

As for recall, Fig. 4(b) shows that ELM-Small had a recall of nearly 90%, ELM-Full has a recall of 77%, while GPLAG, which is also based on PDG, has a recall of only 55%. This is because many plagiarism have methods to disguise, which greatly change the structure of PDG, making GPLAG unable to detect such plagiarism. JPlag can only detect plagiarism without any modification, having an accuracy of around 50%, coinciding with the fact that half of the plagiarism is not disguised.

Performance of Copy Detection. We first compared time for detecting programs with plagiaristic relationship. Fig. 5(a) shows the comparison results. The detecting time mainly depends on the time cost for static analysis tools and the efficiency of identifying programs with plagiaristic relationship. Our approaches and GPLAG costed the same time for static analysis, however, by using ELM, our ELM-based approach ran faster than GPLAG since GPLAG needed to find isomorphic subgraphs, which was time consuming, whereas we used ELM to classify PDGs, which was very efficient.

Fig. 5(b) shows number of generated feature spaces for different number of functions. We can see that pruning isolated control dependence subgraphs can decrease relatively high number of features in each feature space.

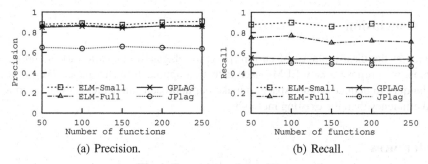

(a) Precision.   (b) Recall.

Fig. 4. Comparison of algorithms

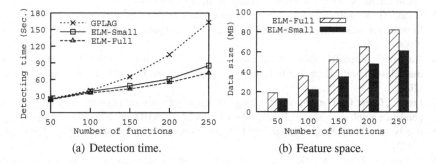

(a) Detection time.   (b) Feature space.

Fig. 5. Performance of copy detection

6   Related Work

The existing work can generally be categorized into the following five groups.

(i) *Text-based method.* Text-based methods directly use source code as intermediate form to compare without utilizing the grammar and syntax features of program [6, 7].

(ii) *Word-based method.* Word-based methods lexically analyze source code at first, and transform source code into a sequence consisting of token, on which the subsequent steps will be conducted [8, 9]. Obviously, word-based methods can better deal with modifications on code fragments like variable rename, pattern adjustment and so on. Compared with text-based methods, they are more robust.

(iii) *Tree-based method.* Tree-based methods conduct lexical and grammatical analysis on the program, transform it into an abstract syntax tree, and then use tree matching or other methods to conduct similarity detection [10–12].

(iv) *Metric-based method*. The basic idea of metric-based methods is to extract some features or indicators from program fragments, and thereby use them to compare instead of directly comparing code or abstract syntax trees [13].

(v) *Semantic-based method*. Semantic-based methods are used to find more veiled similarities among programs [1, 4]. Semantic-based methods conduct static analysis instead of simple grammatical analysis to provide more precise information.

7   Conclusion

Aimed at the disadvantages of current program plagiarism detection tools and algorithms, we propose a new ELM-based detection approach in this paper. We propose techniques to deduce feature space to make our approach can achieve higher accuracy and efficiency than the existing methods.

References

1. Ferrante, J., Ottenstein, K.J., Warren, J.D.: The program dependence graph and its use in optimization. ACM Trans. Program.Lang. Syst. 9(3), 319–349 (1987)
2. Huang, G., Siew, C., Chen, L.: Universal approximation using incremental constructive feed-forward networks with random hidden nodes. IEEE Trans. Neural Netw. 17(4), 879–892 (2006)
3. Huang, G., Zhou, H., Ding, X., Zhang, R.: Extreme learning machine for regression and multiclass classification. IEEE Trans. Syst. Man Cybern. Part B Cybern. 42(2), 513–529 (2012)
4. Liu, C., Chen, C., Han, J., Yu, P.: Gplag: Detection of software plagiarism by program dependence graph analysis. In: Proc. of the 12th ACM SIGKDD, pp. 872–881 (2006)
5. Prechelt, L., Malpohl, G., Phlippsen, M.: Jplag: Finding plagiarisms among a set of programs (2012)
6. Ducasse, S., Rieger, M., Demeyer, S.: A language independent approach for detecting duplicated code. In: Proc. of ICSM, pp. 109–118 (1999)
7. Johnson, J.: Visualizing textual redundancy in legacy source. In: Proc. of the 1994 Conference of the Centre for Advanced Studies on Collaborative Research (CASCON), p. 32 (1994)
8. Kamiya, T., Kusumoto, S., Inoue, K.: CCFinder: A multilinguistic token-based code clone detection system for large scale source code. IEEE Trans. on Software Engineering 28(7), 654–670 (2002)
9. Yan, X., Han, J., Afshar, R.: Clospan: Mining closed sequential patterns in large datasets. In: Proc. SIAM Int'l Conf. Data Mining (May 2003)
10. Baxter, I., Yahin, A., Moura, L., Anna, M.: Clone detection using abstract syntax trees. In: Proc. of ICSM, pp. 368–377 (1998)
11. Falke, R., Koschke, R., Frenzel, P.: Empirical evaluation of clone detection using syntax suffix trees. Empirical Software Engineering 13, 601–643 (2008)
12. Jiang, L., Misherghi, G., Su, Z., Glondu, S.: DECKARD: Scalable and accurate tree-based detection of code clones. In: Proc. of ICSE, pp. 96–105 (2007)
13. Mayrand, J., Leblanc, C., Merlo, E.: Experiment on the automatic detection of function clones in a software system using metrics. In: Proc. of ICSM, pp. 244–253 (1996)

Extreme Learning Machine for Reservoir Parameter Estimation in Heterogeneous Reservoir

Jianhua Cao, Jucheng Yang, and Yan Wang

College of Computer Science and Information Engineering,
Tianjin University of Science and Technology, Tianjin City, China
{Caojh,jcyang,wangy}@tust.edu.cn

Abstract. This study focuses on reservoir parameter estimation using extreme learning machine in heterogeneous sandstone reservoir. The specific aim of work is to obtain accurate porosity and permeability which has proven to be difficult by conventional petrophysical methods in wells without core data. 4950 samples of 15 wells with core data have been used to train the neural network, and robust ELM algorithm provides fast and accurate prediction results. The network model is then applied to estimate porosity and permeability for the remaining wells. Based on the predicted results, reservoir parameter distribution character has been analyzed. Potential zone has been proposed for further research in the survey.

Keywords: Extreme learning machine, reservoir parameter estimation, sandstone reservoir.

1 Introduction

This paper is concerned with reservoir parameter estimation from well log data, and two types of parameters including porosity and permeability are to be discussed and predicted using extreme learning machine.

With the rapid development of neuron science and computer science, a lot of progress have been made in the study of artificial neural networks, and the specific virtues have attracted many engineers and scientists. A steady increase has been witnessed in the literatures concerning the application of neural network models in petroleum engineering, such as sedimentary micro-facies prediction[11], lithology classification[12], reservoir parameters prediction[2,3]. In petrophysical analysis, the neural network models have always acted as a predictor or estimator of deriving petrophysical parameters, such as porosity and permeability where no core data is available[1,9,16,17].

In heterogeneous reservoirs, the geological surroundings are complex, and it is difficult to estimate reservoir parameters precisely using the conventional multilinear and nonlinear statistical technique ways.

Extreme learning machine (ELM) is a single-hidden layer feedforward neural network (SLFN) proposed by Huang[5,6] , which has been widely studied and used in a lot of scientific fields to solve problems of classification[15], pattern recognition[14], and big data mining[7,8]. And in petroleum reservoir prediction, there still has few applications. The ELM approach to training SLFN consists in the random generation of the hidden layer weights, followed by solving a linear system of equations by least-squares for the estimation of the output layer weights. This learning strategy is very fast and gives good prediction accuracy. Theoretically and practically, this algorithm can produce good generalization performance in most cases and can learn thousands of times faster than conventional popular learning algorithms for feed-forward neural networks.

In this paper, we examine the potential of ELM to predict porosity and permeability parameters in a heterogeneous sandstone reservoir. Well logs and core data are paired as training samples, and prediction models for porosity and permeability prediction are set up. The network models are finally applied to interpret reservoir porosity and permeability in the rest wells which have no core data.

2   Geological Background

In this study, 15 wells are selected from YQ survey of Ordos basin, China. All wells have encountered Permian sandstone reservoir. Oil have been discovered in 6 wells, which are marked with red color-filled circles in Fig.1.

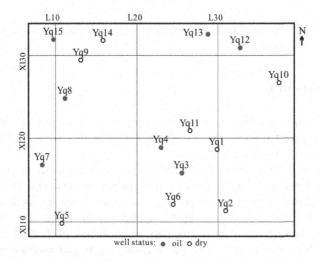

Fig. 1. Well location and basemap of the YQ 3D survey. There are totally 15 wells, with 6 wells encountering oil in Permian sandstone reservoir, while the remaining are dry.

Porous sandstone of Permian is the target reservoir. Owing to the complicated deposition process and variable diagenesis, the sandstone reservoir is heterogeneous and changes fast spatially, which is controlled by the fluvial sedimentary facies. Core analysis shows that the main pore type is intragranular pore(Fig.2a), with proportion more than 55%, and the rest include secondary intergranular pore(Fig.2b), matrix pore and micro crack. And the measured permeability varies greatly both horizontally and vertically.

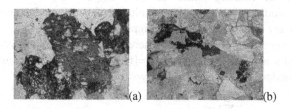

(a)   (b)

Fig. 2. Microscope photo of core samples in YQ survey. Pore spaces are colored and the grey part is the matrix: (a) intragranular pore, (b)intergranular pore

In petrophysical log interpretation, permeability is related to porosity, and this interrelationship is always expressed by the regression equations based on the actual porosity and permeability of core samples(shown as Fig.3). The equation can be recommended for permeability estimation if porosity is determined, but practically, porosity is hard to be accurately calculated using current empirical formula. What's more, the error of porosity calculation is sure to be brought into the permeability estimation process. In this paper, a neural network method is applied to accurately predict reservoir parameters for further reservoir characterization.

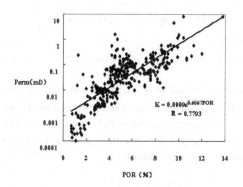

Fig. 3. Porosity vs. permeability(briefly named as Perm). The data are the actual measured results of core samples. Porosity ranges from 2% to 14%, while permeability value has large span from 0.0001md to 10md.

3   Extreme Learning Machine

Extreme Learning Machine(ELM) is a simple supervised learning algorithm proposed by Huang [6] for Single-hidden Layer Feedforward Neural Network(SLFN).

Comparing with the conventional neural networks, ELM has better performance in learning efficiency and universal approximation capability[7]. Different from BP network, the input weights and biases of ELM are randomly assigned and need not to be adjusted within the training phase, and the output weights can be determined analytically by finding the least-square solution. Therefore the neural network is obtained after a few steps with very low computation cost.

Given a training dataset $N=(x_i,t_i)$, where x_i is a $n \times 1$ input vector $x_i=[x_i^1,x_i^2, ...,x_i^n]^T \in R^n$, and t_i is a $m \times 1$ target vector $t_i=[t_i^1,t_i^2,...,t_i^m]^T \in R^m$, the output of a SLFN with M hidden nodes is formulated as follows:

$$\sum_{i=1}^{M} \beta_i g(w_i \cdot x_j + b_i) = t_j , j=1,2,...,N \tag{1}$$

where w_i is the weight vector connecting the ith hidden node with the input nodes, β_i is the weight vector connecting the ith hidden node with the output nodes, and b_i is the threshold of the ith hidden node. $g(.)$ denotes the non-linear activation function of the hidden node. It can be the identity sigmoid or Gaussian function, among a large collection of polynomial functions.

Eq.(1) can be written in a more compact format as follows:

$$H\beta=T \tag{2}$$

where H is the hidden layer output matrix of the network:

$$H = \begin{bmatrix} g(w_1 x_1 + b_1) & \cdots & g(w_N x_1 + b_N) \\ \vdots & \ddots & \vdots \\ g(w_1 x_M + b_1) & \cdots & g(w_N x_M + b_N) \end{bmatrix} \tag{3}$$

β is the matrix of hidden-to-output weights, and T is the target matrix.

In Eq(3), weights (w_i)and biases(b_i) are randomly assigned, and $g(.)$ is known to be selected as sigmoid function, so the output of hidden nodes could be determined, which is H in Eq (2). The remaining problem becomes a set of linear equations, and can be solved by minimum square error estimation:

$$Min_\beta \|H\beta - T\| \tag{4}$$

According to the definition of the Moore-Penrose generalized inverse, the smallest norm least squares solution of (2) is given as:

$$\hat{\beta}=H^{-1}T \tag{5}$$

where H^{-1} is the Moore-Penrose generalized inverse of matrix H.

For Eq.(5), once the H and T is set, it is not difficult to get theβmatrix. The process has proved several advantages: (1) the training error is minimized;(2) the generalization performance is optimal;(3) and the solution is unique.

Totally, the *ELM* algorithm includes[15]: Given a training set $\{(x_i,t_i)|$, $x_i \in R^n, t_i \in R^m, i=1,...,N\}$, an activation function g, and the number of hidden nodes M:

(1)randomly set input weights w_j and biases b_j;
(2)calculate the hidden layer output matrix H;
(3)calculate the output weight matrixβ.

4   Experimental Design

4.1   Data Preparation

Among the 15 wells, 7 wells have core data in Permian formation and most of them have been tested in labs to get the actual reservoir properties, such as porosity and permeability. The total number of tested samples is summed up to 4950, 80% of which will be used as the training data in the ELM network, while the rest 20% as validation samples.

Logs including acoustic slowness(AC), density(DEN), gamma ray(GR), compensated neutron logs(CNL), and deep resistivity (RT) are used as input for the networks, and porosity(POR) and permeability (PERM) measured from cores are to be considered as target. The input logs are paired with the core porosity and permeability as the training samples, and the latter is taken as the target output. Part of the samples are shown as Tab.1.

Table 1. Training data examples from the wells

Well Name	Depth	AC	CNL	DEN	GR	RT	POR	PERM
	meter	us/m	%	g/cc	API	Oh.m	%	md
YQ1	3200	165.52	22.29	2.53	62.75	16.24	12.5	1.58
YQ1	3205.5	159.35	24.36	2.49	55.32	15.39	13.4	18.2
YQ1	3211	185.21	19.52	2.58	71.37	12.53	6.8	1.29
YQ1	3215.5	203.52	16.34	2.64	86.52	8.95	5.1	0.81
YQ1	3350	192.34	20.42	2.55	80.32	9.23	7.6	0.95
YQ1	3350.6	172.56	20.46	2.51	65.85	18.52	10.5	10.2
YQ3	3608.2	175.36	23.54	2.45	62.38	14.62	15.2	10.7
YQ3	3609.5	179.62	21.65	2.50	63.50	13.49	13.4	8.61
YQ3	3710	215.31	15.32	2.68	92.65	7.25	5.8	0.52
YQ3	3715.6	200.45	16.49	2.65	84.67	8.69	8.2	1.23
YQ3	3716.8	192.35	17.59	2.63	75.68	10.52	9.3	1.54
YQ3	3719.2	182.64	19.67	2.57	72.35	13.25	11.4	5.69

Before inputting data into the network, data normalization is necessary. The normalized variable has the following form:

$$X_{new} = \frac{X_{old} - \min X}{\max X - \min X}$$
(6)

where X denotes the logs: GR, AC, DEN, CNL, RT. The new normalized variable X_{new} takes the range from 0 to 1 for all the parameters.

4.2   Network Model

For porosity and permeability estimation, five logs including GR, AC, DEN,CNL and RT are related to petrophysical properties. So these five logs are fed to the ELM network as input with each node denoting one log. Porosity is to be taken as the only network output at the output layer. The network architecture is shown as Fig.4.

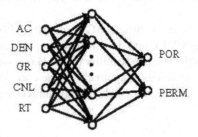

Fig. 4. Schematic of architecture of the ELM network model. Node number of the input layer is 5, and porosity and permeability are the two nodes at the output layer. Number of hidden layer nodes is to be settled by testing.

4.3   Parameter Selection

For ELM network, since weights and biases for the hidden layer are randomly assigned initially, activation function and number of hidden nodes are the only parameters to be determined.

In the study, four types of activation functions have been tested, and they are: Sigmoid function, Radial Basis function, Hardlim function, and Triangular function. At the same time, numbers of hidden nodes are tested accordingly. When selecting one of the activation functions, number of hidden nodes will start from 20, with 20 as incremental step. The test process is to select the optimal activation and hidden node numbers, so that the feasible ELM network for this prediction can be determined and best accuracy result can be obtained for further estimation.

Almost 4000 samples from the wells have been used to train the network model, and the total time lasts about 10 minutes.

Accuracy comparison has been done when using different functions and node numbers. The comparison result is shown as Fig.5.

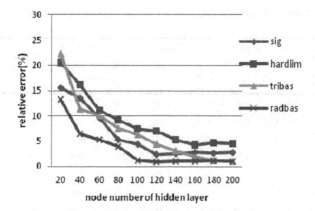

Fig. 5. Relative error percentage comparison using different activation function and node numbers. The error is a measure of misfit between the actual and the predicted based on the network model. The lower the error, the higher the accuracy.

According to the parameters test, the relative error percentage between prediction results and measured properties decreases rapidly as node number of the hidden layer increases, and keeps stable when the number is up to 140. It is also the trend for accuracy of the ELM network. Between the activation functions, the relative prediction error keeps lower than others when using the radial basis function, and when setting the number of hidden layer nodes as 100, accuracy of the network is suitable for the reservoir parameters prediction. Therefore, for the ELM network, radial basis function is set as the activation function and number of the hidden layer nodes as 100.

5   Reservoir Parameters Prediction

5.1   Network Validation

After the neural network parameters are fixed, the prediction network model is then established. The remaining 950 samples are fed to the network for model validation. Results show that the prediction accuracy is satisfactory and the relative error percentage for porosity can be limited to 0.5-1.5%, and for permeability the error is less than 4% (examples shown as Tab.2).

5.2   Reservoir Parameter Prediction

The model is trained and validated, and has been proved to be effective and reliable, so the final step is to run neural network model on the remaining wells in the survey and output the predicted reservoir parameters. Log data of 8 wells without core data have been input the model, and porosity and permeability of the Permian reservoir have been estimated.

Table 2. Validation examples for the network prediction model

Well Name	Depth meter	Core Porosity %	Predicted Porosity %	Absolute Error	Relative error %	Core Perm md	Predicted Perm md	Absolute Error	Relative error %
YQ10	3102	8.96	8.82	-0.14	1.563	4.56	4.67	0.11	2.412
YQ10	3104.5	6.28	6.33	0.05	0.796	0.89	0.92	0.03	3.371
YQ10	3113	11.26	11.15	-0.11	0.977	10.25	10.09	-0.14	1.366
YQ10	3115.5	10.64	10.78	0.14	1.316	15.23	15.39	0.16	1.051
YQ10	3143	7.85	7.75	-0.1	1.274	3.24	3.21	-0.03	0.926
YQ10	3145.6	6.97	7.05	0.08	1.148	0.56	0.52	-0.04	7.143
YQ15	3057.5	13.52	13.64	0.12	0.888	22.39	22.15	0.24	1.072
YQ15	3065.5	10.79	10.92	0.13	1.205	12.54	12.65	0.09	0.718
YQ15	3070.2	8.69	8.79	0.1	1.151	1.25	1.21	-0.04	3.200
YQ15	3075.5	8.29	8.19	-0.1	1.206	0.95	0.91	-0.04	4.211

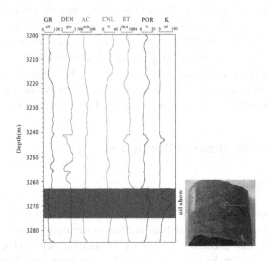

Fig. 6. Reservoir prediction examples for well YQ12.Curves of POR and K(Permeability) are predicted from the network model. Sandstone reservoir from 3263 to 3275 has oil show, and has median porosity but high permeability on the basis of the prediction results. The right photo is the core for YQ12 at the 3270.56m. Small pores can be recognized.

Fig.6 is the plot of original logs and predicted reservoir parameters of well YQ12 from 3200 to 3285m.This interval belongs to the Permian formation, and oil show exists at the depth of 3263-3275m. According to the predicted result, porosity of the oil-bearing interval is about 12.5% and permeability about 32md, which means the sandstone reservoir from 3263m to 3275m has good properties and can be recommended for further evaluation.

Since all wells have been processed using the ELM model, statistical reservoir analysis is then to be carried out in the survey. According to the geological correlation, the top and bottom of target sandstone reservoir in the wells are determined. Therefore under the control of the tops for the reservoir, average porosity and permeability of the target reservoir interval have been calculated for all the 15 wells in the survey. Then flex gridding algorithm is used to interpolate the attribute between wells. Fig.7 is the average porosity map for target reservoir. It is clear that well zones of YQ12 and YQ13 at the northeast part of the survey has porosity higher than 10%, and also for well zones of YQ15,YQ8,YQ4,YQ3 and YQ7 at the western part in the survey. Since there has no direct structural high in the survey, stratigraphic trap is the dominant type of traps. Fluvial porous reservoir is the critical factor for well planning and economic assessment. Therefore these areas are potential for next well-planning selections.

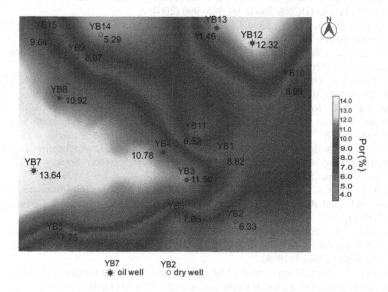

Fig. 7. Average porosity distribution for the target reservoir of YQ survey. The blue number near well symbol is the average porosity based on the ELM network output. Red-yellow color of the colorbar stands for the high porosity.

6   Conclusions

ELM network analysis has been investigated and applied to predict reservoir parameters in heterogeneous sandstone reservoir. Reliable network prediction model has been established in the study, and porosity and permeability are estimated for 15 wells in the survey. West zone in the survey, including wells of YQ15,YQ8, YQ4,YQ3 and YQ7 is considered as favorable area, so as the northeast zone of wells of YQ12 and YQ13. ELM provides new way of artificial intelligence to solve such nonlinear petrophysical problems.

References

1. Ali, M., Chawathe, A.: Using artificial intelligence to predict permeability from petrographic data. Computers & Geosciences 26(8), 915–925 (2000)
2. Chikhi, S., Batouche, M.: Probabilistic neural method combined with radialbias functions applied to reservoir characterization in the Algerian Triassic province. Journal of Geophysics and Engineering 1(2), 134–142 (2004)
3. El Ouahed, A.K., Tiab, D., Mazouzi, A.: Application of artificial intelligence to characterize naturally fractured zones in Hassi Messaoud Oil Field, Algeria. Journal of Petroleum Science and Engineering 49, 122–141 (2005)
4. Gharbi, R.B.C., Mansoori, G.A.: An introduction to artificial intelligence applications in petroleum exploration and production. Journal of Petroleum Science and Engineering 49, 93–96 (2005)
5. Huang, G.-B., Zhu, Q.-Y., Siew, C.-K.: Extreme learning machine: a new learning scheme of feed forward neural networks. In: Proceedings of International Joint Conference on Neural Networks (IJCNN 2004), pp. 985–990 (2004)
6. Huang, G.-B., Zhu, Q.-Y., Siew, C.-K.: Extreme learning machine: theory and applications. Neurocomputing 70(1-3), 489–501 (2006)
7. Huang, G.-B., Ding, X., Zhou, H.: Optimization method based extreme learning machine for classification. Neurocomputing 74, 155–163 (2010)
8. Chorowski, J., Wang, J., Zurada, J.M.: Review and performance comparison of SVM and ELM-based classifiers. Nerocomputing 128(2014), 507–516 (2014)
9. Khoshdel, H., Riabi, M.A.: Multi attribute transform and neural network in porosity estimation of an offshore oil field — A case study. Journal of Petroleum Science and Engineering 78, 740–747 (2011)
10. Kouider El Ouahed, A., Tiab, D., Mazouzi, A., Safraz, A.J.: Application of Artificial Intelligence to Characterize Naturally Fractured Reservoirs. Society of Petroleum Engineers. Paper SPE 84870, pp. 1–14 (2013)
11. Qi, L., Carr, T.R.: Neural network prediction of carbonate lithofacies from well logs, Big Bow and Sand Arroyo Creek fields, Southwest Kansas. Computers & Geosciences 32, 947–964 (2006)
12. Nikravesh, M.: Neural network knowledge-based modelling of rock properties based on well log databases. In: SPE 46206, 1998 SPE Western Regional Meeting, Bakersfield, California, pp. 10–13 (1998)
13. Ouenes, A.: Practical application of fuzzy logic and neural networks to fractured reservoir characterization. Computers & Geosciences 26(8), 953–962 (2010)
14. Minhas, R., Baradarani, A., Seifzadeh, S., et al.: Human action recognition using extreme learning machine based on visual vocabularies. Nerocomputing 73(10), 1906–1917 (2010)
15. Ramon, M., Francesco, C., Amaury, L.: Extreme learning machine for soYQean classification in remote sensing hyperspectral images. Neurocomputing 28, 207–216 (2014)
16. Zazoun, R.S.: Fracture density estimation from core and conventional well logs data using artificial neural networks: The Cambro-Ordovician reservoir of Mesdar oil field, Algeria. Journal of African Earth Sciences 83, 55–73 (2013)
17. Saemi, M., Ahmadi, M., Varjani, A.Y.: Design of neural networks using genetic algorithm for the permeability estimation of the reservoir. Journal of Petroleum Science and Engineering 59, 97–105 (2007)

Multifault Diagnosis for Rolling Element Bearings Based on Extreme Learning Machine

Yuan Lan, Xiaoyan Xiong, Xiaohong Han, and Jiahai Huang

Research Institute of Mechano-Electronic Engineering,
Taiyuan University of Technology,
Xi Kuang Street, Taiyuan, Shanxi, China 030024
{lanyuan,xiongxiaoyan,hanxiaohong,
huangjiahai}@tyut.edu.cn

Abstract. Rolling element bearings constitute the key parts on rotating machinery and their fault diagnosis are of great importance. In this paper, a new intelligent fault diagnosis scheme based on Wavelet Packet Transform (WPT) and Extreme Learning Machine (ELM) is proposed. 16-dimensional wavelet packet node energies were extracted from the original datasets as the feature vector input to the classifiers. A novel classifier, ELM, and its variants, error-minimized ELM (EM-ELM) and online sequential ELM (OS-ELM), were introduced in this study to diagnose the fault on bearings. ELM has been proved to be extremely fast and can provide good generalization performance on many pattern recognition cases. However, preliminary ELM is a batch learning algorithm with a fixed network structure. EM-ELM and OS-ELM are the extends of the preliminary ELM to allow network to grow in the learning process and to learn data sequentially. ELM, EM-ELM and OS-ELM classifiers were evaluated on 13 fault datasets and the empirical results showed that they are really fast and perform well on all the 13 datasets.

Keywords: rolling element bearings fault diagnosis, extreme learning machine, error-minimized extreme learning machine, online sequential extreme learning machine.

1 Introduction

Rolling element bearing is one of the most common yet fragile parts in mechanical equipments and according to statistics, it is responsible for nearly 30% of failures in rotating machines, which leads to significant economic losses and even human casualties. To avoid catastrophe and minimize defective machinery downtime, bearing fault diagnosis is of great industrial importance and attracted attention from researchers and engineering communities. Many fault diagnosis methods [11,16,18,2] have already been used to detect the type and severity of the faults for preventing the rolling element bearings from failures. Generally speaking, a fault diagnosis system includes three crucial parts: data acquisition, feature extraction and fault pattern recognition.

© Springer International Publishing Switzerland 2015   209
J. Cao et al. (eds.), *Proceedings of ELM-2014 Volume 2,*
Proceedings in Adaptation, Learning and Optimization 4, DOI: 10.1007/978-3-319-14066-7_21

Mechanical vibration signal is a typical nonlinear and non-stationary signal due to the strike, velocity chopping, structure transmutation, loading and friction, which is very crucial and has been widely used as the most common and reliable source to extract fault features for rolling element bearing fault diagnosis. In addition, the performance of a fault diagnosis method is highly dependent on the information stored within the extracted fault features. Pan et al. [14] made use of wavelet packet node energies as the input to fuzzy c-means in order to build a health index to track bearing health. Shen et al. [18] employed the statistical parameters of wavelet packet paving and generic support vector regressive classifier to diagnose the fault bearings. Additionally, Du et al. [2] proposed a novel method based on wavelet leaders multifractal features to examine the bearings. In this paper, we utilized the wavelet packet node energies as the feature vector similar to [14].

Essentially, bearing fault diagnosis is a pattern recognition problem. Some intelligent classification methods, such as artificial neural network (ANN) and support vector machine (SVM), have been applied to the fault diagnosis of rotating machinery [15,18,2]. However, these methods share the common shortcomings like slow computational speed and manual intervention requirement. In this paper, a new classifier, extreme learning machine classifier, is introduced and employed to recognize the bearing faults. Extreme learning machine (ELM) [10,9,8,7,6] proposed by Huang et al. is developed for generalized single hidden layer feedforward networks (SLFNs) with a wide variety of hidden nodes. Different from other learning algorithms, ELM randomly selects all the hidden node parameters, after which the network can be represented as a linear system and the output weights can be computed analytically. It is extremely fast and tends to obtain the smallest training error and the smallest norm of weights that lead to good generalization performance. In addition, ELM has many variants including error-minimized ELM (EM-ELM) [3] and online sequential ELM (OS-ELM) [12] etc. EM-ELM extends preliminary ELM such that hidden node(s) can be added in the network sequentially with one-by-one or group-by-group mode, with which it becomes easier to evaluate the significance of each hidden node or group of hidden nodes and we can obtain the suitable network size with a predetermined threshold. In many industrial applications, the training data could not be collected at once. In such case, OS-ELM could learn the training data one-by-one or chunk-by-chunk instead of conducting re-training using the past data with the new collected data. ELM and its variants could provide good generalization performance for many regression and classification applications and they may run much faster than SVM classifiers [9].

The rest of paper is organized as follows. Section 2 provides a brief review of ELM and its variants. Section 3 describes the data sets of bearing fault diagnosis and feature extraction. Section 4 presents the results of bearing fault diagnosis by using ELM classifiers and its variants. Finally, section 5 is the conclusion of the paper.

2 A Brief Review of Extreme Learning Machine and Its Variants

2.1 Preliminary Extreme Learning Machine (ELM)

ELM [10,9,8,7,6] is a learning algorithm that was developed for generalized SLFNs with wide variety of hidden nodes. Consider N arbitrary distinct samples $(\mathbf{x}_i, \mathbf{t}_i) \in \mathbf{R}^n \times \mathbf{R}^m$. If a SLFN with L hidden nodes can approximate these N samples with zero error, it then implies that there exist β_i, $\mathbf{a}_i$ and b_i such that

$$f_L(\mathbf{x}_j) = \sum_{i=1}^{L} \beta_i G(\mathbf{a}_i, b_i, \mathbf{x}_j) = \mathbf{t}_j, \ \ j = 1, \cdots, N. \tag{1}$$

where $\mathbf{a}_i$ and b_i are the learning parameters of the hidden nodes, β_i is the output weight, and $G(\mathbf{a}_i, b_i, \mathbf{x})$ denotes the output of the ith hidden node with respect to the input $\mathbf{x}$.

Equation (1) can be written compactly as:

$$\mathbf{H}\beta = \mathbf{T} \tag{2}$$

where

$$\mathbf{H}(\mathbf{a}_1, \cdots, \mathbf{a}_L, b_1, \cdots, b_L, \mathbf{x}_1, \cdots, \mathbf{x}_N)$$
$$= \begin{bmatrix} G(\mathbf{a}_1, b_1, \mathbf{x}_1) & \cdots & G(\mathbf{a}_L, b_L, \mathbf{x}_1) \\ \vdots & \cdots & \vdots \\ G(\mathbf{a}_1, b_1, \mathbf{x}_N) & \cdots & G(\mathbf{a}_L, b_L, \mathbf{x}_N) \end{bmatrix}_{N \times L} \tag{3}$$

$$\beta = \begin{bmatrix} \beta_1^T \\ \vdots \\ \beta_L{}^T \end{bmatrix}_{L \times m} \quad \text{and} \quad \mathbf{T} = \begin{bmatrix} \mathbf{t}_1^T \\ \vdots \\ \mathbf{t}_N^T \end{bmatrix}_{N \times m} \tag{4}$$

$\mathbf{H}$ is called the hidden layer output matrix of the network [5].

According to ELM theories [10,9,8,7,6] all the hidden nodes $(\mathbf{a}_i, b_i)$ can be randomly assigned instead of being tuned. The solution of equation (2) is estimated as:

$$\hat{\beta} = \mathbf{H}^\dagger \mathbf{T} \tag{5}$$

where $\mathbf{H}^\dagger$ is the Moore-Penrose generalized inverse [17] of the hidden layer output matrix $\mathbf{H}$. The ELM algorithm[1] which only consists of three steps can then be summarized as:

ELM Algorithm: Given a training set $\aleph = \{(\mathbf{x}_i, \mathbf{t}_i) | \mathbf{x}_i \in \mathbf{R}^n, \mathbf{t}_i \in \mathbf{R}^m, i = 1, \cdots, N\}$, activation function $g(x)$, and hidden node number L,

---

[1] The source codes of ELM can be downloaded from
http://www.ntu.edu.sg/home/egbhuang/

1. Assign random hidden nodes by randomly generating parameters $(\mathbf{a}_i, b_i)$, $i = 1, \cdots, L$.
2. Calculate the hidden layer output matrix $\mathbf{H}$.
3. Calculate the output weight β: $\beta = \mathbf{H}^\dagger \mathbf{T}$

Remark 1: Universal approximation capability of ELM has been analyzed by Huang, et al [8,7] using incremental methods and it has been shown that single SLFNs with randomly generated (variety of) hidden nodes can universally approximate any continuous target functions.

Remark 2: ELM is a batch learning algorithm with a fixed network structure. The network structure of ELM is usually determined by a tedious trial and error method before pattern recognition process.

2.2 Error-Minimized Extreme Learning Machine (EM-ELM)

EM-ELM is an error minimization based method in which the number of hidden nodes can grow one-by-one or group-by-group until optimal. The approach can significantly reduce the computational complexity and its convergence was proved as well.

Assume we have a set of training data $\{(\mathbf{x}_i, t_i)\}_{i=1}^N$, the maximum number of hidden nodes L_{max}, and the expected learning accuracy ϵ. There are two phases in EM-ELM algorithm.

Initialization Phase

1. Randomly generate a hidden node $(\mathbf{a}_1, b_1)$, where $\mathbf{a}$ and b are the parameters of hidden node. And set $j = 0$ and $L_0 = 1$.
2. Calculating the hidden layer output matrix: $\mathbf{h}_0$

$$\mathbf{h}_0 = \begin{bmatrix} G(\mathbf{a}_1, b_1, \mathbf{x}_1) \cdots G(\mathbf{a}_1, b_1, \mathbf{x}_N) \end{bmatrix}^T \tag{6}$$

where $G(\mathbf{x})$ is the activation function and $G(\mathbf{a}_1, b_1, \mathbf{x}_i)$ denotes the output of the hidden node w.r.t input $\mathbf{x}_i$.
3. Calculating the corresponding output error $E(\mathbf{h}_0) = \|\mathbf{h}_0\mathbf{h}_0^\dagger \mathbf{t} - \mathbf{t}\|$.

Recursively Growing Phase
while $L_j < L_{max}$ and $E(\mathbf{H}_j) > \epsilon$

1. Randomly add a hidden node to the existing SLFNs. The number of hidden nodes $L_{j+1} = L_j + 1$ and the corresponding hidden layer output matrix $\mathbf{H}_{j+1} = [\mathbf{H}_j \quad \delta\mathbf{h}_j]$, where $\delta\mathbf{h}_j$ is shown below.

$$\delta\mathbf{h}_j = \begin{bmatrix} G(\mathbf{a}_{L_j+1}, b_{L_j+1}, \mathbf{x}_1) \cdots G(\mathbf{a}_{L_j+1}, b_{L_j+1}, \mathbf{x}_N) \end{bmatrix}^T \tag{7}$$

2. Updating the output weight β [1]

$$\mathbf{D}_j = \frac{\delta \mathbf{h}_j^T (\mathbf{I} - \mathbf{H}_j \mathbf{H}_j^\dagger)}{\delta \mathbf{h}_j^T (\mathbf{I} - \mathbf{H}_j \mathbf{H}_j^\dagger) \delta \mathbf{h}_j}$$

$$\mathbf{U}_j = \mathbf{H}_j^\dagger - \mathbf{H}_j^\dagger \delta \mathbf{h}_j \mathbf{D}_j$$

$$\beta^{(j+1)} = \mathbf{H}_{j+1}^\dagger \mathbf{t} = \begin{bmatrix} \mathbf{U}_j \\ \mathbf{D}_j \end{bmatrix} \mathbf{t}$$

(8)

3. $j = j + 1$.

endwhile

2.3 Online Sequential Extreme Learning Machine (OS-ELM)

Assume the network has L hidden nodes and the data $\aleph = \{(\mathbf{x}_i, \mathbf{t}_i) | \mathbf{x}_i \in \mathbf{R}^n, \mathbf{t}_i \in \mathbf{R}^m, i = 1, \ldots, N\}$ presents to the network sequentially (one by one or chunk by chunk). There are two phases in OS-ELM algorithm, an initialization phase and a sequential phase. In the initialization phase, it is required by the algorithm that $rank(\mathbf{H}_0) = L$. (where $\mathbf{H}_0$ denotes the hidden layer output matrix for initialization phase), which means the number of training data required in the initialization phase N_0 is equal to or greater than L, ie. $N_0 \geq L$. If the first L training data are not distinct, more training data may be needed. The most significant improvement of OS-ELM is presented in the sequential phase. As a property of sequential learning system, the data used to train the OS-ELM network would be discarded once it is used and never retrieved.

Initialization Phase. A small chunk of training data is used to initialize the learning, $\aleph_0 = \{(\mathbf{x}_i, \mathbf{t}_i)\}_{i=1}^{N_0}$ from the given training set $\aleph = \{(\mathbf{x}_i, \mathbf{t}_i) | \mathbf{x}_i \in \mathbf{R}^n, \mathbf{t}_i \in \mathbf{R}^m, i = 1, \ldots, N\}$, where N_0 denotes the number of observations used in initialization phase, $N_0 \geq L$.

1. Randomly assign the input parameters: for additive hidden nodes, parameters are input weights $\mathbf{a}_i$ and bias b_i; for RBF hidden nodes, parameters are center $\mathbf{a}_i$ and impact factor b_i; $i = 1, \ldots, L$.
2. Calculate the initial hidden layer output matrix $\mathbf{H}_0$

$$\mathbf{H}_0 = \begin{bmatrix} G(\mathbf{a}_1, b_1, \mathbf{x}_1) & \cdots & G(\mathbf{a}_L, b_L, \mathbf{x}_1) \\ \vdots & \cdots & \vdots \\ G(\mathbf{a}_1, b_1, \mathbf{x}_{N_0}) & \cdots & G(\mathbf{a}_L, b_L, \mathbf{x}_{N_0}) \end{bmatrix}_{N_0 \times L}$$

(9)

3. Estimate the initial output weight $\beta^{(0)}$
We have

$$\mathbf{T}_0 = \begin{bmatrix} \mathbf{t}_1 \cdots \mathbf{t}_{N_0} \end{bmatrix}_{N_0 \times m}^T$$

(10)

then, the problem is to minimize $\|\mathbf{H}_0 \beta - \mathbf{T}_0\|$, and from [12], we know that $\mathbf{H}^\dagger = (\mathbf{H}^T \mathbf{H})^{-1} \mathbf{H}^T$. The solution to minimize $\|\mathbf{H}_0 \beta - \mathbf{T}_0\|$ is given by $\beta^{(0)} = \mathbf{P}_0 \mathbf{H}_0^T \mathbf{T}_0$, where $\mathbf{P}_0 = (\mathbf{H}_0^T \mathbf{H}_0)^{-1}$, and $\mathbf{K}_0 = \mathbf{H}_0^T \mathbf{H}_0 = \mathbf{P}_0^{-1}$.
4. Set $k = 0$.

Sequential Learning Phase. Present the $(k+1)$th chunk of new observations, $\aleph_{k+1} = \{(\mathbf{x}_i, \mathbf{t}_i)\}_{i=(\sum_{j=0}^{k} N_j)+1}^{\sum_{j=0}^{k+1} N_j}$, and N_{k+1} denotes the number of observations in the $(k+1)$th chunk.

1. Compute the partial hidden layer output matrix $\mathbf{H}_{k+1}$, which is shown in Equation (11).

$$\mathbf{H}_{k+1} = \begin{bmatrix} G(\mathbf{a}_1, b_1, \mathbf{x}_{(\sum_{j=0}^{k} N_j)+1}) & \cdots & G(\mathbf{a}_L, b_L, \mathbf{x}_{(\sum_{j=0}^{k} N_j)+1}) \\ \vdots & \cdots & \vdots \\ G(\mathbf{a}_1, b_1, \mathbf{x}_{\sum_{j=0}^{k+1} N_j}) & \cdots & G(\mathbf{a}_L, b_L, \mathbf{x}_{\sum_{j=0}^{k+1} N_j}) \end{bmatrix}_{N_{k+1} \times L} \tag{11}$$

2. Calculate the output weight $\beta^{(k+1)}$
   We have

$$\mathbf{T}_{k+1} = \begin{bmatrix} \mathbf{t}_{(\sum_{j=0}^{k} N_j)+1} & \cdots & \mathbf{t}_{\sum_{j=0}^{k+1} N_j} \end{bmatrix}_{N_{k+1} \times m}^{T} \tag{12}$$

and

$$\mathbf{K}_{k+1} = \mathbf{K}_k + \mathbf{H}_{k+1}^T \mathbf{H}_{k+1} \tag{13}$$

$$\beta^{(k+1)} = \beta^{(k)} + \mathbf{K}_{k+1}^{-1} \mathbf{H}_{k+1}^T (\mathbf{T}_{k+1} - \mathbf{H}_{k+1} \beta^{(k)}) \tag{14}$$

From equation (14), we find that $\mathbf{K}_{k+1}^{-1}$ is used to compute $\beta^{(k+1)}$. In order to avoid calculating inverse in the recursive process, the Woodbury formula [4] is applied to modify the equations.

$$\begin{aligned} \mathbf{K}_{k+1}^{-1} &= (\mathbf{K}_k + \mathbf{H}_{k+1}^T \mathbf{H}_{k+1})^{-1} \\ &= \mathbf{K}_k^{-1} - \mathbf{K}_k^{-1} \mathbf{H}_{k+1}^T (\mathbf{I} + \mathbf{H}_{k+1} \mathbf{K}_k^{-1} \mathbf{H}_{k+1}^T)^{-1} \mathbf{H}_{k+1} \mathbf{K}_k^{-1} \end{aligned} \tag{15}$$

And $\mathbf{P}_{k+1} = \mathbf{K}_{k+1}^{-1}$, we modify the equations (13)(14) using (15):

$$\mathbf{P}_{k+1} = \mathbf{P}_k - \mathbf{P}_k \mathbf{H}_{k+1}^T (\mathbf{I} + \mathbf{H}_{k+1} \mathbf{P}_k \mathbf{H}_{k+1}^T)^{-1} \mathbf{H}_{k+1} \mathbf{P}_k \tag{16}$$

$$\beta^{(k+1)} = \beta^{(k)} + \mathbf{P}_{k+1} \mathbf{H}_{k+1}^T (\mathbf{T}_{k+1} - \mathbf{H}_{k+1} \beta^{(k)}) \tag{17}$$

3. Set $k = k + 1$. Go to 1) in this sequential learning phase.

3   Multifault Diagnosis for Rolling Element Bearings

Generally speaking, an intelligent machine fault diagnosis scheme includes three parts: data acquisition, fault feature extraction and fault pattern recognition. In this paper, we intend to apply the ELM and its variants on fault diagnosis of rolling element bearing. The experiment data are obtained from the Bearing Data Center of Case Western Reserve University (CWRU),which has been widely considered as a benchmark fault diagnosis data [13]. The data are first processed by wavelet packet transform (WPT) at 4th decomposition depths to enhance the signal-to-noise ratio. And the energies of $2^4 = 16$ frequency ranges are used as the desired features and input to the pattern recognizer. The status of the rolling element bearing is determined by the output of the classifier. More details are presented as follows.

3.1   Data Sets

The experiment data are kindly provided by the Bearing Data Center of Case Western Reserve University (CWRU),which has been widely considered as a benchmark fault diagnosis data [13]. The data provided by the Bearing Data Center are huge and we only use part of them. The test stand of the experimental system mainly includes three parts, a $2hp$ motor, a torque transducer and a dynamometer. The motor shaft is supported by bearing with the type of $6205 - 2RSJEMSKF$. Single point faults were introduced to the test bearings using electro-discharge machining with fault diameters of $7mils$, $14mils$, and $21mils$. An accelerometer is attached to the housing with magnetic base at the drive end to acquire the vibration data from the test bearings. Vibration data were collected using a 16 channel DAT recorder with a sample frequency of $12kHz$, and each dataset contents 480,000 points. All the bearing data used in this paper is under $0hp$ motor load and the motor speed is approximate 1797rpm. The bearing is tested under four different situations (normal condition, inner race fault , ball fault and outer race fault) and three different fault degrees ($7mils$ - slight, $14mils$ - medium, and $21mils$ - serious).

In order to evaluate the ELM classifier and its variants, we separate the experiment data sets with $0hp$ load into 13 fault data sets with different fault condition combinations. Each sample in the data sets contents 2048 points. The data specification is presented in Table 1. For $DALL$ data set, each fault location type includes data from three different fault degrees. Data sets from $D070707$ to $D210714$ are the different combinations of fault degree for three fault locations with the normal data. Data sets DNI, DNB, and DNO consider one fault at a time with the normal situation. In addition, data sets DI, DB, and DO are prepared for recognizing the fault degree on each fault location.

3.2   Feature Extraction

As mentioned in the previous section, after preprocessing the data, each sample in the data sets contents 2048 points. In order to make the data from different fault type and fault degree comparable, we conducted normalization for each element in each sample. Let one sample be $\mathbf{x} = [x_1, x_2, ..., x_{2048}]$. The normalization has been done according to the equation below.

$$x_i' = \frac{x_i - \mu}{\sigma} \quad i = 1, 2, ..., 2048 \tag{18}$$

Table 1. Data specification of fault diagnosis of rolling element bearings

Dataset	Training data	Testing data	Fault diameter (mil)	Fault type	Classification label
DALL	180	60	0	N	1
			7 14 21	I	2
			7 14 21	B	3
			7 14 21	N	4
D070707	150	50	0 7 7 7	NIBO	1 2 3 4
D141414	150	50	0 14 14 14	NIBO	1 2 3 4
D212121	150	50	0 21 21 21	NIBO	1 2 3 4
D071421	150	50	0 7 14 21	NIBO	1 2 3 4
D142107	150	50	0 14 21 7	NIBO	1 2 3 4
D210714	150	50	0 21 7 14	NIBO	1 2 3 4
DNI	120	60	0	N	1
			7 14 21	I	2
DNB	120	60	0	N	1
			7 14 21	B	2
DNO	120	60	0	N	1
			7 14 21	O	2
DI	100	50	7 14 21	I	1 2 3
DB	100	50	7 14 21	B	1 2 3
DO	100	50	7 14 21	O	1 2 3

D - data, N - normal, I - inner race fault, B - ball fault, O - outer race fault

where μ, σ are the average value and standard deviation respectively and x_i' is the ith element of $\mathbf{x}$ after normalization. Fig. 1 shows a original sample data with the value range of $[-0.3, 0.3]$ and the same sample data after normalization with the value range of $[-3, 3]$.

In this paper, we applied wavelet packet transform (WPT) to remove the noise as well as extract the useful features. As a widely used wavelet basis function for machine fault diagnosis, $db2$ wavelet was employed for WPT step in this study. After WPT, 16 signals were extracted from the 16 nodes at the $4th$ decomposition level. Take one sample from the normal data set for an example. Fig. 2 presents the WPT tree and decomposition signal from node $(4, 0)$ of the selected sample. Calculate the energy value of each decomposition signal, we have

$$E_i = \sum_{j=1}^{L} |y_{ij}|^2, \quad i = 0, 1, ..., 15 \tag{19}$$

where y_{ij} is the jth element of the ith decomposition signal, L is the length of the decomposition signal. Hence, we have obtained one 16-dimension feature vector, $V = [E_0, E_1, ..., E_{15}]$. Normally, the value of E_i is big, so we normalize the feature vector using

$$V' = \left[\frac{E_0}{R} \ \frac{E_1}{R} \ \cdots \ \frac{E_{15}}{R} \right] \tag{20}$$

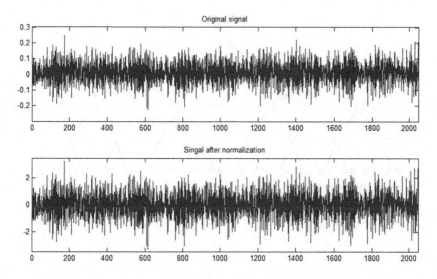

Fig. 1. Original sample data(up) and data after normalization (down)

where $R = \sqrt{\sum_{i=0}^{15} |E_i|^2}$, and V' is the feature vector after normalization. All the samples in the datasets described in Table 1 were processed following the procedures above and the resultant 16-dimensional data were used for evaluating the ELM classifier and its variants.

4   Experiments and Results

In this section, the performance of preliminary ELM and its variants were evaluated on the manipulated datasets indicated in Table 1. The source code of preliminary ELM and its variants could be found on [19]. All the evaluations were carried out in the Matlab R2010b environment running on a desktop with Pentium(R) Core(TM)i5 CPU 3.1GHz and 8GB of RAM. Cross-validation was used for evaluating the three classifiers, namely preliminary ELM,EM-ELM and OS-ELM. In each round of simulation, the target dataset was randomly separated into the values given in Table 1. 20 round of simulations were conducted and the average values were recorded as the final results. We compared the results obtained by preliminary ELM, EM-ELM and OS-ELM, and the details are shown in this section. The activation function used in all the classifiers were sigmoidal function. For EM-ELM classifier applied on any dataset, the initial number of hidden node in the network $L_0 = 1$, the number of hidden nodes added into the network in each step $L_i = 1$, the maximum number of hidden nodes allowed in the hidden layer $L_{max} = 200$, and the expected training accuracy was set to $\epsilon = 99\%$. For OS-ELM classifier applied on any dataset, the chunk of training data used to initialize the learning is $N_0 = 50$ (for datasets DI, DB, and DO) and $N_0 = 100$ (for the remaining datasets), the number of

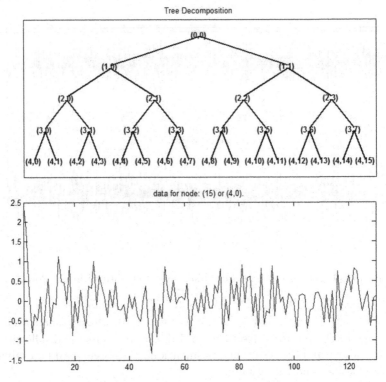

Fig. 2. Four level wavelet packet decomposition: WPT tree (up), decomposition signal from node $(4, 0)$ (down)

instances learnt in each step in the sequential learning phase was set to $N_i = 5$ for all the datasets.

4.1 Performance Comparison of ELM, EM-ELM and OS-ELM on *DALL* Dataset

According to Table 1, dataset *DALL* contains instances from normal data, inner race fault data of all three fault degrees, ball fault data from all three fault degrees and outer race fault data from all three fault degrees. The fault diagnosis results are shown in Table 2. The testing accuracy of all the classifiers is quite high and above 96%. Preliminary ELM is the fastest; EM-ELM could obtain high accuracy with the smallest network structure; and although OS-ELM could outperform ELM and EM-ELM, it needs more time and more hidden nodes.

Table 2. Performance comparison of ELM, EM-ELM and OS-ELM on DALL dataset

Dataset	Algorithm	Time (s)	Accuracy(%)		Testing Std	#Node
			Training	Testing		
DALL	ELM	0.0148	99.94	98.58	0.0124	50
	EM-ELM	0.0242	99.64	96.17	0.0265	25.5
	OS-ELM	0.0758	100	99.5	0.0134	160

4.2 Performance Comparison of ELM, EM-ELM and OS-ELM on Multi-location and Multi-degree Case

It is very common that multiply faults on rolling element bearings occur simultaneously with different fault degree. In this paper, the bearing is tested under four different situations (normal condition, inner race fault , ball fault and outer race fault) and three different fault degrees ($7mils$ - slight, $14mils$ - medium, and $21mils$ - serious). Hence, the number of all the possible fault combinations is $3 \times 3 \times 3 = 27$. We conducted the evaluations for all the fault combinations and 6 of them are presented in this section in Table 3. All the three classifiers perform well on these fault combinations.

Table 3. Performance comparison of ELM, EM-ELM and OS-ELM on multi-location and multi-degree case

Dataset	Algorithm	Time (s)	Accuracy(%)		Testing Std	#Node
			Training	Testing		
D070707	ELM	0.0001	100	99.9	0.0045	16
	EM-ELM	0.0102	99.7	99.4	0.0131	7.15
	OS-ELM	0.0008	100	100	0	15
D141414	ELM	0.0001	100	100	0	20
	EM-ELM	0.007	99.57	98.4	0.0239	9.9
	OS-ELM	0.0617	100	100	0	50
D212121	ELM	0.0086	100	99.6	0.0082	40
	EM-ELM	0.0055	9.57	98.5	0.0182	13.25
	OS-ELM	0.0484	100	99.7	0.0073	45
D071421	ELM	0.0001	100	100	0	22
	EM-ELM	0.0055	99.7	98.3	0.0227	11.4
	OS-ELM	0.0289	100	100	0	25
D142107	ELM	0.0008	100	100	0	18
	EM-ELM	0.0063	99.7	98.3	0.0208	9.2
	OS-ELM	0.0094	100	100	0	25
D210714	ELM	0.0001	100	100	0	12
	EM-ELM	0.0047	99.6	99.3	0.0098	8.2
	OS-ELM	0.0188	100	100	0	20

4.3 Performance Comparison of ELM, EM-ELM and OS-ELM on *DNI*, *DNB*, and *DNO* Datasets

In this subsection, we separate the faults of the rolling element bearing and consider only one fault at a time. The diagnosis results are shown in Table 4. It is obvious that the fault can be fully identified with ELM algorithm and its variants. At the same time, it implies that the features extracted from the original data are of great significance.

Table 4. Performance comparison of ELM, EM-ELM and OS-ELM on DNI, DNB, and DNO datasets

Dataset	Algorithm	Time (s)	Accuracy(%) Training	Testing	Testing Std	#Node
DNI	ELM	0.0001	100	100	0	10
	EM-ELM	0.0023	99.88	99.33	0.0113	4.25
	OS-ELM	0.0008	100	100	0	8
DNB	ELM	0.0001	100	100	0	8
	EM-ELM	0.0016	99.8	99.25	0.0127	4.1
	OS-ELM	0.0008	100	100	0	9
DNO	ELM	0.0001	100	100	0	8
	EM-ELM	0.0008	99.96	99.5	0.0134	4
	OS-ELM	0.0001	100	100	0	8

D - data, N - normal, I - inner race fault, B - ball fault, O - outer race fault

4.4 Performance Comparison of ELM, EM-ELM and OS-ELM on Each Fault Location

It is also very interesting to know whether the extracted features are distinct for different fault degrees on any fault location. Therefore, in this section, we

Table 5. Performance comparison of ELM, EM-ELM and OS-ELM on each fault location

Dataset	Algorithm	Time (s)	Accuracy(%) Training	Testing	Testing Std	#Node
DI	ELM	0.0001	100	100	0	16
	EM-ELM	0.007	100	98.7	0.0187	8.85
	OS-ELM	0.0016	100	100	0	16
DB	ELM	0.0031	99.65	97.7	0.0134	26
	EM-ELM	0.0133	100	97.3	0.0208	23.4
	OS-ELM	0.0023	99.45	97.3	0.0187	18
DO	ELM	0.0016	100	100	0	16
	EM-ELM	0.0016	100	99	0.0152	8.45
	OS-ELM	0.0023	100	100	0	20

D - data, N - normal, I - inner race fault, B - ball fault, O - outer race fault conducted the evaluations considering only faulty data on each fault location and the results are shown in Table 5. The performance of all the three classifiers is quite good, which implies the good performance in subsection 4.2.

5   Conclusion

Rolling element bearings constitute the key parts on rotating machinery and their fault diagnosis are of great importance. In this paper, ELM and its variants, EM-ELM and OS-ELM, are introduced on rolling element bearings fault diagnosis. ELM has been proposed for generalized single hidden layer feedforward networks with a wide variety of hidden nodes. They are extremely fast in learning and perform well on many artificial and real regression and classification applications. However, preliminary ELM is a batch learning algorithm with a fixed network structure. EM-ELM and OS-ELM are the extends of the preliminary ELM to allow network to grow in the learning process and to learn data sequentially. Empirical studies in this paper have shown that ELM classifiers and its variants could perform fast and very well on bearing faults diagnosis.

Acknowledgments. We wish to thank the Western Reserve University for providing us with the dataset and several helpful explanations. In addition, the work described in this paper is supported by the International Scientific and Technological Cooperation Projects of Shanxi Province (no.2014081030), the Natural Science Foundation of Shanxi Province (no.2014021024-1,2014011021-1), and the Taiyuan University of Technology Group Fund (no.1205-04020102).

References

1. Cline, R.E.: Representations for the generalized inverse of a partitioned matrix. Journal of the Society for Industrial and Applied Mathematics 12(3), 588–600 (1964)
2. Du, W., Tao, J., Li, Y., Liu, C.: Wavelet leaders multifractal features based fault diagnosis of rotating mechanism. Mechanical Systems and Signal Processing 43, 57–75 (2014)
3. Feng, G., Huang, G.B., Lin, Q., Gay, R.: Error minimized extreme learning machine with growth of hidden nodes and incremental learning. IEEE Transactions on Neural Networks 20(8), 1352–1357 (2009)
4. Golub, G.H., Loan, C.F.V.: Matrix Computations, 3rd edn. The Johns Hopkins University Press, Baltimore (1996)
5. Huang, G.B.: Learning capability and storage capacity of two-hidden-layer feedforward networks. IEEE Transactions on Neural Networks 14(2), 274–281 (2003)
6. Huang, G.B., Chen, L.: Convex incremental extreme learning machine. Neurocomputing 70, 3056–3062 (2007)
7. Huang, G.B., Chen, L.: Enhanced random search based incremental extreme learning machine. Neurocomputing 71, 3060–3068 (2008)
8. Huang, G.B., Chen, L., Siew, C.K.: Universal approximation using incremental constructive feedforward networks with random hidden nodes. IEEE Transactions on Neural Networks 17(4), 879–892 (2006)

9. Huang, G.B., Zhu, Q.Y., Siew, C.K.: Extreme learning machine: Theory and applications. Neurocomputing 70, 489–501 (2006)
10. Huang, G.B., Zhu, Q.Y., Siew, C.K.: Extreme learning machine: A new learning scheme of feedforward neural networks. In: Proceedings of International Joint Conference on Neural Networks (IJCNN 2004), Budapest, Hungary, July 25-29, vol. 2, pp. 985–990 (2004)
11. Lei, Y., He, Z., Zi, Y., Hu, Q.: Fault diagnosis of rotating machinery based on multiple anfis combination with gas. Mechanical Systems and Signal Processing 21, 2280–2294 (2007)
12. Liang, N.Y., Huang, G.B., Saratchandran, P., Sundararajan, N.: A fast and accurate online sequential learning algorithm for feedforward networks. IEEE Transactions on Neural Networks 17(6), 1411–1423 (2006)
13. Loparo, K.A.: Bearings vibration data set. Case Western Reserve University (2014), http://csegroups.case.edu/bearingdatacenter/home
14. Pan, Y.N., Chen, J., Li, X.L.: Bearing performance degradation assessment based on lifting wavelet packet decomposition and fuzzy c-means. Mechanical Systems and Signal Processing 24, 559–566 (2010)
15. Paya, B.A., Esat, I.I., Badi, M.N.M.: Artificial nerual network based fault diagnostics of rotating machinery using wavelet transforms as a preprocessor. Mechanical Systems and Signal Processing 11(5), 751–765 (1997)
16. Randall, R.B., Antoni, J.: Rolling element bearing diagnostics - a tutorial. Mechanical Systems and Signal Processing 25, 485–520 (2011)
17. Serre, D.: Matrices: Theory and Applications. Springer-Verlag New York, Inc. (2002)
18. Shen, C., Wang, D., Kong, F., Tse, P.W.: Fault diagnosis of rotating machinery based on the statstical paramets of wavelet paving and a generic support vector regressive classifier. Measurement 46, 1551–1564 (2013)
19. Zhu, Q.Y., Huang, G.B.: Source codes of ELM algorithm. School of Electrical and Electronic Engineering, Nanyang Technological University, Singapore (2004), http://www.ntu.edu.sg/home/egbhuang/

Gradient-Based No-Reference Image Blur Assessment Using Extreme Learning Machine*

Baojun Zhao[1], Shuigen Wang[1], Chenwei Deng[1,**],
Guang-Bin Huang[2], and Baoxian Wang[1]

[1] School of Information and Electronics, Beijing Institute of Technology, China
[2] School of EEE, Nanyang Technological University, Singapore
{zbj,sgwang,cwdeng}@bit.edu.cnm, EGBHuang@ntu.edu.sg, wbx1025@bit.edu.cn

Abstract. In this paper, we propose a perceptual no-reference image blur evaluation method using a new machine learning technique, i.e., extreme learning machine (ELM). The new model, Blind Image Blur quality Evaluator (BIBE), exploits scene statistics of gradient magnitudes to model the properties of blur images, and then the underlying features are derived from the gradient magnitudes distribution. The resultant feature is finally mapped to an associated quality score using ELM. Experimental results on public databases show that the proposed BIBE correlates well with human perceived blurriness, and outperforms the state-of-the-art no-reference blur evaluation metrics as well as generic no-reference image quality assessment methods.

Keywords: No-reference blur metric, Extreme learning machine, Gradient magnitude, Generalized Gaussian distribution.

1 Introduction

With the rapid development of digital techniques, multimedia content has become a very popular means of entertainment and communication, such as high definition television (HDTV), streaming Internet protocol TV (IPTV), and websites like Youtube, Facebook and Flickr, etc. An enormous amount of visual data is making its way to consumers. This phenomenon has resulted in great advancements in multimedia acquisition, compression, transmission, enhancement, and reproduction, etc. However, impairments generally exist along the visual signal processing and communication. The visibility of these impairments has a drastic effect on the Quality of Experience (QoE) of the consumers. Hence, image quality assessment (IQA) techniques are in great demand for measuring the perceived quality of the multimedia content [1].

* This work was supported by the National Natural Science Foundation of China under Grant 61301090, the Beijing Excellent Talent Fund under Grant 2013D009011000001, and in part by the Excellent Young Scholars Research Fund of Beijing Institute of Technology under Grant 2013YR0508.
** Corresponding author.

Since human visual system (HVS) is the ultimate receiver of sensory information in many cases, subjective quality assessment is the most reliable way to measure image quality. Nevertheless, subjective quality evaluation suffers from high cost, heavy complexity, and infeasibility to be used in real-time applications. Thus, it is necessary to develop objective quality metrics.

According to the availability of reference information, there is a general agreement that objective quality assessment metrics can be categorized into full-reference (FR), reduced-reference (RR) and no-reference (NR) methods. To evaluate the quality of distorted images, FR metrics utilize the information of original reference image, RR ones use some features extracted from the original images, and NR methods do not require any reference information, and are the most useful techniques in applications where the original image is not available. However, on the other hand, NR metrics designing is quite challenging as the corresponding reference information cannot be used for the gauging of distorted image quality.

Unlike FR IQA, where a reference image is available to be used for the evaluation of most distortion types, NR IQA approaches generally aim to capture one or few distortions, e.g., blur, white noise, blockiness, since various distortions have distinct properties and it is difficult to use the same features to model all distortion types. In this paper, we mainly concentrate on NR blur assessment, which is one of the most important issues needed to be tackled in many applications, such as image acquisition and compression [2].

A number of NR blur evaluation models have been proposed in literature, including pixel-based ones, statistical properties based ones, transform-based and gradient-based ones. In [3][4], local kurtosis is used in the frequency domain for measuring sharpness, since the kurtosis is inversely proportional to the sharpness of edge. Ong et al. [5] and Marziliano et al. [6] measured blurriness by analyzing the spread of edges. DCT coefficients and Wavelet coefficients are employed for blur estimation in [7-10]. In [11], Ferzli et al. analyzed that the aforementioned existing blur metrics cannot well predict the blurriness in images with different contents, and they proposed a blur assessment method by integrating the concept of just noticeable blur into a probability summation model. The metric proposed in [11] is able to predict the amount of blurriness in images with different content, but it does not correlate well with images having nonuniform saliency content. Sadaka et al. [12] tried to improve the performance of [11] by incorporating a visual attention model, such that the areas in the images which are most likely noticed by humans are given more weight than the others. Hassen et al. [13] proposed a multiscale sharpness metric based on the local phase coherence (LPC) of complex wavelet coefficients. In [14], Karam et al. developed a probabilistic model to assess image blur by adopting the cumulative probability of blur detection. Liang et al. [15] and Yan et al. [16] developed their blur algorithms based on the histogram of gradient profile variance and gradient profile sharpness, respectively.

In this paper, we exploit the statistical distribution of gradient for blur evaluation and propose a Blind Image quality Blur Evaluator (BIBE). A generalized

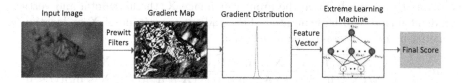

Fig. 1. The diagram of the proposed metric

Gaussian model is utilized to model the distribution of gradient. More specifically, gradient magnitudes are computed for each distorted image, and the parameters of the statistical distribution of gradient magnitudes are then estimated to form the underlying features. Extreme learning machine (ELM) [17] is employed to map the resulting features into the corresponding quality score. Experimental results demonstrate that the proposed method correlates highly consistent with human subjective scores in the tested four publicly available datasets.

2   Proposed Algorithm

In this section, we are to present a detailed description of the proposed algorithm. The algorithm diagram is shown in Fig. 1, we can see that the gradients of input image are first computed by a Prewitt filter, and the resultant gradient distribution can be modeled by a generalized Gaussian function. The parameters of the gradient distribution are then obtained, and they can be used to form a feature vector to represent the properties of image content. The feature vector is finally mapped into a quality score using the ELM.

2.1   Gradient Magnitude Computation

As demonstrated in Fig. 1, gradients are first computed for input images. For digital images, the gradient magnitude is defined as the root mean square of image directional gradients along two orthogonal directions (e.g., horizontal and vertical). Mathematically, the gradient is usually computed by convolving an image with a linear filter such as the classic Roberts, Sobel, Scharr and Prewitt filters or some task-specific ones [18-20]. For computational simplicity, Prewitt filter is utilized to calculate the image gradient using 3×3 templates in Eq. (1). Noted that by adopting other filters such as the Sobel and Scharr filters, the proposed method would have similar IQA results. The Prewitt filters along horizontal (x) and vertical (y) directions are defined as:

$$\mathbf{h}_x = \begin{bmatrix} 1/3 & 0 & -1/3 \\ 1/3 & 0 & -1/3 \\ 1/3 & 0 & -1/3 \end{bmatrix}, \mathbf{h}_y = \begin{bmatrix} 1/3 & 1/3 & 1/3 \\ 0 & 0 & 0 \\ -1/3 & -1/3 & -1/3 \end{bmatrix} \tag{1}$$

Convolving $\mathbf{h}_x$ and $\mathbf{h}_y$ with the input test image $\mathbf{X}$, the horizontal and vertical gradient images can be generated. The gradient magnitude of $\mathbf{X}$ at location i, denoted by $\mathbf{m}(i)$, is computed as follows:

$$\mathbf{m}(i) = \begin{cases} \sqrt{(\mathbf{X} \otimes \mathbf{h}_x)^2(i) + (\mathbf{X} \otimes \mathbf{h}_y)^2(i)}, & (\mathbf{X} \otimes \mathbf{h}_x)(i) \times (\mathbf{X} \otimes \mathbf{h}_y)(i) \geq 0 \\ -\sqrt{(\mathbf{X} \otimes \mathbf{h}_x)^2(i) + (\mathbf{X} \otimes \mathbf{h}_y)^2(i)}, & (\mathbf{X} \otimes \mathbf{h}_x)(i) \times (\mathbf{X} \otimes \mathbf{h}_y)(i) < 0 \end{cases}$$
(2)

where symbol "$\otimes$" denotes the convolution operation. The distribution of the gradient magnitude $\mathbf{m}$ of $\mathbf{X}$ is similar to the distributions in Fig. 2 (e) with different shapes, which will be further analyzed in the following section.

2.2 Gradient Distribution Modeling

It has been widely agreed that blur largely distorts image edges from sharpness to smoothness, and this results in smaller image gradient magnitudes. As shown in Fig. 2, we can see that the blurred image has steep/sharp distribution of gradient, while its corresponding original natural image has flat distribution with heavy tail. Therefore, these distributions can be modeled by a generalized Gaussian distribution whose coefficients change as the degree of blurriness changing.

Moreover, previous studies [21-25] on natural scene statistics based IQA demonstrated that generalized Gaussian can effectively capture the distributions of DCT/wavelet coefficients, and normalized luminance, etc. In this paper, we follow the similar concept of these works, but use the generalized Gaussian to model the gradient distribution.

The generalized Gaussian distribution (GGD) with zero mean is given as follows:

$$f(x; \alpha, \beta) = \frac{\alpha}{2\beta\Gamma(1/\alpha)} \exp\left(-\left(\frac{|x|}{\beta}\right)^\alpha\right)$$
(3)

Note that the GGD model requires symmetric distributions. In real applications, original natural images have fairly regular gradient distributions as shown in Fig. 2 (e) (the red solid line). However, the blur distortion disturbs this correlation structure, which may results in asymmetric distribution. In this case, the GGD model mentioned above is not suitable. Practically, the deviation of asymmetric distribution can be captured by analyzing the products of adjacent coefficients computed along horizontal, vertical and diagonal orientations: $\mathbf{m}(i,j)\mathbf{m}(i,j+1)$, $\mathbf{m}(i,j)\mathbf{m}(i+1,j)$, $\mathbf{m}(i,j)\mathbf{m}(i+1,j+1)$ and $\mathbf{m}(i,j)\mathbf{m}(i+1,j-1)$ for $i \in i, 2, \ldots, M$ and $j \in i, 2, \ldots, N$, supposed that gradient $\mathbf{m}$ is $M \times N$ [23]. The products of neighboring coefficients can be well modeled using a zero mode asymmetric generalized Gaussian distribution (AGGD) [26]:

$$f(x; \gamma, \beta_l, \beta_r) = \begin{cases} \frac{\gamma}{(\beta_l+\beta_r)\Gamma(\frac{1}{\gamma})} \exp\left(-\left(\frac{-x}{\beta_l}\right)^\gamma\right) & \forall x < 0 \\ \frac{\gamma}{(\beta_l+\beta_r)\Gamma(\frac{1}{\gamma})} \exp\left(-\left(\frac{x}{\beta_r}\right)^\gamma\right) & \forall x \geq 0 \end{cases}$$
(4)

where

$$\beta_l = \sigma_l \sqrt{\frac{\Gamma(\frac{1}{\gamma})}{\Gamma(\frac{3}{\gamma})}}, \quad \beta_r = \sigma_r \sqrt{\frac{\Gamma(\frac{1}{\gamma})}{\Gamma(\frac{3}{\gamma})}}$$
(5)

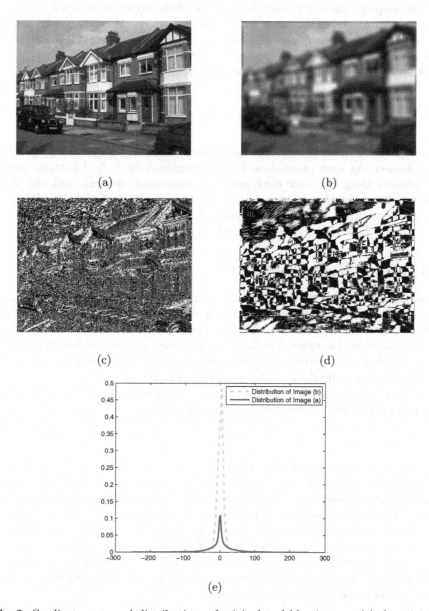

Fig. 2. Gradient maps and distributions of original and blur images. (a) the original distortion-free image, (b) the corresponding blur image; (c) and (d) are the corresponding gradient maps of (a) and (b), respectively; (e) is the gradient distributions of (a) and (b). The red dashed line represents the distribution of original distortion-free image (a) and green dashed line represents the distribution of blur image (b).

The shape parameter γ controls the shape of the distribution, while β_l and β_r are scale parameters that control the spread on each side of the mode, respectively. The parameters of the AGGD(γ,β_l,β_r) can be efficiently estimated using the moment-matching based approach in [26].

Another useful coefficient, i.e., the mean of the distribution is defined as

$$\eta = (\beta_r - \beta_l)\frac{\Gamma(\frac{2}{\gamma})}{\Gamma(\frac{1}{\gamma})} \tag{6}$$

The four parameters $(\eta, \gamma, \beta_l, \beta_r)$ can efficiently model blurred image gradient distribution. As each orientation has 4 parameters $(\eta, \gamma, \beta_l, \beta_r)$, there are 16 parameters along the four directions, i.e., horizontal, vertical, and two diagonal orientations. Hence, each image has 16 parameters for the modeling of image distribution, and these 16 parameters can form a feature vector C = $\{\eta_h, \gamma_h, \beta_{l_h}, \beta_{r_h}, \ldots, \eta_v, \gamma_v, \beta_{l_v}, \beta_{r_v}\}$, which is used for blur assessment.

2.3   Feature Mapping with ELM

From the analysis of the proposed algorithm in Subsection 2.2, we know that the gradient distribution of a test image generates a 16 dimensional feature. Such features can separate out different gradient distributions of different blur images. In order to obtain the final quality score, a mapping function should be established from extracted features to image quality scores. In the proposed metric, ELM is utilized for learning a feature mapping model.

Given one image feature vector C consisting 16 parameters, our goal is to find a function of C to represent the quality score Q as

$$Q = f(C) \tag{7}$$

where f is a function mapping C to the final score Q and Q can be further normalized to [0 1]. However, f is generally difficult to be solved in practice due to the limited knowledge and complexity of the HVS. To estimate f, in this paper, ELM [17] is adopted by predicting the underlying complex relationship between C and subjective quality scores. Here subjective quality scores are used for training the parameters of the function f. Since the subjective scores are given by human observers, the human vision knowledge toward IQA is inexplicitly incorporated into the trained model and thus the resultant objective quality score is consistent with human perception.

3   Performance Evaluation

In this section, extensive experimental results are presented to evaluate the overall accuracy of the proposed IQA metric (BIBE), in comparison with the existing relevant state-of-the-art no-reference metrics, including the specialized blur quality metrics, i.e., QS-SVM [2], kurtosis based [3], Ong et al. [5], Marziliano et al. [6], Marichal et al. [7], Ferzli et al. [11], LPC [13], CPBD [14], Liang et al. [15], and Yan et al. [16], and the generic no-reference quality methods, i.e., DIIVINE [21], BLIINDS-II [22], BRISQUE [23], and NIQE [24].

In this paper, four publicly available and subject-rated benchmark databases are used, including TID2008 [27], CSIQ [28], LIVE [29], IVC [30]. ELM is employed for mapping extracted features into quality scores. The k-fold cross validation (CV) strategy [31] is employed for each database. In each train-test sequence, 80% of the whole blur image dataset is chosen for training, and the remaining 20% for testing. The same training and testing process is repeated for 1000 times, and the average value is taken as the final quality result.

Objective experimental results of BIBE on the four databases are shown in Table 1. The high SRCC and PLCC values demonstrate that BIBE is highly consistent with human perception.

To further demonstrate the good performance of BIBE, comparisons are presented among BIBE and the existing no-reference blur metrics and generic no-reference IQA approaches. Table 2 shows the SRCC and PLCC results of all the metrics on the blur images of LIVE database and the highest SRCC and PLCC values are highlighted in boldface. We can see that BIBE achieves the best performances on both SRCC and PLCC for the LIVE database. Compared with the well-known generic no-reference IQA metrics, e.g., DIIVINE [21], BLIINDS-II [22], BRISQUE [23], and NIQE [24], BIBE outperforms them as well.

The SRCC value comparisons of the TID2008, CSIQ and IVC databases are provided in Tables 3-5, respectively. Best performances are highlighted in boldface. It can be seen that the proposed BIBE is much better than other approaches. On the other hand, we can note that BIBE is more robust than the compared no-reference blur metrics and generic metrics across all the databases. BIBE can achieve 0.9260 of SRCC on the TID2008, while the SRCC results of other metrics are smaller than 0.9.

Table 1. Evaluation Performance of BIBE for Blur Distorted Images

Databases	SRCC	PLCC	RMSE
TID2008	0.9260	0.9220	0.4545
CSIQ	0.9594	0.9604	0.0799
LIVE	0.9665	0.9651	4.9406
IVC	0.9384	0.9628	0.2616

Table 2. SRCC and PLCC Comparisons for Blur Distorted Images on the LIVE Database

IQA Metrics	SRCC	PLCC
BIBE	**0.9665**	**0.9651**
QS-SVM [2]	0.9352	0.9326
kurtosis based [3]	0.7500	0.7200
Ong et al. [5]	0.8230	0.8720
Marziliano et al. [6]	0.8840	0.8940
Marichal et al. [7]	0.8930	0.8040
Ferzli et al. [11]	0.9360	0.9320
LPC [13]	0.9368	0.9239
CPBD [14]	0.9437	0.9107
Liang et al. [15]	0.9430	0.9260
Yan et al. [16]	0.9230	0.9340
DIIVINE [25]	0.9373	0.9370
BLIINDS-II [26]	0.8912	0.8994
BRISQUE [27]	0.9511	0.9506
NIQE [28]	0.9341	0.9525

Table 3. SRCC Comparison for Blur Distorted Images on the TID2008 Database

	BIBE	Marziliano et al.	Ferzli et al.	LPC	CPBD	DIIVINE	BLIINDS-II	BRISQUE
SRCC	**0.9260**	0.7165	0.7045	0.8030	0.8406	0.8620	0.8500	0.881

Table 4. SRCC Comparison for Blur Distorted Images on the CSIQ Database

	BIBE	Marziliano et al. [6]	Ferzli et al. [11]	LPC [13]	CPBD [14]
SRCC	**0.9594**	0.7562	0.7488	0.8795	0.8453

Table 5. SRCC Comparison for Blur Distorted Images on the IVC Database

	BIBE	Marziliano et al. [6]	Ferzli et al. [11]	LPC [13]	CPBD [14]
SRCC	**0.9384**	0.7896	0.7722	0.9022	0.8404

4   Conclusion

In this paper, a perceptual no-reference blur metric (BIBE) has been derived based on the statistical distribution property of image gradient and extreme learning machine. Gradient magnitudes are computed for each input image, and the generalized Gaussian distribution or asymmetric generalized Gaussian distribution is exploited to model the distribution of gradient magnitudes. A 16-dimensional feature is thus obtained for the modeling of image content. Then extreme learning machine is utilized for mapping the resulting feature into a quality score. Experimental results have shown that the proposed blur evaluation metric exhibits high consistency with subjective viewing scores and has better performance and robustness than the existing no-reference blur approaches and generic no-reference metrics.

References

1. Karam, L.J., Ebrahimi, T., Hemami, S.S., Pappas, T.N., Safranek, R.J., Wang, Z., Watson, A.B.: Introduction to the issue on visual media quality assessment. IEEE J. Sel. Topics Signal Process. 3, 189–192 (2009)
2. Chen, M.J., Bovik, A.C.: No-reference image blur assessment using multiscale gradient. J. Img. Video Proc. 3, 189–192 (2011)
3. Zhang, N., Vladar, A., Postek, M., Larrabee, B.: A kurtosis-based statistitcal measure for two-dimensional processes and its application to image sharpness. In: Proc. Section of Physical and Engineering Sciences of American Statistical Society, pp. 4730–4736 (2003)
4. Caviedes, J., Oberti, F.: A new sharpness metric based on local kurtosis, edge and energy information. Signal Process.: Image Commun. 19, 147–161 (2004)
5. Ong, E.P., Lin, W.S., Lu, Z.K., Yao, S.S., Yang, X.K., Jiang, L.F.: No-reference quality metric for measuring image blur. In: Proc. IEEE Int. Conf. Image Processing, pp. 469–472 (2003)
6. Marziliano, P., Dufaux, F., Winkler, S., Ebrahimi, T.: Perceptual blur and ringing metrics: application to JPEG2000. Signal Process.: Image Commun. 19, 163–172 (2004)
7. Marichal, X., Ma, W.Y., Zhang, H.J.: Blur determination in the compressed domain using DCT information. In: IEEE International Conference on Image Processing (1999)
8. Tong, H., Li, M.J., Zhang, H.J., Zhang, C.S.: Blur detection for digital images using wavelet transform. In: IEEE International Conference on Multimedia and EXPO, vol. 1, pp. 17–20 (2004)
9. Ciancio, A., Targino, A.N., Silva, E.D., Said, A., Obrador, P., Samadani, R.: Object no-reference image quality metric based on local phase coherence. IET Electron Letters 45, 1162–1163 (2009)
10. Yao, S., Ong, E., Loke, M.H.: Perceptual distortion metric based on wavelet frequency sensitivity and multiple visual fixations. In: Proceedings of the IEEE International Symposium on Circuits and Systems, pp. 408–411 (2008)
11. Ferzli, R., Karam, L.J.: A no-reference objective image sharpness metric based on the notion of just noticeable blur (JNB). IEEE Trans. Image Process. 18, 717–728 (2009)
12. Sadaka, N.G., Karam, L.J., Ferzli, R., Abousleman, G.P.: A no-reference perceptual image sharpness metric based on saliency-weighted foveal pooling. In: Proc. IEEE Int. Conf. Image Process., pp. 369–372 (2008)

13. Hassen, R., Wang, Z., Salama, M.: No-reference image sharpness assessment based on local phase coherence measurement. In: Proc. Int. Conf. Acoust., Speech, Signal Process., pp. 2434–2437 (2010)
14. Narvekar, N.D., Karam, L.J.: A no-reference image blur metric based on the cumulative probability of blur detection (CPBD). IEEE Trans. Image Process. 20, 2678–2683 (2011)
15. Liang, L.H., Wang, S.Q., Chen, J.H., Ma, S.W., Zhao, D.B., Gao, W.: No-reference perceptual image quality metric using gradient profiles for JPEG2000. Signal Processing: Image Communications 25, 502–516 (2010)
16. Yan, Q., Xu, Y., Yang, X.K.: No-reference image blur assessment based on gradient profile sharpness. In: 2013 IEEE International Symposium on Broadband Multimedia Systems and Broadcasting (BMSB), pp. 1–4 (2013)
17. Huang, G.-B., Zhu, Q.Y., Siew, C.K.: Extreme learning machine: Theory and applications. Neurocomputing 70, 489–501 (2006)
18. Neuenschwander, A.L., Crawford, M.M., Magruder, L.A., Weed, C.A., Cannata, R., Fried, D., Knowlton, R., Heinrichs, R.: Classification of LADAR data over Haitian urban environments using a lower envelope follower and adaptive gradient operator. In: Proc. SPIE (2010)
19. Coleman, S.A., Scotney, B.W., Suganthan, S.: Multi-scale edge detection on range and intensity images. Pattern Recognition 44, 821–838 (2011)
20. Nezhadarya, E., Rabab, K.W.: An efficient method for robust gradient estimation of RGB color images. In: 16th IEEE International Conference on Image Processing, ICIP (2009)
21. Moorthy, A.K., Bovik, A.C.: Blind Image Quality Assessment: From Natural Scene Statistics to Perceptual Quality. IEEE Trans. Image Process. 20, 3350–3364 (2011)
22. Saad, M.A., Bovik, A.C., Charrier, C.: Blind image quality assessment: A natural scene statistics approach in the dct domain. IEEE Trans. Image Process. 21, 3339–3352 (2012)
23. Mittal, A., Moorthy, A.K., Bovik, A.C.: No-reference image quality assessment in the spatial domain. IEEE Trans. Image Process. 21, 4695–4708 (2012)
24. Mittal, A., Soundararajan, R., Bovik, A.C.: Making a completely blind image quality analyzer. IEEE Signal processing Letters 22, 209–212 (2013)
25. Wang, Z., Simoncelli, E.P.: Reduced-reference image quality assessment using a wavelet-domain natural image statistic model. In: SPIE, vol. 5666, pp. 149–159 (2005)
26. Lasmar, N.-E., Stitou, Y., Berthoumieu, Y.: Multiscale skewed heavy tailed model for texture analysis. In: 16th IEEE International Conference on Image Processing, ICIP (2009)
27. Ponomarenko, N.: Tampere image database, Tid 2008 (2008), http://www.ponomarenko.info/tid2008.htm
28. Larson, E.C., Chandler, D.M.: Categorical image quality (csiq) database, http://vision.okstate.edu/csiq
29. Sheikh, H.R., Seshadrinathan, K., Moorthy, A.K., Wang, Z., Bovik, A.C., Cormack, L.K.: Mict image and video quality assessment research at live, http://live.ece.utexas.edu/research/quality/
30. Ninassi, A., Callet, P.L., Autrusseau, F.: Pseudo non-reference image quality metric using perceptual data hiding. In: SPIE Human Vision and Electronic Imaging (2006)
31. Bartlett, P., Boucheron, S., Lugosi, G.: Model selection and error estimation. Journal of Machine Learning 48, 85–113 (2002)

OS-ELM Based Real-Time RFID Indoor Positioning System for Shop-Floor Management

Zhixin Yang[*], Lei Chen, and Pengbo Zhang

Department of Electromechanical Engineering
Faculty of Science and Technology
University of Macau, Macau SAR, China
{Zxyang,mb15550,mb35515}@umac.mo

Abstract. Shop-floor management is featured with dynamic and mixed-product assembly lines, where the real-time positions of manufacturing objects are critical for effective information interaction and management decision making. This paper proposes to adopt RFID technology to constantly acquire wireless signal sent from tags mounted on manufacturing objects. To build the mapping mechanism between RFID signals and object positions, online sequential extreme learning machine (OS-ELM) is applied for training and testing. Besides extremely fast learning speed and high generalization performance, OS-ELM could avoid retraining for new arrived objects and disturbance existed in dynamic shop-floor environment. The verification through experiments demonstrate that the proposed OS-ELM based RFID indoor positioning system is superior than other prevailing indoor positioning approaches in terms of accuracy, efficiency and robustness.

Keywords: online sequential extreme learning machine (OS-ELM), RFID, indoor positioning, shop-floor management, Manufacturing Execution System.

1   Introduction

Manufacturing enterprises located in some hot industrial zones, like Singapore and the Pearl River Delta region in southern China, are facing severe competitions from their global counterparts. The ability to quickly respond to market and diverse customer requirements has been well adopted as the key to survive, which, however, increases the complexity in both material flow and information flow across various manufacturing systems [1]. One product is typically composed of a considerable large number of components, each of which involves different setup requirements and process plans. Accordingly, the production resource management needs to coordinate all the relevant manufacturing objects, including Work-In-Progress (WIP) materials, key components, workstations, working staff, and important tools, etc., in a dynamic production and mixed-model assembly lines environment. The positions and moving paths of these manufacturing objects are constantly changing and hard to be estimated. Such uncertainty tends to disturb the normal production plan and schedules [2]

---

[*] Corresponding author.

© Springer International Publishing Switzerland 2015   233
J. Cao et al. (eds.), *Proceedings of ELM-2014 Volume 2*,
Proceedings in Adaptation, Learning and Optimization 4, DOI: 10.1007/978-3-319-14066-7_23

Manufacturing Execution System (MES) is recently introduced to act as the most important bridge between the physical shop-floor objects and the management oriented Enterprise Resource Planning [3, 4]. MES is designed to manage shop-floor operations such as scheduling as well as its execution and control, timely monitoring the positions and running status of relevant manufacturing objects, material delivery and consumption, as well as manufacturing progress [5]. MES has been widely adapted and adopted since its tangible and intangible benefits [6]. However, the current deployed MES in shop-floor suffers the following three shortcomings: (1) inefficient information acquisition method (paper-based shop-floor data collection, and manual recording of material situation through physical interaction), (2) lack of real-time positioning and monitoring of WIP, staff and workstations, (3) difficulty in dynamical control of manufacturing objects for rescheduled plan. Many efforts, such as bar-coding technique or visualization system like CCTV, have been proposed to address these shortcomings, however the problem still not been well unsolved. Real-time indoor positioning method is very important as it could quickly generate a set of points along the object trajectory, which could then be used to approximate the traveling path of the target object for determining whether or not to adjust its production plan.

RFID (Radio Frequency Identification) uses wireless communication networking technology which improves the information interaction mechanism from physical contact manner to wireless retrieving and processing level. Many researchers have deployed RFID in manufacturing industries to leverage MES for fitting the requirement of the next generation advanced manufacturing technology - wireless manufacturing. RFID has been reported of its capabilities to improve the efficiency of information acquisition from various objects [7]. RFID technology could be simply deployed on virtually all manufacturing objects, by mounting one RFID tag on the targeted object. Once the manufacturing objects are tagged, they become traceable and are called as smart objects (SOs). The data carried by them can be collected and updated when their locations or status change from time to time [8]. Under RFID-enabled MES, tagged object can transmit information to receiver, so manufacturing process, workstation statue, raw material consumption and workforce position will be collected accurately and timely. RFID facilitated real-time data collection will synchronize the information flow and material flow [9, 10]. As RFID could constantly capture the dynamic wireless signals sent from the tags mounted on manufacturing objects in complex manufacturing environment, this paper will adopt RFID as information acquisition mechanism of MES. The architecture of RFID aided MES is depicted in Fig. 1.

The Extreme Learning Machine (ELM) proposed by Huang et al. [11] is a novel learning scheme for single-hidden layer feedforward networks, which is with extremely fast learning speed and high generalization performance for both classification and regression problems [12-14]. Therefore, ELM provides the great potential to perform real-time positioning prediction for MES. Moreover, Zou, et al. [15] has applied ELM into indoor positioning problem and the experiment results approve that ELM based positioning system could achieve 55% of accuracy improvement than that of LANDMARC method.

After fulfilling the requirements of indoor positioning problem in terms of accurately prediction and extremely fast testing capability by ELM, can we directly deploy it for MES? The job-shop environment is more complex and the dynamic manufacturing with mixed-production brings extra challenges for its indoor positioning application. Disturbance is unavoidable in shop-floor as it could be caused by layout changes and object movement, on the other hand, the signal fluctuation property of RFID make one tag in same location be read as different RSS by one fixed reader. The manufacturing objects change their positions from time to time in various moving frequencies, which will impact positioning accuracy. Therefore, ELM based positioning system faces difficulty to maintain high prediction accuracy for real-time positioning in dynamic MES environment. Unfortunately, traditional sequential learning method handles data one-by-one only and cannot handle data on a chunk by chunk. Online sequential extreme learning machine (OS-ELM) [23] developed by Liang et al, can learn training data at varied length which is not only one-by-one but also chunk-by-chunk. After training the OS-ELM model with a set of initial samples, the newly arrived data could be directly input the model and generate its position prediction directly. It can be applied into dynamic MES environment and maintain high prediction accuracy for real-time positioning purpose. The experimental results verify that OS-ELM based method outperforms than other prevailing indoor positioning approaches.

2   The Proposed RT-MES Architecture

In modern manufacturing shop-floor with mixed-production lines, where the manufacturing objects (MOs) including raw materials, components, and staff, arrive at different batch requirements. So positioning and tracking these manufacturing objects are essential to management of labor, material, and equipment, and production scheduling. This paper designs one type of real-time manufacturing execution system (RT-MES) based on RFID technology as shown in Fig. 1. This architecture consists of four layers: RFID enabled manufacturing workshop layer; object perception layer; information integration and application service application layer.

RFID enabled manufacturing workshop layer is configured to collect RSS signals from multi-resource manufacturing objects (MOs) including WIP materials, workstations, and staff. RFID technology could be simply deployed on virtually all manufacturing objects, by mounting one RFID tag on the targeted object. Once the manufacturing objects are tagged, they become traceable and are called as smart objects (SOs). When one WIP moves along its production line that is connected with different set of workstations, the RFID signal transmitted from the tag mounted on that WIP will be received by the preset readers differently. There are huge amount of RSS data being continuously acquired by RFID readers, and sent for positioning estimation in the second object perception layer.

Object perception layer is responsible for quickly and accurately processing the acquired RFID signals and mapping them to the positioning predictions. At initial step, tag registration enable each MO mounted with a RFID tag be identifiable.

Each tag encodes its status, batch, consumption rate, and staff personal information. The signals transmitted by readers will be sent to server through wireless network which ensures the real-time, accurate and reliable interconnection. The captured RFID RSS signals, together with their registration information, will be input to the proposed OS-ELM based positioning system for real-time coordinate approximation.

Information integration layer is used to convert the scattered positioning data into one continuous path curve. It will facilitate the visible monitoring of specific objects as well as provide standard production information to make sequencing, scheduling even to managerial level for decision-making. Object positioning path information will help production planning and scheduling.

Application service layer serves for transmitting the integrated information of shop-floor level to different functional modules in enterprise level. Multi-resource information will be integrated according to the requirements raised from different departments.

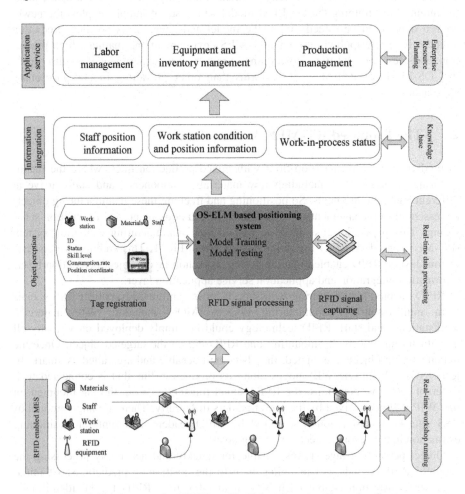

Fig. 1. The framework of RFID enabled Manufacturing Execution System

3   The Methodology of OS-ELM Based Positioning

The flowchart of the proposed OS-ELM based real-time indoor positioning system for the MES is shown in Fig. 2. The proposed real-time positioning system consists of three models, namely, (1) real-time shop-floor signal capturing system; (2) OS-ELM model construction; (3) real-time position identification output.

The real-time data acquisition system uses RFID devices which are deployed in the shop-floor to capture signal strength of tagged objects. The captured RFID RSS signals, together with their registration information, will be continuously transmitted to the proposed OS-ELM based positioning system for real-time coordinate approximation The real-timely collected dataset is spilt two sub-datasets, the training dataset $D_{Training}$ and the testing dataset $D_{Testing}$. $D_{Training}$ is employed to train the OS-ELM based positioning regression model.

The proposed data pre-processing approaches are utilized both for $D_{Training}$ and $D_{Testing}$. The minor difference in physical structure of RFID device tends to cause disturbance of signal strength from even same location. The physical difference elimination methods are employed to reduce the disturbance from tags to increase the positioning accuracy. In order to avoid the domination of large feature values, both datasets will be normalized into [0, 1].

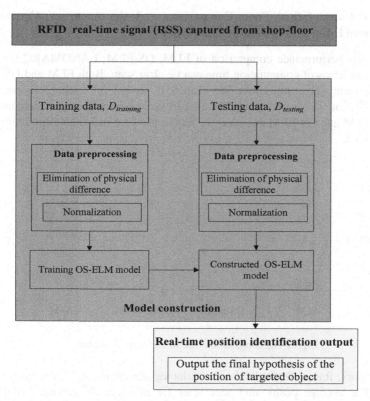

Fig. 2. Framework of OS-ELM based indoor positioning system

4   Experiment Evaluation and Result Discussion

4.1   Experiment Scenario

In our proposed positioning system, fixed RFID readers, a set of reference tags and target tags are deployed to collect signal strength. RFID reader (M250) with working frequency at 433MHz is used, which could identify 1,400 tags simultaneously. RFID tags used in our system are active tag (battery-powered) which transmits not only RSS value but also the temperature of experiment environment. Each tag is registered with a unique ID which can be identified by reader. During the experiment process, tags transmit their information to reader in every 10 second. Each reader captures all reachable tag signals and sends them via internet to Object Perception layer as described in Fig. 1. In this experimental case study is conducted in Innovation Design and Integrated Manufacturing Laboratory of University of Macau, at a room with $36m^2$. The size of the test bed is a square 6m x 6m. 9 reference tags and 3 tracking tags are used in the experiment. The interval of every two adjacent reference tags is 2m and all reference tag covers the whole test bed. Three readers are set on the corner of the test bed to capture the RSS values of the tags.

4.2   The Computation Time Comparison and the Accuracy Comparison among Different Positioning Algorithms

In Fig. 3, the performance comparison of ELM, OS-ELM, LANDMARC [16] and AFM [17] in terms of computation time can be clear seen. Both ELM and OS-ELM based positioning algorithms take less than 0.05 second for testing, comparing to LANDMARC at 0.2 second and AFM at 0.36 second respectively. It demonstrates that both ELM and OS-ELM can meet the requirement for real-time estimation of the target position.

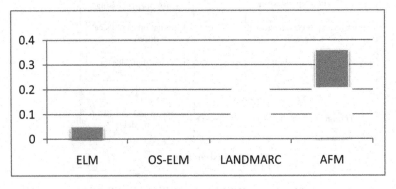

Fig. 3. The computation time of different algorithms

In this work, four different kinds of positioning algorithms are evaluated for comparison. The average positioning accuracies for different algorithms, LANDARC (with 4 neighbor reference tags), AFM, ELM (RBF with 50 hidden nodes), and

OS-ELM (RBF with 20 hidden nodes), are shown in Table.4. The accuracy of LANDMARC, AFM, ELM and OS-ELM is 1.7m, 1.86m, 1.36m and 1.02m respectively. In complex and dynamic shop-floor environment, traditional algorithms cannot obtain ideal positioning accuracy which is vulnerable to the disturbance caused by random movements and fluctuation of signal strength. Both LANDMARC and AFM methods require extensive searching/ranking of reference tags, and thus be inefficient. ELM and OS-ELM positioning algorithms can overcome the random disturbance problem in complex workshop environment. The experiment demonstrates that OS-ELM always performs best than among the four algorithms, as shown in Fig. 14. OS-ELM can effectively handle real-time positioning problem in continuous changing environment.

Table 1. Positioning accuracy comparison among different algorithms

Average positioning accuracy	LANDMARC	AFM	ELM	OS-ELM
	1.7m	1.86m	1.36m	1.02m

5   Conclusion and Discussion

This paper proposes a RFID enabled indoor positioning for real-time manufacturing execution system (RT-MES) based on OS-ELM. Real-time tracking of relevant manufacturing objects will support adaptive production scheduling and optimization of workstation/workforce utilizations in modern manufacturing enterprises. RT-MES will compensate the deficiency of ERP system through synchronization of information flow and material flow.

We design the RFID based manufacturing information acquisition and processing architecture for RT-MES. Virtually all manufacturing objects including WIPs, workstations, and staff, could be deployed with RFID technology by mounting them with registered RFID tags, which make them be able to continuously transmit RSS signal to MES in high frequency and therefore be traceable. Such RFID based model is much more efficient and specific to record manufacturing objects information comparing to traditionally manual or video recording methods.

Modern manufacturing environment poses urged demand on robust indoor positioning method. Regular movement of manufacturing objects and random disturbance in shop-floor will bring fluctuation of radio signal which will make object positioning be difficult. Traditional positioning algorithms cannot solve the above problem accurately and efficiently. The core positioning principle of LANDMARC relies on the position information provided by reference tags. Theoretically, the more reference tags used in shop-floor, the higher accuracy will be achieved. However, with the increasing of tag's density in shop-floor, the model tends to suffer another problem regarding signal interference. Unreliable signal acquisition will seriously downgrade the positioning accuracy. AFM is independent on the quantity of reference tags, but the dynamic environment also forms constraint on its accuracy. Scatted displacement of manufacturing objects tends to cause fluctuation on the signal propagation curve.

Machine learning provides different perspective to handle this problem. However, many practices that apply various machine learning methodologies fail to meet the accuracy and efficiency requirements of RT-MES. Moreover, dynamic environment challenge the application of machine learning for positioning because offline trained model cannot fit the continuous changing environment. Sequential learning method can solve the online learning problem in dynamic environment but its learning rate still cannot satisfy requirements in MES. OS-ELM is a kind of learning method that not only handles the online learning problem but also can reduce the model training time, indeed, it make real-time positioning in non-stationary environment possible. OS-ELM avoids the retraining for new arrived objects and is not sensitive to disturbance existed in dynamic shop-floor environment. The performance of OS-ELM based indoor positioning has been evaluated by error comparison among several different localization algorithms. The experiment result demonstrates the proposed positioning system can achieve the best accuracy and efficiency among all the four methods. One of the main contributions of this work is that OS-ELM is firstly introduced to RFID positioning system in MES and could effectively solve the synchronization issue of information and material flow. The proposed model implements real-time object positioning with robust performance in complex working environment, which enables RT-MES being deployed in reality.

Acknowledgment. The authors would like to thank the funding support by the University of Macau, Grant numbers: MYRG079(Y1-L2)-FST13-YZX.

References

1. Yang, Z., Wong, P.-K., Vong, C.-M., Lo, K.-M.: Constraint-based adaptive shape deformation technology for customised product development. International Journal of Materials and Product Technology 44, 1–16 (2012)
2. Dietrich, A.J., Kirn, S., Sugumaran, V.: A service-oriented architecture for mass customization - A shoe industry case study. IEEE Transactions on Engineering Management 54, 190–204 (2007)
3. Morel, G., Panetto, H., Zaremba, M., Mayer, F.: Manufacturing Enterprise Control and Management System Engineering: paradigms and open issues. Annual Reviews in Control 27, 199–209 (2003)
4. Simão, J.M., Stadzisz, P.C., Morel, G.: Manufacturing execution systems for customized production. Journal of Materials Processing Technology 179, 268–275 (2006)
5. Blanc, P., Demongodin, I., Castagna, P.: A holonic approach for manufacturing execution system design: An industrial application. Engineering Applications of Artificial Intelligence 21, 315–330 (2008)
6. Dai, Q., Zhong, R., Huang, G.Q., Qu, T., Zhang, T., Luo, T.Y.: Radio frequency identification-enabled real-time manufacturing execution system: a case study in an automotive part manufacturer. International Journal of Computer Integrated Manufacturing 25, 51–65 (2012)

7. Zhu, X., Mukhopadhyay, S.K., Kurata, H.: A review of RFID technology and its managerial applications in different industries. Journal of Engineering and Technology Management 29, 152–167 (2012)
8. Huang, G.Q., Zhang, Y.F., Jiang, P.Y.: RFID-based wireless manufacturing for walking-worker assembly islands with fixed-position layouts. Robotics and Computer-Integrated Manufacturing 23, 469–477 (2007)
9. Trappey, A.J.C., Lu, T.-H., Fu, L.-D.: Development of an intelligent agent system for collaborative mold production with RFID technology. Robotics and Computer-Integrated Manufacturing 25, 42–56 (2009)
10. Osman, M.S., Ram, B., Stanfield, P., Samanlioglu, F., Davis, L., Bhadury, J.: Radio frequency identification system optimisation models for lifecycle of a durable product. International Journal of Production Research 48, 2699–2721 (2010)
11. Huang, G.-B., Zhu, Q.-Y., Siew, C.-K.: Extreme learning machine: Theory and applications. Neurocomputing 70, 489–501 (2006)
12. Fu, H., Vong, C.-M., Wong, P.-K., Yang, Z.: Fast detection of impact location using kernel extreme learning machine. Neural Computing and Application (2014)
13. Sun, Y., Yuan, Y., Wang, G.: An OS-ELM based distributed ensemble classification framework in P2P networks. Neurocomputing 74, 2438–2443 (2011)
14. Yap, K.S., Yap, H.J.: Daily maximum load forecasting of consecutive national holidays using OSELM-based multi-agents system with weighted average strategy. Neurocomputing 81, 108–112 (2012)
15. Zou, H., Wang, H., Xie, L., Jia, Q.-S.: An RFID indoor positioning system by using weighted path loss and extreme learning machine. In: 2013 IEEE 1st Inter. Conf. on Cyber-Physical Systems, Networks, and Applications, Taiwan, pp. 66–71 (August 2013)
16. Lionel, M.N., Liu, Y., Lau, Y.C., Patil, A.P.: LANDMARC: Indoor location sensing using active RFID. In: First IEEE International Conference on Pervasive Computing and Communications (PerCom 2003), pp. 407–415 (2003)
17. Yang, Z., Chen, L.: Adaptive Fitting Reference Frame for 2-D Indoor Localization Based on RFID. International Journal of Engineering and Technology 6(2), 114–118 (2014)

An Online Sequential Extreme Learning Machine for Tidal Prediction Based on Improved Gath-Geva Fuzzy Segmentation

Jianchuan Yin[1] and Nini Wang[2,3]

[1] Navigation College, Dalian Maritime University, 116026 Dalian, China
[2] Institute of Geographic Science and Natural Resources Research,
Chinese Academy of Sciences, 100101 Beijing, China
[3] Department of Mathematics, Dalian Maritime University, 116026 Dalian, China

Abstract. A novel sampling optimization scheme is proposed for the on-line sequential extreme learning machine (OS-ELM) based on improved Gath-Geva (IGG) fuzzy segmentation. As is known that most of the complex systems are time-varying in nature whose dynamics varies with changes of internal and environmental factors. When OS-ELM is implemented to identify the time-varying dynamics, it select samples in the sampling pool for a certain period of time. Under such circumstance, samples representing system dynamics of different time span are mixed together thus the online representing ability of OS-ELM for current dynamics is deteriorated. To construct optimal sample pool for OS-ELM thus improve its representing ability for current system dynamics, in this study, time series of system output and relevant factors are online segmented by an IGG fuzzy segmentation approach. Time series are segmented as per dynamics characteristics such as mean value and variance. The changing points split up the time series into several segments and the changing points themselves represent changes in system dynamics. Samples within the same segment are considered as possessing similar characteristics. The OS-ELM selects samples from sampling pool which consists of samples with better representing ability for current dynamics. Furthermore, the membership degree and novelty degree are both employed to decide the importance of samples thus improve the generalization ability of OS-ELM . To achieve accurate tidal prediction, the IGG-based sampling scheme is applied in online tidal prediction for representing the influences of environmental factors. In the meantime, conventional harmonic analysis is applied to represent the influences of celestial bodies and coastal topology. The harmonic method and improved OS-ELM are combined together and the resulted modular prediction scheme is applied for online tidal level prediction of port of King Point. Simulation results demonstrates the feasibility and effectiveness of the proposed sampling scheme and the modular tidal prediction approach.

Keywords: Online Sequential Extreme Learning Machine, Gath-Geva Fuzzy Segmentation, Tidal Prediction, Modular prediction.

© Springer International Publishing Switzerland 2015
J. Cao et al. (eds.), *Proceedings of ELM-2014 Volume 2,*
Proceedings in Adaptation, Learning and Optimization 4, DOI: 10.1007/978-3-319-14066-7_24

1   Introduction

Tidal prediction is an important issue in areas of marine safety and efficiency [1]. Among various tidal prediction approaches, the conventional harmonic analysis method is the most commonly used one and is still the basis for long-term tidal prediction [2]. Harmonic analysis expresses the tide as superposition of several sinusoidal constituents. However, tidal level is not only influenced by celestial bodies and coastal topography but also by meteorological factors such as atmospheric pressure, wind and rainfall, *etc* [3,4]. Furthermore, these factors are time-varying in nature, whose influences on tidal change are complex and hard to be represented by strictly founded model.

To represent time-varying system dynamics, adaptive learning strategies are needed for representing time-varying dynamics. One type of sequential learning mode is online sequential extreme learning machine (OS-ELM) [5], which is capable of handling samples which arrives one by one or chunk by chunk. OS-ELM is derived from the novel theory of extreme learning machine (ELM) proposed by Huang [6,7]. OS-ELM has been applied in varies areas and simulation results indicate that it produces satisfying generalization performance as well as faster training speed [5]. However, the samples are received and processed passively, and there needs an optimal sample selection strategy for OS-ELM to improve its representation and generalization abilities.

In this paper, a new sampling method is presented for OS-ELM based on the online time series segmentation algorithm referred to as improved GathCGeva clustering (IGG) algorithm [8]. IGG is aimed for online detecting the changing points of time-varying dynamics automatically thus deciding the sampling pool with similar characteristics. Attributing to the fact that the changes of some complex nonlinear time series are usually vague and do not suddenly happen on any particular time point, IGG realized the fuzzy segmentation of time series. The efficiency of the IGG has been validated via adaptive segmentation of hydrometeorological time series [8].

The OS-ELM based on IGG segmentation is implemented in tidal online prediction for representing the time-varying tidal changes caused by environmental factors. Conventional harmonic method is also implemented to represent influences of celestial movements and coastal topography simultaneously. The parametric method and nonparametric is combined together and the resulted modular structure can achieve better prediction accuracy and satisfying stability [9]. The OS-ELM in modular model is adjusted adaptively based on IGG segmentation and makes predictions consequently at each step. Finally, the proposed modular prediction model was applied to real-time tidal level prediction at Port of King Point to validate its feasibility and effectiveness.

2   Online Sequential Extreme Learning Machine

In this part, we skip the rigorous proof for ELM [6,10,11]. The main idea of ELM is that for N arbitrary distinct samples $(\mathbf{x}_k, t_k)$, in order to obtain arbitrarily small non-zero training error, one may randomly generate $\tilde{N}(\leq N)$ hidden nodes (with random parameters $\mathbf{a}_i$ and b_i). Under this assumption, $\mathbf{H}$ is completely defined. Then, output weights $\boldsymbol{\omega}$ are estimated as

$$\hat{\boldsymbol{\omega}} = \mathbf{H}^\dagger \mathbf{T} = (\mathbf{H}^T \mathbf{H})^{-1} \mathbf{H}^T \mathbf{T}, \tag{1}$$

where $\mathbf{H}^\dagger$ is the Moore-Penrose generalized inverse of the hidden layer output matrix $\mathbf{H}$. Calculation of the output weights can be done in a single step. This avoids any lengthy training procedure to choose control parameters (learning rate and learning epochs, etc.) thus enables its extreme processing speed. Universal approximation capability of ELM has been analyzed in [6], which indicated that SLFNs with randomly generated additive or radial basis function (RBF) nodes can universally approximate any continuous target function on any compact subspace of $\mathbf{R}_n$.

As training data may be presented one-by-one or chunk-by-chunk, the ELM is modified so as to make it suitable for online sequential computation [5]. Suppose a new chunk of data is given, it results in a problem of minimizing

$$\left\| \begin{bmatrix} \mathbf{H}_0 \\ \mathbf{H}_1 \end{bmatrix} \omega - \begin{bmatrix} \mathbf{T}_0 \\ \mathbf{T}_1 \end{bmatrix} \right\|. \tag{2}$$

When a new sample arrives or a chunk of samples arrive, the connecting weight $\omega^{(k+1)}$ can be achieved iteratively. When the $(k+1)$-th new chunk of data arrives, the recursive method is implemented for acquiring the updated solution. $\omega^{(k+1)}$ can be updated by

$$\omega^{(k+1)} = \omega^{(k)} + \mathbf{K}_{k+1}^{-1} \mathbf{K}_{k+1}^T (\mathbf{T}_{k+1} - \mathbf{H}_{k+1} \omega^{(k)}). \tag{3}$$

with

$$\mathbf{K}_{k+1}^{-1} = \mathbf{K}_k^{-1} - \mathbf{K}_k^{-1} \mathbf{H}_{k+1}^T (I + \mathbf{H}_{k+1} \mathbf{K}_k^{-1} \mathbf{H}_{k+1}^T)^{-1} \times \mathbf{H}_{k+1} \mathbf{K}_k^{-1}. \tag{4}$$

3   Improved GG Clustering-Based Time Series Segmentation

Given a time-series, time-series segmentation problem is to partition the time-series into several internally homogeneous subseries[12]. Time series segmentation is important for time series dynamics analysis and prediction. While in practice, changes of system is not always abrupt but vague under some conditions, so the fuzzy clustering technique is used in segmentation. The close relationships between fuzzy clustering techniques and probability models indicate that it is possible to adopt probability models to modify fuzzy approaches [13].

To couple segmentation order selection and model parameter estimation in a unified framework, a reformulation of CEM^2 algorithm [14] for GG clustering-based time series segmentation is employed. The algorithm starts with a large number of clusters all over the space, which makes the algorithm robust with respect to initialization. And all that has to be done is to select the necessary clusters in a top-down manner. Let

$$\boldsymbol{\eta}(l) = \{\boldsymbol{\eta}_i(l), \alpha_i(l)|i = 1, \ldots, c_{nz}\}$$
$$= \{\boldsymbol{\eta}_i^t(l), \boldsymbol{\eta}_i^x(l), \alpha_i(l)|i = 1, \ldots, c_{nz}\}$$
$$= \{v_i^t(l), \sigma_{i,t}^2(l), \mathbf{v}_i^x(l), \mathbf{F}_i^x(l), \alpha_i(l)|i = 1, \ldots, c_{nz}\}$$

be the vector of parameters at the current iteration. The *E-step* updates the fuzzy partition matrix $\mathbf{U}(l) = \left[\mu_{i,k}^{(l)}\right]_{c_{nz} \times n}$, where $\mu_{i,k}^{(l)}$ is the membership degree of the kth data point to the ith cluster:

$$
\mu_{i,k}^{(l)} = \frac{1}{\sum_{j=1}^{c_{nz}} \left(D(\mathbf{z}_k, \boldsymbol{\eta}_i(l))/D(\mathbf{z}_k, \boldsymbol{\eta}_j(l))\right)^{2/m-1}}
$$
$$
= \frac{\left(\alpha_i(l)p(t_k|\boldsymbol{\eta}_i^t(l))p(\mathbf{x}_k|\boldsymbol{\eta}_i^x(l))\right)^{m-1}}{\sum_{j=1}^{c_{nz}} \left(\alpha_j(l)p(t_k|\boldsymbol{\eta}_j^t(l))p(\mathbf{x}_k|\boldsymbol{\eta}_j^x(l))\right)^{m-1}}. \tag{5}
$$

Then, the *M-step* involves simultaneously updating the vector of the parameters $\boldsymbol{\eta}(l)$ and annihilating clusters with vanishing mixing coefficients to reduce the segmentation order. For the ith cluster, the mixing coefficient $\alpha_i(l + 1)$ can be calculated by [14,15]:

$$
\alpha_i(l + 1) = \frac{\max\left\{0, \sum_{k=1}^n \mu_{i,k}^{(l)} - \frac{N}{2}\right\}}{\sum_{j=1}^{c_{nz}} \max\left\{0, \sum_{k=1}^n \mu_{j,k}^{(l)} - \frac{N}{2}\right\}}. \tag{6}
$$

The normalized mixing coefficient vector $\mathcal{W}(l + 1)$ is defined as

$$
\mathcal{W}(l + 1) = \{\alpha_1(l + 1), \ldots, \alpha_{c_{nz}}(l + 1)\} = \frac{\{\alpha_1(l + 1), \ldots, \alpha_{c_{nz}}(l + 1)\}}{\sum_{j=1}^{c_{nz}} \alpha_j(l + 1)}. \tag{7}
$$

Then, the ith cluster corresponding to $\alpha_i(l + 1) = 0$ is annihilated, else the parameters of the ith cluster $\boldsymbol{\eta}_i(l + 1)$ is updated by

$$
v_i^t(l + 1) = \left(\sum_{k=1}^n \left(\mu_{i,k}^{(l)}\right)^m\right)^{-1} \sum_{k=1}^n \left(\mu_{i,k}^{(l)}\right)^m t_k,
$$
$$
\mathbf{v}_i^x(l + 1) = \left(\sum_{k=1}^n \left(\mu_{i,k}^{(l)}\right)^m\right)^{-1} \sum_{k=1}^n \left(\mu_{i,k}^{(l)}\right)^m \mathbf{x}_k,
$$
$$
\sigma_{i,t}^2(l + 1)
$$
$$
= \frac{\sum_{k=1}^n \left(\mu_{i,k}^{(l)}\right)^m \left(t_k - v_i^t(l + 1)\right) \left(t_k - v_i^t(l + 1)\right)^T}{\sum_{k=1}^n \left(\mu_{i,k}^{(l)}\right)^m}, \tag{8}
$$
$$
\mathbf{F}_i^x(l + 1)
$$
$$
= \frac{\sum_{k=1}^n \left(\mu_{i,k}^{(l)}\right)^m \left(\mathbf{x}_k - \mathbf{v}_i^x(l + 1)\right) \left(\mathbf{x}_k - \mathbf{v}_i^x(l + 1)\right)^T}{\sum_{k=1}^n \left(\mu_{i,k}^{(l)}\right)^m}.
$$

It is to be noted that if the initial number of clusters c_{nz} is too large, it may happen that no cluster has enough initial support ($\sum_{k=1}^{n} \mu_{i,k}^{(l)} < \frac{N}{2}$, for $i = 1, \ldots, c_{nz}$) [14]. To sidestep this difficulty, CEM^2 with the *M-step* is adopted to sequentially update parameters $\boldsymbol{\eta}(l)$ at the lth iteration: update $\alpha_1(l+1)$ by Eq. 6 and $\boldsymbol{\eta}_1(l+1)$ by Eq. 8, recompute $\mathcal{W}(l+1)$ by Eq. 7, update $\alpha_2(l+1)$ and $\boldsymbol{\eta}_2(l+1)$, recompute $\mathcal{W}(l+1)$, and so on.

After convergence, i.e., when the relative decrease between the fuzzy partition matrices of the previous and the current iteration $\|\mathbf{U}(l) - \mathbf{U}(l-1)\|$ falls below a threshold ε, there is no guarantee that we have found a minimum of MML criterion. Following the method of [14], we check if smaller values of MML criterion are achieved by annihilating the least probable cluster (with smallest α_i).

4   OSELM-Based Modular Tidal Level Prediction Simulation Experiments

4.1   Structure of Modular Prediction Model

Two steps are conducted during identification process. Firstly, the harmonic method is implemented to describe the periodical tidal change , and the prediction of y is denoted as y_{M}. y_{R} denotes the residual between $y(t)$ and y_{M}. The OS-ELM is then used for online prediction of y_{R} based on nonlinear autoregressive with exogenous inputs (NARX) model:

$$y(t) = f(y(t-1), \ldots, y(t-n_y), u(t-1), \ldots, u(t-n_u)), \tag{9}$$

where the y and u are system output and input, n_y and n_u denote order of y and u, respectively. In this study, u in (9) contains water temperature T, air pressure P and wind speed V, that is, $u(t-1), \ldots, u(t-n_u)$ contains:

$$T(t-1), \ldots, T(t-n_T), P(t-1), \ldots, P(t-n_P), V(t-1), \ldots, V(t-n_V), \tag{10}$$

with n_T, n_P and n_V are orders of the T, P and V in prediction model, respectively. The residual information is considered as the effects of time-varying environmental changes. For m-steps-ahead prediction, the processes of identification and prediction are expressed as follows:

$$y(t) = f(y(t-m), \ldots, y(t-m-n_y+1), u(t-m), \ldots, u(t-m-n_u+1)), \tag{11}$$

and

$$y(t+m) = f(y(t), \ldots, y(t-n_y+1), u(t-m), \ldots, u(t-m-n_u+1)), \tag{12}$$

After the OS-ELM is constructed by learning data pairs of y_{R} in current selected pool, the modular prediction model is achieved by combing the OS-ELM with the harmonic method and implemented for prediction. The configuration of the prediction process in the modular model is illustrated as Fig. 1.

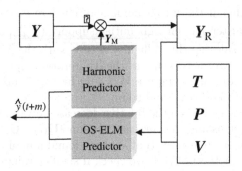

Fig. 1. Prediction process of the modular model based on OS-ELM

Once the identification is completed, historical information of y, P, T and V are then set as input according to (12) and the $y_R(t + m)$ would be the m-steps-ahead prediction of the influence of environment to the tidal levels. Predictions generated by the two modules are summed together to form the final prediction model. That is, the two identification module is combined in series connection to form one tidal forecast model and get the final tidal prediction result.

4.2 Tidal Prediction Simulation

In order to verify the feasibility and efficiency of the proposed sample selection scheme for OS-ELM, online tidal prediction is performed. The study takes use of the measured hourly tidal data of American Port of King Point in New York, United States. The data are measured from GMT0000 April 1 to GMT2300 April 30, 2014, 720 samples in total. All the measurement data of King Point and the values of harmonic constants in this study are achieved from web site of American National Oceanic and Atmospheric Administration: http://co-ops.nos.noaa.gov.

As the harmonic method only takes consider of the period influences of celestial bodies whereas ignores the influence of the environmental changes, there exists time-varying errors in the prediction results. To depict the time-varying error more clearly, the curve of prediction error is shown in Fig. 2.

It can be shown from Fig. 2 that the maximum prediction error reaches almost 0.7m, voyage plan based on such prediction may cause danger to the ship and seafarers when sails through shallow water or under bridge. Therefore, there is a practical need to reduce the error and give accurate tidal predictions. The index of root mean square error (RMSE) is utilized to evaluate the errors of identification and prediction by $RMSE_I$ and $RMSE_P$, respectively.

The RMSE for prediction based on harmonic method is 0.181680m. In this study, prediction error of harmonic method is identified and predicted by OS-ELM based on IGG segmentation. The changes of air pressure, water temperature and wind speed are time-varying in nature will impose influences to tidal level changes. Therefore, in this study, the time span which have largest correlation to the current changes will be determined and the corresponding samples

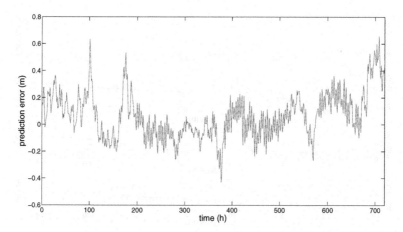

Fig. 2. Prediction error by using harmonic method

are the candidate sampling pool. This will help to avoid unfavorable effects of samples which have little or contrary effects to the changes of system dynamics.

Before the process of segmentation, the value of variable is normalized to $[-1 \quad 1]$. The optimal membership degree for P, T, V and y_R are obtained and shown in Fig. 3.

The corresponding crisp segmentation result is shown in Fig. 4. It can also be noticed that some of the crisp changing points between different variables are correlated and this phenomenon can be further studied to investigate the latent correlation between system dynamics changes and relevant environmental factors. The crisp changing point for the 4 variables are listed in Table 1.

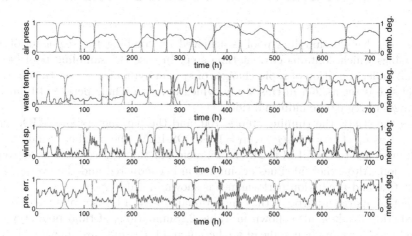

Fig. 3. Optimal membership degree for P, T, V and y_R

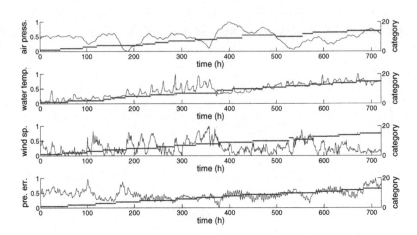

Fig. 4. Crisp segmentation result

Table 1. Segmentation results for the variables

Factor	Change points
Air pressure	1, 44, 92, 122, 219, 245, 273, 328, 374,428,499,556,578,650
Water temperature	1, 60, 137, 152, 183, 235, 285, 288, 372,374,383,387,407,471,521,571,620,669
Wind speed	1, 47, 100, 134, 185, 247, 248, 249, 285, 329, 377,380,381,449,527,536,537,539,579,630,631,633,671,672,676
Residual error	1, 59, 118, 162, 211, 212, 213, 287, 289, 290, 331, 387,427,504,505,507,583,586,587,636,637,639,683

Altogether 26 sampling pool is achieved by combing the segments of each variables which contains current time step. In case the sampling pool are too small under condition of rapid changes of system dynamics, we set the minimum pool size is set as 25h in this study, which covers a whole period of tidal level changes of regular semidiurnal tide.

In the study, the running step is 700, and the parameters for NARX model are set as n_y=6, n_P=1, n_T=1, n_V=1. For OS-ELM, the number of hidden neurons assigned to the ELM is 24, and Sigmoid function was selected as activation function. Altogether 50 times of simulation is conducted and the average identification and prediction performance is shown in Table 2.

For comparison purpose, the conventional OS-ELM is also conducted and the prediction result is also shown in Table 2. Simulations of tidal predictions are conducted for more hours ahead prediction and the results are shown in Table 2. Simulation results show that the prediction error has reduced and the prediction results are more stable with less abrupt changes. And the RMSE for OS-ELM

Table 2. Simulation results of online ship roll motion prediction

Algorithms	Improved OS-ELM			OS-ELM		
Results	$RMSE_I$(m)	$RMSE_P$(m)	Time(s)	$RMSE_I$(m)	$RMSE_P$(m)	Time(s)
1-hour-ahead	0.0109	0.0229	0.0085	0.0123	0.0286	0.0057
3-hours-ahead	0.0206	0.0294	0.0079	0.0592	0.0362	0.0056
6-hours-ahead	0.0213	0.0373	0.0079	0.1168	0.1515	0.0064
12-hours-ahead	0.0352	0.0401	0.0088	0.1910	0.2314	0.0062

prediction based on IGG model is0.0109m, which is less than 0.0123m by using conventional OS-ELM , also less than 0.1816801m by using harmonic method.

It is shown in Table 2 that, by combining the prediction of harmonic method and OS-ELM based on IGG, the resulted prediction scheme can achieve better generalization accuracy with small loss of processing speed. With the increase of prediction domain, the prediction error of both approaches decreases, but the predicted approach still remains relative high prediction accuracy. The simulation results indicates that the optimization of sampling pool for OS-ELM can improve its representing ability for system dynamics especially the dynamics of time-varying dynamics, with small damage to the processing speed of OS-ELM.

5   Conclusions

A OS-ELM is used for tidal prediction based on improved Gath-Geva fuzzy segmentation approach. The approach selects sampling pool consciously and improves the generalization performance of OS-ELM with little loss of processing speed. The sampling pool exhibits better real-time representing ability for changes of time-varying system dynamics. The improved OS-ELM module are combined with conventional harmonic method and the resulted modular prediction scheme possesses satisfying prediction accuracy. The novel sampling pool selection can also be implemented to other occasions of online applications.

Acknowledgments. This work is supported by the Applied Basic Research Fund of the Chinese Ministry of Transport (Grant No. 2014329225010) and Fundamental Research Funds for the Central Universities (Grant No. 3132014028).

References

1. Fang, G.H., Zheng, W.Z., Chen, Z.Y.: Analysis and Prediction of Tide and Tidal Current. Ocean Press, Beijing (1986)
2. Lee, T.: Back-propagation neural network for long-term tidal predictions. Ocean Eng. 31, 225–238 (2004)
3. Liang, S., Li, M., Sun, Z.: Prediction models for tidal level including strong meteorologic effects using a neural network. Ocean Eng. 35, 666–675 (2008)
4. Yin, J.C., Zou, Z.J., Xu, F.: Sequential learning radial basis function network for real-time tidal level predictions. Ocean Eng. 57, 49–55 (2013)

5. Liang, N.Y., Huang, G.B., Saratchandran, P., Sundararajan, N.: A fast and accurate online sequential learning algorithm for feedforward networks. IEEE Trans. Neur. Netw. 17(6), 1411–1423 (2006)
6. Huang, G.B., Zhu, Q.Y., Siew, C.K.: Extreme learning machine: theory and applications. Neurocomputing 70, 489–501 (2006)
7. Huang, G.B., Wang, D.H., Lan, Y.: Extreme learning machines: a survey. Int. J. Mach. Learn. Cyb. 2(2), 107–122 (2011)
8. Wang, N.N., Liu, X.D., Yin, J.C.: Improved Gath-Geva clustering for fuzzy segmentation of hydrometeorological time series. Stoch. Env. Res. Risk A. 26(1), 139–155 (2012)
9. Jiang, N., Zhao, Z.Y., Ren, L.Q.: Design of structural modular neural networks with genetic algorithm. Adv. Eng. Softw. 34, 17–24 (2003)
10. Huang, G.B., Ding, X.J., Zhou, H.M.: Optimization method based extreme learning machine for classification. Neurocomputing 74, 155–163 (2010)
11. Huang, G.B., Zhou, H.M., Ding, X.J., Zhang, R.: Extreme learning machine for regression and multiclass classification. IEEE Trans. Syst. Man Cy. B 42(2), 513–529 (2012)
12. Abonyi, J., Feil, B., Nemeth, S., Arva, P.: Fuzzy clustering based segmentation of time-series. In: Berthold, M., Lenz, H.-J., Bradley, E., Kruse, R., Borgelt, C. (eds.) IDA 2003. LNCS, vol. 2810, pp. 275–285. Springer, Heidelberg (2003)
13. Povinelli, R., Johnson, M., Lindgren, A., Ye, J.: Time series classification using Gaussian mixture models of reconstructed phase spaces. IEEE Trans. Knowl. Data Eng. 16(6), 779–783 (2004)
14. Figueiredo, M., Jain, A.K.: Unsupervised learning of finite mixture models. IEEE Trans. Pattern Anal. Mach. Intell. 24(3), 381–396 (2002)
15. Fu, Z., Robles-Kelly, A., Zhou, J.: Mixing linear SVMs for nonlinear classification. IEEE Trans. Neural Netw. 21, 1963–1975 (2010)

Recognition of Human Stair Ascent and Descent Activities Based on Extreme Learning Machine

Wendong Xiao[1,*], Yingjie Lu[1], Jian Cui[2], and Lianying Ji[3]

[1] School of Automation and Electrical Engineering
University of Science and Technology Beijing, Beijing, China
[2] School of Instrumentation Science and Opto-electronics Engineering
BeiHang University, Beijing, China
[3] School of Electronic, Electrical and Communication Engineering
University of Chinese Academy of Sciences, Beijing, China
wdxiao@ustb.edu.cn

Abstract. Recently, people pay much attention to physical activity recognition using wearable sensors, mostly by accelerometers. In recognition of various activities, distinguishing between stair ascent and descent is difficult, as the difference between the acceleration signals of the two activities is not so significant. Also the activity characteristics of the people are different from person to person. This paper will be focused on the development of highly-efficient algorithm for distinguishing between human stair ascent and descent by a triaxial accelerometer attached to the subject's chest. In the experiment, a number of features are extracted from the raw accelerometer signal based on the tilt angle, the signal magnitude area, and the wavelet transform. Extreme Learning Machine (ELM) is proposed as a novel stair ascent and descent recognition method. Experimental results show that ELM classifier can achieve superior recognition performance with higher accuracy as well as faster learning speed than BP classifier.

Keywords: Activity Recognition, Feature Extraction, Extreme Learning Machine (ELM).

1 Introduction

Physical activity recognition plays an important role in many fields such as physical training and health care. Especially nowadays, faced with the aging problem, a growing number of old people live alone, and urgently demand advanced solutions for their health monitoring, including their physical activity recognition.

Human physical activity recognition is usually based on the video analysis and the wearable sensor signal analysis. The video analysis approach needs to obtain the position and attitude information from a series of body image sequences. Due to the complexity and variability of the human physical activity, the accuracy and efficiency of video recognition cannot completely satisfy the practical requirements in related industries, it is also weak for privacy of the monitored person [1][2]. With the

© Springer International Publishing Switzerland 2015
J. Cao et al. (eds.), *Proceedings of ELM-2014 Volume 2*,
Proceedings in Adaptation, Learning and Optimization 4, DOI: 10.1007/978-3-319-14066-7_25

development of Micro-Electro-Mechanical System (MEMS) and wearable sensor network (WSN) technologies, activity recognition based on the wearable sensors, especially accelerometers, has received increasing interests due to its advantages that can be implemented easily anywhere anytime. In previous studies, many researchers use multiple sensors for physical activity recognition [3]-[6]. But the hardware would obstruct the movements of human, and is not practical for long-term wearing. Also the cost of system will increase with the number of sensors. Therefore more researchers are seeking activity recognition approaches by using only one accelerometer sensor for collecting the acceleration signal [7]-[9].

For the recognition of physical activities, the present studies mainly focus on sitting, standing, walking, running and other basic activities [7]. To distinguish activities such as stair ascent and stair descent is much difficult, as the difference between the acceleration signals of the two activities is not so significant. Also the motion characteristics of the people are different from person to person, likely to cause misrecognition. This paper will be focused on the development of highly-efficient algorithm for distinguishing between human stair ascent and stair descent by using a triaxial accelerometer put on the human chest.

Feature extraction is crucial for activity recognition. A range of different approaches has been used to obtain features from accelerometer data. The common methods of feature extraction include the time domain analysis, the frequency domain analysis and the time-frequency domain analysis. The time domain method extracts features, such as the mean [10][11], the signal magnitude area (SMA)[7][12], and the correlation coefficient [7][10][13], from the collected acceleration signal directly, usually it requires small sample size for calculation. The frequency domain method extracts features from the frequency-domain signal, such as FFT coefficient [14][15]. More recently, wavelet analysis, which can combines time and frequency information, has been used to derive the so-called time-frequency analysis [3][16][17].

As for the recognition techniques, a large number of classification methods have been investigated, including Bayesian classifier, Neural Network classifier, and SVM classifier. These classifiers often suffer from slow learning speed and poor generalization performance, and are not suitable for the real-time activity recognition scenarios requesting high accuracy [18]. Therefore novel Extreme Learning Machine (ELM) [18]-[22] classifier will be proposed in this paper as a highly-efficient and accurate activity recognition approach.

The paper is organized as follows: The proposed approach is presented in Section 2, including the general description of the approach, feature extraction method, and the proposed ELM classifier. Experimental results are reported in Section 3 to give, the comparison between ELM and BP recognition methods. Finally, conclusions and the future work are given in Section 4.

2   Proposed Approach

The proposed activity recognition approach uses a triaxial accelerometer to collect data. As shown in Fig. 1, the overall approach consists of the following steps: data acquisition, preprocessing, feature extraction, ELM learning in the training phase and ELM based classification in the real-time implementation, followed by the performance evaluation of the classifier.

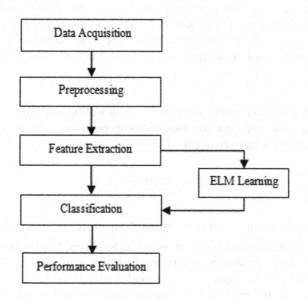

Fig. 1. The architecture of the proposed activity recognition approach

2.1   Data Acquisition

The accelerometer used in this paper is the ADXL345 triaxial accelerometer of the AD Company. The sampling frequency is set at 50Hz. The acceleration signal is sent to a mobile phone via Bluetooth, and is stored in the SD card of the phone. The subject wears the accelerometer on the chest to collect the acceleration signal. When the body wears the accelerometer vertically, the X-axis represents the acceleration in the lateral direction, the Y-axis represents the acceleration in the longitudinal direction, and the Z-axis represents the acceleration in the vertical direction. In order to facilitate observations, the value of each axis was divided by the gravitational acceleration (taken as 9.8 m/s^2), so the data are the multiples of the gravitational acceleration.

In the testing, 15subjects' data are collected, 10 female and 5 male. There are 12 steps of every stair. Each subject is required to repeat 5 times, using his/her normal movement style for stair ascent and stair descent without specific requirements. As the apparatus will be influenced by the gravity, the data are filtered by a high-pass filter with cut-off frequency of 0.5Hz. Also as frequency of human daily activities is not too high, a low-pass filter with cut-off frequency of 20Hz is used to filter out high frequency noise. In addition, the data are smoothed by a median filter to eliminate the independent noise point.

2.2   Feature Extraction

The feature extraction is based on the sliding time window. This paper uses the tilt angle, the signal magnitude area (SMA), and the wavelet analysis to extract the features. As the use of a 50% overlap between successive sliding windows has been shown to be effective in previous studies for activity classification [23], a rectangular window, with the length of 128 sample points (2.56s) and 50% overlap, is used in this paper. The features used in the paper will be detailed in the following paragraphs.

Tilt Angle

No matter what the state of the apparatus is, there will be vertical gravity. When tilting the device, there will be a gravitational component in the Z-axis. The tilt angle is calculated from the acceleration in the Z-axis according to

$$\theta = \arccos \frac{a_z}{g}, \tag{1}$$

where, a_z is the gravity component in the Z-axis. It refers to the relative tilt of the body in space. The gravitational component can be obtained by low-pass filter. Fig. 2 shows the tilt angle distribution of different people stepping up and down stairs after the accelerometer bias correction. We can find that the body will lean forward during stairs ascent, and the angle is not perpendicular to the horizontal direction. While during stair descent the angle between the body and the horizontal direction is close to 90 degree, and the body is close to upright.

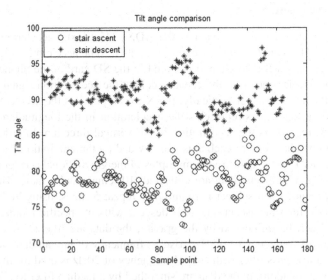

Fig. 2. Tilt angle distribution of stair ascent and stair descent

SMA

Through the analysis, we can know that during stair descent the acceleration fluctuation is greater in Y-axis and Z-axis. This hints us to use acceleration amplitudes of different axes to distinguish between stair ascent and stair descent. In this paper, we adopt the SMA (Signal Magnitude Area) to extract a feature quantity according to

$$SMA = \sum_{i=1}^{N} \left(|x(i)| + |y(i)| + |z(i)| \right),$$

(2)

where $x(i), y(i), z(i)$ indicate the values of X, Y, Z-axis acceleration signals at the i th sample point after preprocessing, N is the length of the sliding time window. SMA can indicate the fluctuation degree of the acceleration signal, the higher the value is, the more violent the fluctuation is. Fig. 3(a) shows the SMA comparison of one person. It can distinguish the stair ascent and stair descent well. For different people, the states of activity are not the same. As shown in Fig. 3(b), the SMA values of three subjects during stair ascent and stair descent may be overlaping. Therefore SMA cannot be used alone for distinguishing stair ascent and stair descent.

Wavelet Energy

Acceleration signal in the vertical direction can distinguish the stair ascent and the stair descent better. Convert the Y-axis acceleration signal to wavelet transform decomposition. During stair descent, the human body is swing relatively greater and the acceleration changes larger, resulting in the larger detail coefficients in wavelet decomposition. Although SMA also makes use of this feature, the Wavelet can reflect the geometry of the waveform, i.e., different waveforms will correspond to different coefficients. This paper uses db5 as the mother wavelet. Decomposing the vertical components to 5 layers, the wavelet energy is calculated as the sum of the squared detail coefficients at levels 4 and 5, according to

$$WE = \sum_{j=4}^{5} CD_j^{2},$$

(3)

where, CD_j are the detail coefficients of Y-axis acceleration signal. Fig. 4 shows the wavelet energy distribution of the stair ascent and stair descent of different subjects. The wavelet energy of the stair descent is larger. From the figure, the wavelet energy can generally distinguish stair ascent and stair descent, but there may be overlapping which can be distinguished incorrectly.

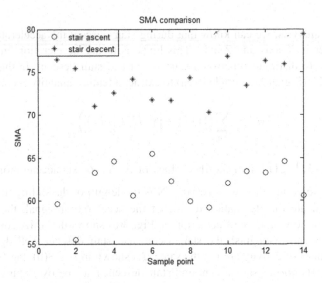

(a) SMA comparison of one person

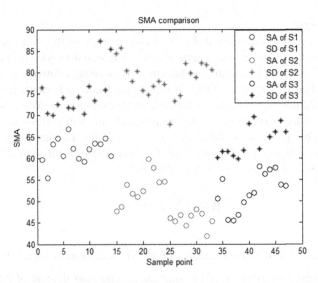

(b) SMA comparison of three subjects S1, S2, S3, SA indicates stair ascent, SD indicates stair descent

Fig. 3. SMA distribution of stair ascent and stair descent

In summary, as the activities of different people differ, a single feature is difficult for accurate determination of the status of the human activity, so a combination of these three features will be used in this paper. Fig. 5 shows the three features extracted from the collected samples, where the features are used as X, Y and Z axis. It is clear that the features can distinguish stair ascent and stair descent.

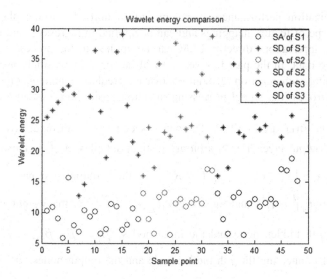

Fig. 4. Wavelet energy distribution of stair ascent and stair descent of three subjects S1, S2, S3, SA indicates stair ascent, SD indicates stair descent

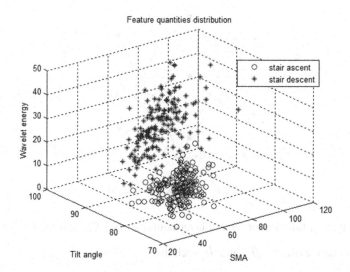

Fig. 5. Feature quantities distribution of stair ascent and stair descent

2.3   ELM Classifier

ELM was proposed by G. B. Huang [18]-[22], originally proposed for standard Single-hidden Layer Feed-forward Neural Network (SLFN), with random hidden nodes, and has recently been extended to kernel learning as well [24]. Compared with traditional training methods, ELM has the advantages of high accuracy, fast learning speed and good generalization performance. It can provide a unified learning platform with widespread type of feature mappings and can be applied in regression and multi-class classification applications directly. ELM can be applied to the processing of sparse data and large data. In the past few years, ELM has been increasingly used in many fields, including human action recognition, disease prediction, human computer interface, image processing, signal processing, location positioning system and network security.

Fig. 6 is the structure of ELM. The network consists of an input layer, a hidden layer and an output layer. For N arbitrary distinct samples (x_i, t_i), the input signal vector is $x_i = [x_{i1}, x_{i2}; \cdots, x_{im}]^T \in R^m$, the output signal vector is $t_i = [t_{i1}, t_{i2}, \cdots, t_{in}]^T \in R^n$, and $w_j = [w_{j1}, w_{j2}, \cdots, w_{jm}]^T$ is the weight vector connecting the jth hidden node and the input nodes, $\beta_j = [\beta_{j1}, \beta_{j2}, \cdots, \beta_{jn}]^T$ is the weight vector connecting the jth hidden node and the output nodes, b_j is the threshold of the jth hidden node.

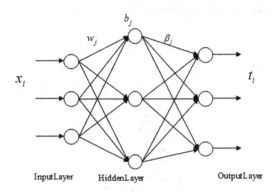

Fig. 6. The structure of ELM

The activation function of hidden layer neurons is $g(x)$, and the hidden layer has L neurons, there exist w_j, β_j and b_j such that

$$\sum_{j=1}^{L} \beta_j g(w_j x_i + b_j) = t_i, i = 1, \cdots N.$$ (4)

The output of the network is

$$T = \begin{bmatrix} t_1^T \\ \vdots \\ t_N^T \end{bmatrix}_{N \times n}. \qquad (5)$$

So the above equations can be written as

$$H\beta = T, \qquad (6)$$

where, H is the hidden layer output matrix of the ELM,

$$H(w_1, w_2, \cdots, w_L, b_1, b_2, \cdots, b_L, x_1, x_2, \cdots, x_N) =$$
$$\begin{bmatrix} g(w_1 x_1 + b_1) & g(w_2 x_1 + b_2) & \cdots & g(w_L x_1 + b_L) \\ g(w_1 x_2 + b_1) & g(w_2 x_2 + b_2) & \cdots & g(w_L x_2 + b_L) \\ \vdots & \vdots & \ddots & \vdots \\ g(w_1 x_N + b_1) & g(w_2 x_N + b_2) & \cdots & g(w_L x_N + b_L) \end{bmatrix}_{N \times L}, \qquad (7)$$

and

$$\beta = \begin{bmatrix} \beta_1^T \\ \vdots \\ \beta_L^T \end{bmatrix}_{L \times n}. \qquad (8)$$

According to the theorem proven by G. B. Huang, when the activation function is infinitely differentiable, and $L \leq N$, the parameters of SLFN do not need all to be adjusted. w and b can be selected randomly before training, and remains constant during the training process. To train an SLFN is simply equivalent to finding a least-squares solution $\hat{\beta}$ of the linear system $H\beta = T$, i.e.,

$$\left\| H\hat{\beta} - T \right\| = \min_{\beta} \| H\beta - T \|, \qquad (9)$$

and can be derived as

$$\hat{\beta} = H^\dagger T, \qquad (10)$$

where $H^\dagger$ the Moore-Penrose generalized inverse of the matrix H. Then we can calculate the output value for prediction for a given input.

3   Experimental Results

The collected data are randomly divided into two sets, namely the training set and the testing set. The classification process consists of two parts, generating the model by the training set and testing the performance of the model by the testing set. 50 trials were conducted for all the algorithms and the classification results are averaged over them.

In this section, we use the combination of the tilt angle, SMA, and wavelet energy features and the ELM classifier to classify the samples, and give the comparison between the performance of the ELM and BP classifiers.

3.1   Classification Results

Before training the samples, we need to determine the number of hidden layer nodes. To prevent the influence of some large data, and to shorten the run time, the data is normalized to the range [0, 1]. We normalize the data before training and testing, and set the number of nodes in the hidden layer as 40. Fig. 7 is the testing set classification results of ELM classifier. For a sample, the coincidence of the predicted value and the true value indicates a correct classification, while the incoincidence indicates an error classification.

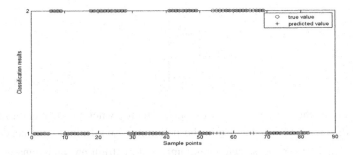

Fig. 7. Classification results of ELM classifier

3.2   Comparison of Classifiers

We compared the classification performance of ELM classifier with BP classifier. We use two classifiers respectively for training with the training set and testing with the testing set, each repeating 50 times. Take the average accuracy rate and the average time as a basis for comparison. The classification results are shown in Table 1. As shown in the table, for both classifiers, their classification results are similar. But BP classifier takes longer time for training and testing, while the ELM classifier can classify samples better and faster.

We also use RMSE and ROC [25] curve to evaluate the two classifiers. RMSE also called standard error, the result is shown in Table2. It is clear that RMSE of ELM classifier is smaller than BP classifier. So ELM's classification accuracy is higher and more stable.

Table 1. Classification results of ELM classifier and BP neural network classifier

Classifier	Time(s)		Accuracy rate(%)	
	Training	Testing	Training	Testing
ELM	0.016	$<10^{-4}$	99.080	96.098
BP	0.832	0.039	98.199	95.365

Table 2. Comparison of two classifiers' RMSE

	ELM	BP
RMSE	0.0244	0.3347

ROC curve (Receiver Operating Characteristic curve) analysis [25] is the common method for evaluating the performance of classifier. ROC curve takes full advantage of the predicted probability value, and can take into account the sensitivity and specificity requirement. ROC curve uses the false positive rate as horizontal axis, and the truth positive rate as the vertical axis. For a perfect classifier, TPR=1 and FPR=0.

Fig. 8 shows the two classifiers' ROC curves. As shown in the figure, compared with BP neural network, ELM's ROC curve is nearer to the upper left corner, with a better classification performance.

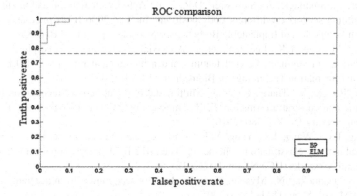

Fig. 8. ROC curves of two classifiers

4   Conclusion

This paper aims to develop a highly-efficient approach for distinguishing between human stair ascent and descent by a triaxial accelerometer. A number of features, including the tilt angle, SMA and wavelet energy, are defined and extracted from the accelerometer signals. Then the Extreme Learning Machine (ELM) classifier is proposed to classify the stair ascent and stair descent activities. Experimental results show that ELM can achieve superior recognition performance compared with the BP classifier.

In the future, we would like to analyze more complex physical activity, and apply ELM to solve more complex classification problems.

Acknowledgment. We thank Professor Jiankang Wu for his valuable comments and Yuejie Lu for his work related to this research issue.

References

1. Li, R.F., Wang, L.L., Wang, K.: A survey of human body action recognition. Pattern Recognition and Artificial Intelligence 27, 35–48 (2014)
2. Sun, Q.R., Wang, W.M., Liu, H.: Study of human action representation in video sequences. CAAI Transaction on Intelligent Systems 8, 189–198 (2013)
3. Mantyjarvi, J., Himberg, J., Seppanen, T.: Recognizing human motion with multiple acceleration sensors. In: The 2001 IEEE International Conference on Systems, Man, and Cybernetics, vol. 2, pp. 747–752 (2001)
4. Foerster, F., Fahrenberg, J.: Motion pattern and posture: correctly assessed by calibrated accelerometers. Behavior Research Methods, Instruments, & Computers: A Journal of the Psychonomic Society, Inc. 32, 450–457 (2000)
5. Kern, N., Schiele, B., Schmidt, A.: Multi-sensor activity context detection for wearable computing. In: Aarts, E., Collier, R.W., van Loenen, E., de Ruyter, B. (eds.) EUSAI 2003. LNCS, vol. 2875, pp. 220–232. Springer, Heidelberg (2003)
6. Maurer, U., Smailagic, A., Siewiorek, D.P., et al.: Activity recognition and monitoring using multiple sensors on different body positions. In: The 2006 IEEE International Workshop on Wearable and Implantable Body Sensor Networks, pp. 113–116 (2006)
7. Khan, A.M., Lee, Y.K., Lee, S.Y., et al.: A triaxial accelerometer-based physical-activity recognition via augmented-signal features and a hierarchical recognizer. IEEE Transactions on Information Technology in Biomedicine 14, 1166–1172 (2010)
8. Li, M., Rozgic, V., Thatte, G., et al.: Multimodal physical activity recognition by fusing temporal and cepstral information. IEEE Transactions on Neural Systems and Rehabilitation Engineering 18, 369–380 (2010)
9. Chuang, F.C., Wang, J.S., Yang, Y.T., et al.: A wearable activity sensor system and its physical activity classification scheme. In: The 2012 IEEE International Joint Conference on Neural Networks (IJCNN), pp. 1–6 (2012)
10. Ravi, N., Dandekar, N., Mysore, P., et al.: Activity recognition from accelerometer data. In: AAAI, vol. 5, pp. 1541–1546 (2005)
11. Pirttikangas, S., Fujinami, K., Nakajima, T.: Feature selection and activity recognition from wearable sensors. In: Youn, H.Y., Kim, M., Morikawa, H. (eds.) UCS 2006. LNCS, vol. 4239, pp. 516–527. Springer, Heidelberg (2006)
12. Karantonis, D.M., Narayanan, M.R., Mathie, M., et al.: Implementation of a real-time human movement classifier using a triaxial accelerometer for ambulatory monitoring. IEEE Transactions on Information Technology in Biomedicine 10, 156–167 (2006)

13. Bao, L., Intille, S.S.: Activity recognition from user-annotated acceleration data. In: Ferscha, A., Mattern, F. (eds.) PERVASIVE 2004. LNCS, vol. 3001, pp. 1–17. Springer, Heidelberg (2004)

14. Sugimoto, A., Hara, Y., Findley, T.W., et al.: A useful method for measuring daily physical activity by a three-direction monitor. Scandinavian Journal of Rehabilitation Medicine 29, 37–42 (1997)

15. Ermes, M., Pärkka, J., Mantyjarvi, J., et al.: Detection of daily activities and sports with wearable sensors in controlled and uncontrolled conditions. IEEE Transactions on Information Technology in Biomedicine 12, 20–26 (2008)

16. Nyan, M.N., Tay, F.E.H., Seah, K.H.W., et al.: Classification of gait patterns in the time–frequency domain. Journal of Biomechanics 39, 2647–2656 (2006)

17. Wang, N., Ambikairajah, E., Lovell, N.H., et al.: Accelerometry based classification of walking patterns using time-frequency analysis. In: The 29th Annual International Conference of the IEEE Engineering in Medicine and Biology Society, pp. 4899–4902 (2007)

18. Huang, G.B., Zhu, Q.Y., Siew, C.K.: Extreme learning machine: A new learning scheme of feedforward neural networks. In: Proc. IJCNN, Budapest, Hungary, vol. 2, pp. 985–990 (2004)

19. Huang, G.B., Zhu, Q.Y., Siew, C.K.: Extreme learning machine: Theory and applications. Neurocomputing 70, 489–501 (2006)

20. Huang, G.B., Chen, L., Siew, C.K.: Universal approximation using incremental constructive feedforward networks with random hidden nodes. IEEE Trans. Neural Netw. 17, 879–892 (2006)

21. Huang, G.B., Chen, L.: Convex incremental extreme learning machine. Neurocomputing 70, 3056–3062 (2007)

22. Huang, G.B., Chen, L.: Enhanced random search based incremental extreme learning machine. Neurocomputing 71, 3460–3468 (2008)

23. Preece, S.J., Goulermas, J.Y., Kenney, L.P.J., et al.: A comparison of feature extraction methods for the classification of dynamic activities from accelerometer data. IEEE Transactions on Biomedical Engineering 56, 871–879 (2009)

24. Huang, G.B., Zhou, H., Ding, X., et al.: Extreme learning machine for regression and multiclass classification. IEEE Transactions on Systems, Man, and Cybernetics, Part B: Cybernetics 42, 513–529 (2012)

ELM Based Dynamic Modeling for Online Prediction of Molten Iron Silicon Content in Blast Furnace

Ping Zhou[*], Meng Yuan, and Hong Wang

State Key Laboratory of Synthetical Automation for Process Industries,
Northeastern University, Shenyang 110819, China
zhouping@mail.neu.edu.cn

Abstract. Silicon content ([Si]) of the molten metal is an important index reflecting the product quality and thermal status of the whole blast furnace (BF) ironmaking process. Since the direct online measure on this index is difficult and larger time lag exists in the offline assay procedure, quality modeling is required to achieve online estimation of [Si], which is an open problem for realizing BF automation. Focusing on this practical problem, this paper proposes a data-driven dynamic modeling method for [Si] prediction using extreme learning machine (ELM) with the help of principle component analysis (PCA). First, data-driven PCA is introduced to pick out the most pivotal variables from multitudinous factors that influence [Si] to serve as the secondary variables of modeling. Second, since this BF metallurgical process is nonlinearity dynamic system with severe time-varying characteristic, dynamic ELM modeling technology with good generalization performance and strong nonlinear mapping capability is proposed by applying the self-feedback structure on traditional ELM. The self-feedback connection enables ELM to overcome the static mapping limitation of its feedforward network structure so as can cope with dynamic time-series prediction problems very well. At last, industrial experiments and compared studies demonstrate that the constructed model has a better modeling and estimating accuracy as well as a faster learning speed when compared with different modeling method and different model structure.

Keywords: extreme learning machine (ELM), silicon content, dynamic modeling, principle component analysis (PCA), blast furnace.

1 Introduction

Blast furnace (BF) is a giant countercurrent reactor and heat exchanger in metallurgical industry, and is the first step towards the production of steel [1~3]. As one of the most complex industrial reactors, the BF has received broad interests both theoretically and experimentally due to its complexity and the key role of iron and steel industry on the national economy. However, it is true that the operation and control of an industrial BF is a serious problem, and still relies on the manual operation of foremen

---

[*] Corresponding author.

experientially [1, 2]. So far, there remains some open problems both in metallurgical fields and engineering control fields, such as the closed-loop control or operational optimization for the whole BF ironmaking process [4~6].

Undoubtedly, the most crucial obstacle for closed-loop control of BF is that the current regular instruments do not have the ability to feed the need of online measure for molten iron quality, such as the silicon content ([Si]) in the final hot metal. In the past decades, through continuous efforts and attempts, a great deal of models and algorithms have been developed trying to tackle the modeling problem for [Si] prediction. These existing methods including linear model based methods like ARX and ARMAX models [6~8], partial least squares based methods [9], and nonlinear intelligent based methods like artificial neural network (ANN) model [10~12], and support vector machine (SVM) model [1, 2, 13, 14]. Though these existing methods have made some achievements in practical application, most of these studies are only focused on the static modeling for [Si] prediction while little attention has been paid to dynamical modeling of this quality parameter.

The BF ironmaking process is a complicated dynamic system with many influential factors and large time-lag. To capture the system dynamics, the time series and time delays of the relevant input and output variables should be took into account during the process modeling. This also means that the existing static prediction models cannot capture the process nonlinear dynamics very well, thus do not provide much accuracy estimation. Therefore, the self-feedback structure which can construct a dynamic system may appear more important for the BF system with serious nonlinear dynamics and large time lag. Moreover, most of the existing prediction models are trained by gradient-based algorithms such as back propagation (BP) algorithm and its variants. It is clear that the learning speed of such intelligent models is insufficiently fast as larger number of training data may be required. Moreover, the BP-like algorithm usually suffers from high computational burden, poor generalization ability, and local optima and overweighting problems [15].

On the other hand, a new machine learning approach that is termed as the extreme learning machine (ELM) has been recently proposed by Huang et al. in [15~18], and verified on a number of benchmark and real-world problems including pattern classification and prediction modeling [16~25]. The ELM and its variants have been considered as a promising learning algorithm in contrast with other algorithms such as BP NN and SVM. This is because ELM has the following advantages: 1) much faster learning speed; 2) higher generalization performance in comparison with BP NN and SVM; and 3) no extra parameters need to be tuned except the predefined network architecture [15~18, 22~25]. In this paper, a data-driven dynamic modeling method to predict molten iron silicon content using ELM with the help of principle component analysis (PCA) [26,27] is proposed. In the design of this predictive model, data-driven PCA for reducing the input variables space of ELM has been constructed. Moreover, output self-feedback architecture has been introduced to establish a dynamic ELM model for practical BF dynamic system. This self-feedback structure enable ELM to overcome the static mapping limitation of its feedforward network structure so as can cope with dynamic time-series prediction problems very well. Lastly, performance of the proposed dynamic ELM based prediction model is compared with other well-known modeling algorithms by industrial experiments on $2^{\#}$ BF in Liuzhou Iron & Steel Group Co. of China.

2   Description of BF Ironmaking System and Its Quality Index

The BF ironmaking is a continuous production process conducted in a closed vertical furnace where materials reduction from iron ore to molten iron takes place every time using carbon coke and gas in high temperature and high pressure environment. When a BF ironmaking system runs, the solid raw materials consisting of coke and fresh ore are charged layer by layer with definite quantities from the top, while the preheated compressed air, together with pulverized coal, is introduced at the bottom through tuyeres, entering just above the hearth, which is a crucial region of BF where the final molten metal product gathers. The hot air at approximately 1200℃ passes upward through the charge and reacts with the descending coke and the supplementary injected oil to generate carbon dioxide, which then changes to CO and H_2 at high temperature. A lot of heat energy is released during this period that can heat up the hearth as high as 2000℃. The generated CO and H_2 further reduces the descending iron ore to form hot metal accumulating in the hearth, and some unreduced impurities (mainly SiO_2) form the slag (mainly $CaSiO_3$) floating on the hot metal being lighter. The liquid hot metal and slag are periodically tapped out by opening a clay-lined tapholes for the subsequent processing. Generally, it will take 6~8 hours for each period of BF ironmaking [28].

For a practical BF production process, silicon content ([Si]) is an important index indicating the chemical heat of molten iron. High [Si] means a large quantity of slag, and this would be easier to wipe off the phosphorus and sulphur in the hot metal. However, excessive [Si] will make cast iron become stiff and brittle, even lead lower yield of metal and easier splashing. In addition, high [Si] will result in a corresponding increase of SiO_2 in the slag, thereby influencing slagging speed of calclime, extending converting time and intensifying corrosion to furnace lining. From an energy point of view, it would be desired to operate the BF process at low molten metal silicon content, still avoiding the risk of cooling the hearth which may result in chilled hearth. Generally, the content of silicon should be controlled in 0.5%~0.7%.

Nowadays, it is still an insoluble dilemma to realize the closed-loop control of molten iron quality in ironmaking BF. The main bottleneck is that the directly online-measurement on [Si] parameter is difficult to be realized with the existing conventional measuring means. Moreover, the offline assaying process for this index takes a long lag time, usually more than 1 hour. Therefore, online prediction based [Si] modeling must be established. Effective online prediciton or estimation for [Si] not only can offer useful information for operators to judge the inner smelting state and operational condition, but also plays a key role in realizing closed-loop control and operational optimization as well as energy-saving and cost-reducing.

3   Modeling Strategy

The proposed data-driven modeling strategy for [Si] prediction is shown in Fig.1. First, data-driven PCA technology with a strong ability to handle strong high-dimensional nonlinear correlated data is introduced to pick a few key factors as the input variables of model so as to reduce the dimension and difficulty for prediction modeling. Then, ELM with better nonlinear mapping and fast process capability modeling technology is brought in this paper. In the meantime, output self-feedback structure is put into use on the basis of traditional ELM in this method, and the output variables derived from previous time are feed back to the network input layer. This feedback outputs together with input variables at different time constitutes a dynamic ELM structure which has a storage capacity and has the ability to tackle data in different time, thus overcoming the limitation of static modeling of traditional ELM.

Remark 1: The proposed modeling strategy has two advantages: 1) The dynamic property of time series and time delays is considered by feeding the output and inputs in previous time through a self-feedback structure. This self-feedback connection enables ELM to overcome the static mapping limitation of its feedforward network structure. Thus the improved version of ELM can capture the process nonlinear dynamics very well by remembering prior input and output states and using both the prior and current states to calculate new output value; 2) Different from the BP-like modeling algorithm usually suffering from high computational burden, poor generalization ability, and local optima and overweighting problems, the ELM based modeling profits from much faster learning speed, higher generalization performance, and easy of implantation and use (no extra parameters need to be tuned except the predefined network architecture).   □

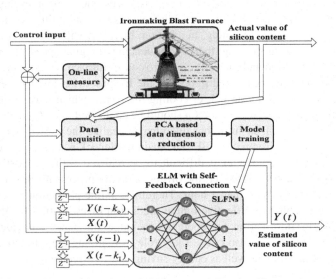

Fig. 1. Strategy diagram of nonlinear intelligent modeling for silicon content prediction

4   Modeling Algorithm

4.1   Selection of Secondary Variables by PCA-Based Dimension Reduction

PCA is a kind of method trying to grasp the main contradiction part in statistical analysis process and analyze the main influencing factors from multiple objects in order to simplify the complex problems. Actually, the principle components conducted by PCA are the combination of column vectors picked by varimax from input matrix. Since correlations and noises are always existed in practical industrial data, principle components with a small variance are usually some noisy information. Abandoning this data will not cause a crucial information loss and can even achieve de-noising in some extent.

Consider the following data set

$$u_i = \mathbf{X}v_i \tag{1}$$

where $\mathbf{X}_{n \times m}$ is the measured n data array on m variables, u_i is the score vector, v_i is the characteristic unit vector of covariance matrix $\mathbf{X}^T\mathbf{X}$, named load vector. The variance of u_i is λ_i which is also the eigenvalue of $\mathbf{X}^T\mathbf{X}$, and satisfies $\text{Var}(t_i) = \lambda_i$, $\lambda_1 \geq \cdots \geq \lambda_m \geq 0$. PCA is also a procedure used to explain the variance in a single data matrix. The principal component decomposition of $\mathbf{X}$ can be represented as follows:

$$\mathbf{X} = \mathbf{U}\mathbf{V}^T = \sum_{i=1}^{m} u_i v_i^T + \mathbf{E} \tag{2}$$

In Eq.(2) $u_i v_i^T$ is the ith principal component, and $\mathbf{E}$ is a matrix of residuals. It is to be noted that the score vectors are orthogonal and so are the loading vectors which are of unit length. Eq.(2) indicates that a rank n matrix $\mathbf{X}$ can be decomposed as the sum of n rank 1 principal components. The number of principle component kept in Eq.(2) is determined by the total variance. The variance contribution and the total variance of principal component can be represented as follows:

$$\eta_k = \lambda_k / (\sum_{j=1}^{m} \lambda_j) \tag{3}$$

$$C\eta_k = \sum_{i=1}^{k} \eta_i = \sum_{i=1}^{k} \lambda_i / \sum_{j=1}^{p} \lambda_j \tag{4}$$

where η_k is the kth principle component variance contribution; $C\eta_k$ is the total variance of the first k terms. Usually, the total variance varies should be larger than 85%. Only in this case, the data dimension can be reduced on the premise of not losing useful information.

Remark 2: A problem of the PCA-based dimension reduction is that the conducted principle components are comprehensive representation of the original higher-dimension physical variables. However, by computing the *component matrix* which contains the correlations between the principle component and the original physical variable, one can obtained the lower-dimension physical variables which related to the principle components mostly, according to some specific requirements.   □

4.2   ELM with Self-feedback Connection

Extreme learning machine (ELM) is an algorithm for single hidden layer feedforward networks (SLFNs) with additive or radial basis function (RBF) hidden nodes whose learning speed can be thousands of times faster than conventional feedforward network learning algorithm like BP algorithm while reaching better approximation performance. The procedure of the ELM algorithm used here can be summarized as follows: For N arbitrary distinct samples $(\mathbf{X}_i, \mathbf{Y}_i)$, where $\mathbf{X}_i = [x_{i1}, x_{i2}, \cdots, x_{in}]^T \in \mathbf{R}^n$ and $\mathbf{Y}_i = [y_{i1}, y_{i2}, \cdots, y_{im}]^T \in \mathbf{R}^m$, the output of a SLFN with $\tilde{N}$ hidden nodes can be represented by

$$f_{\tilde{N}}(\mathbf{X}) = \sum_{i=1}^{\tilde{N}} \beta_i G(\mathbf{a}_i, b_i, \mathbf{X}), \quad \mathbf{X} \in \mathbf{R}^n, \mathbf{a}_i \in \mathbf{R}^n \tag{5}$$

where $\mathbf{a}_i$ and b_i are the learning parameters of hidden nodes, β_i is the output weigh, and $G(\mathbf{a}_i, b_i, \mathbf{X})$ is the output of the ith hidden node with respect to the input data $\mathbf{X}$.

In supervised batch learning, the learning algorithms use a finite number of input-output samples for training. For N arbitrary distinct samples $(\mathbf{X}_i, \mathbf{Y}_i)$, if an SLEN with $\tilde{N}$ additive hidden nodes can approximate these N samples with zero error, it then implies that there exist $\mathbf{a}_i, b_i$ and β_i such that

$$f_{\tilde{N}}(\mathbf{X}_j) = \sum_{i=1}^{\tilde{N}} \beta_i G(\mathbf{a}_i, b_i, \mathbf{X}) = \mathbf{Y}_j, \quad J = 1, \cdots, N \tag{6}$$

Eq.(6) can be written compactly as:

$$\mathbf{H}\beta = \mathbf{Y} \tag{7}$$

where

$$\mathbf{H}(\mathbf{a}_1, \cdots, \mathbf{a}_{\tilde{N}}, b_1, \cdots, b_{\tilde{N}}, \mathbf{x}_1, \cdots, \mathbf{x}_N) = \begin{bmatrix} g(\mathbf{a}_1 \odot \mathbf{x}_1 + b_1) & \cdots & g(\mathbf{a}_{\tilde{N}} \odot \mathbf{x}_1 + b_{\tilde{N}}) \\ \vdots & \ddots & \vdots \\ g(\mathbf{a}_1 \odot \mathbf{x}_N + b_1) & \cdots & g(\mathbf{a}_{\tilde{N}} \odot \mathbf{x}_N + b_{\tilde{N}}) \end{bmatrix}_{N \times \tilde{N}} \tag{8}$$

$\beta = \begin{bmatrix} \beta_1^T \cdots \beta_{\tilde{N}}^T \end{bmatrix}_{\tilde{N} \times m}^T$, $\mathbf{Y} = \begin{bmatrix} y_1^T \cdots y_N^T \end{bmatrix}_{N \times m}^T$, and $\mathbf{a}_i \odot \mathbf{X}$ denotes the inner product o f vector $\mathbf{a}_i$ and $\mathbf{X}$ in $\mathbf{R}^n$.

The purpose of ELM is training the net to find a least-squares solution $\widehat{\beta}$ of the linear system $\mathbf{H}\beta = \mathbf{Y}$

$$\left\| \mathbf{H}(a_1, \cdots, a_{\widetilde{N}}, b_1, \cdots, b_{\widetilde{N}})\widehat{\beta} - \mathbf{Y} \right\| = \min_{\beta} \left\| \mathbf{H}(a_1, \cdots, a_{\widetilde{N}}, b_1, \cdots, b_{\widetilde{N}})\beta - \mathbf{Y} \right\| \quad (9)$$

And the solution of the above linear system can be solved by the inverse of matrix β by the Moore-Penrose method, which is

$$\widehat{\beta} = \mathbf{H}^{\dagger}\mathbf{Y} \tag{10}$$

where $\mathbf{H}^{\dagger}$ is the Moore-Penrose generalized inverse of $\mathbf{H}$ [15~17].

Remark 3: For the simplicity of the paper, the prediction modeling process based on ELM with additive hidden node is summarized as follows: Giving a training set $\mathbf{Z} = \{(\mathbf{X}_i, \mathbf{Y}_i) \mid \mathbf{X}_i \in \mathbf{R}^n, \mathbf{Y}_i \in \mathbf{R}^m, i = 1, \cdots\}$ for prediction modeling, and hidden neuron number $\widetilde{N}$, the input weight $\mathbf{a}_i$ and bias b_i can be assigned arbitrarily to calculate the output matrix $\mathbf{H}$ of hidden layer by using Eq.(8). After that, the output weight β can be calculated by Eq.(10), which is essential for estimating output only based on estimating inputs.   □

Remark 4: The hidden node number $\widetilde{N}$ is the only parameter need to be predefined in the presented modeling method. In order to achieve optimal approximation ability of training and realize fast convergence aiming complex industrial data, a proper (maybe optimal) $\widetilde{N}$ can be determined as the one which results in the lowest validation error through several trainings and validations.   □

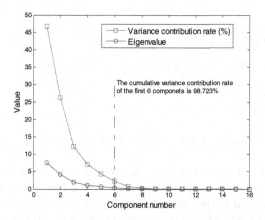

Fig. 2. Eigenvalue and variance contribution rate of each component

5   Industrial Experiments

In this section, a medium-sized blast furnace with the working volume of 2000 m³ in Liuzhou Iron & Steel Group Co. is chosen to perform the validation of silicon content prediction model. On the foundation of process mechanism and existing monitoring instruments status, 16 measurable parameters influencing [Si] are determined as blast temperature (℃), blast pressure (kPa), blast humidity (g/m³), and so on. Considering that the impact of strong correlation between the selected 16 input variables, PCA is used to determine the key input variables that influence the molten iron silicon content mostly. According to Eq.(3) and Eq.(4), the eigenvalue and the variance contribution rate of each component can be calculated as shown in Fig.2. It can be summarized that the cumulative variance contribution rate of the first 6 terms is 98.723%>98%. This means these 6 principal components are sufficient to describe the major variances in the data. Then, by computing the component matrix of principle components, 6 process variables can be determined as the secondary input of the [Si] prediction model. These secondary variables include hot blast pressure x_1 (kPa), hot blast temperature (℃), oxygen enrichment percentage (%), volume of coal injection (Kg/t), blast humidity (g/m³), and gas volume of bosh (m³/min).

The optimal number of hidden units is selected as the one which results in the lowest validation error. Through experiments analysis, the optimal number of hidden nodes with sigmoidal function is set as $\tilde{N} = 25$. The corresponding modeling result of the developed ELM model with self-feedback structure is shown in Fig.3, where the good modeling accuracy with practical data has been demonstrated.

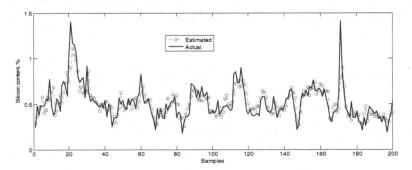

Fig. 3. Modeling results with proposed method

The developed ELM prediction model has been test on $2^{\#}$ blast furnace in Liuzhou Steel of China for quite a long time. Fig.4 shows the estimated results using the proposed modeling method for predicting [Si], where the figure compares the predicted trend with the actual one. Moreover, in order to show the superiority of the proposed method more intuitively, comparisons with various popular prediction models have been made. Here, BP NN without self-feedback (SFB) connection, BP NN with SFB connection, and traditional ELM without SFB connection have been chosen to conduct the prediction comparison on the same observations. From Fig.4, it can be seen that the proposed model has the best estimation performance among all the developed prediction models. For example, it results in the best estimation trend and accuracy, and the shape of the estimated curve values match the measured ones very well and better than that with other three methods.

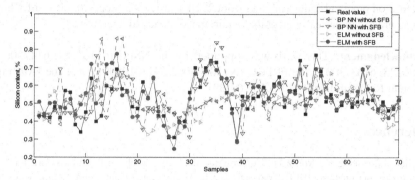

Fig. 4. Estimation results of molten iron silicon content with different models

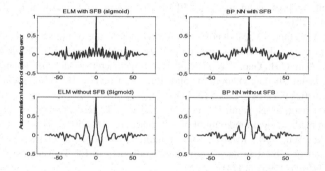

Fig. 5. Autocorrelation function of estimating error of different models

It is well known that a good model should have its estimated error autocorrelation close to a white noise. So in this text, we draw the autocorrelation function of estimating error of different models as shown in Fig.5. It can be seen that the autocorrelation results of algorithm like BP NN without SFB connection, and ELM without SFB connection is much worse than that with a SFB structure, respectively. Although one can obtain that the measuring error autocorrelation of the proposed ELM with SFB connection and BP NN with SFB connection are all satisfactory and close to the shape of the white noise here, the above estimation result confirmed the effectiveness and superiority of the proposed method in predicting accuracy.

6   Conclusions

This paper proposed a data-driven modeling for prediction of molten iron silicon content using PCA and ELM with self-feedback structure. Performance of the proposed ELM based prediction model is compared with BP algorithm and different model structure on practical industrial data from $2^{\#}$ BF in Liuzhou Steel Company of China. The accuracy can basically meet the requirements of actual operation.

Acknowledgment. This work was supported by the NSFC (61104084, 614730646, 61290323, 6133 3007), the FRFCU (N130508002, N130108001), and the IAPI Fundamental Research Funds (2013ZCX02-09).

References

1. Jian, L., Gao, C.H., Xia, Z.H.: Constructing multiple kernel learning framework for blast furnace automation. IEEE Transactions on Automation Science and Engineering 9(4), 763–777 (2012)
2. Gao, C.H., Jian, L., Luo, S.H.: Modeling of the thermal state change of blast furnace hearth with support vector machines. IEEE Transactions on Industrial Electronics 59(2), 1134–1145 (2012)
3. Brik, W., Marklund, O., Medvedev, A.: Video monitoring of pulverized coal injection in the blast furnace. IEEE Transactions on Industrial Applications 38(2), 571–576 (2002)
4. Saxen, H., Gao, C.H., Gao, Z.W.: Data-driven time discrete models for dynamic prediction of the hot metal silicon content in the blast furnace-A review. IEEE Transactions on Industrial Informatics 9(4), 2213–2225 (2013)
5. Ueda, S., Natsui, S., Nogami, H., Yagi, J., Ariyama, T.: Recent progress and future perspective on mathematical modeling of blast furnace. Int. ISIJ 50(7), 914–923 (2010)
6. Phadke, M., Wu, S.M.: Identification of multi input-multi output transfer function and noise model of a blast furnace from closed-loop data. IEEE Transactions on Automatic Control 19(6), 944–951 (1974)
7. Castore, M., Gandolfi, G., Palella, S., Taspedini, G.: Dynamic model for hot-metal Si prediction in blast-furnace control. In: Proc. Developments Ironmaking Practice, Iron and Steel Institute, London, U.K., pp. 152–159 (1972)
8. Chao, Y.C., Su, C.W., Huang, H.P.: The adaptive autoregressive models for the system dynamics and prediction of blast-furnace. Chem. Eng. Commun. 44, 309–330 (1986)
9. Bhattacharaya, T.: Prediction of silicon content in blast furnace hot metal using partial least squares (PLS). Int. ISIJ 45(1), 1943–1945 (2005)
10. Jiménez, J., Mochón, J., de Ayala, J.S., Obeso, F.: Blast furnace hot metal temperature prediction through neural networks-based models. Int. ISIJ 44, 573–580 (2004)
11. Saxén, H., Pettersson, F.: Nonlinear prediction of the hot metal silicon content in the blast furnace. Int. ISIJ 47(12), 1732–1737 (2007)
12. Chen, J.: A predictive system for blast furnaces by integrating a neural network with qualitative analysis. Engineering Applications of Artificial Intelligence 14(1), 77–85 (2001)
13. Tang, X., Zhuang, L., Jiang, C.: Prediction of silicon content in hot metal using support vector regression based on chaos particle swarm optimization. Expert Syst. Appl. 36(9), 11853–11857 (2009)

14. Gao, C.H., Ge, Q.H., Jian, L.: Rule extraction from fuzzy-based blast furnace SVM multic-lassifier for decision-making. IEEE Transactions on Fuzzy Systems 22(3), 586–596 (2014)
15. Liang, N.Y., Huang, G.B., Saratchandran, P., Sundararajan, N.: A fast and accurate online sequential learning algorithm for feedforward networks. IEEE Transactions on Neural Networks 17(6), 1411–1423 (2006)
16. Huang, G.B., Zhu, Q.Y., Siew, C.K.: Extreme learning machine: a new learning scheme of feedforward neural networks. In: 2004 IEEE International Joint Conference Neural Networks, vol. 2, pp. 985–990 (2004)
17. Huang, G.B., Zhu, Q.Y., Siew, C.K.: Extreme learning machine: Theory and applications. Neurocomputing 70(1-3), 489–501 (2006)
18. Huang, G.B., Babri, H.A.: Upper bounds on the number of hidden neurons in feedforward networks with arbitrary bounded nonlinear activation functions. IEEE Trans. on Neural Networks 9(1), 224–229 (1998)
19. Huang, G.B., Zhou, H.M., Ding, X.J., Zhang, R.: Extreme learning machine for regression and multiclass classification. IEEE Transactions on Systems, Man, and Cybernetics, Part B: Cybernetics 42(2), 513–529 (2012)
20. Lan, Y., Soh, Y.C., Huang, G.B.: Ensemble of online sequential extreme learning machine. Neurocomputing 72(13-15), 3391–3395 (2009)
21. Moreno, R., Corona, F., Lendasse, A., Graña, M., Galvão, S.: Extreme learning machines for soybean classification in remote sensing hyperspectral images. Neurocomputing 128, 207–216 (2014)
22. Sun, Y.J., Yuan, Y., Wang, G.R.: Extreme learning machine for classification over uncertain data. Neurocomputing 128, 500–506 (2014)
23. Savitha, R., Suresh, S., Sundararajan, N.: Fast learning circular complex-valued extreme learning machine (CC-ELM) for real-valued classification problems. Information Sciences 187, 277–290 (2012)
24. Yu, Q., Miche, Y., Séverin, E., Lendasse, A.: Bankruptcy prediction using extreme learning machine and financial expertise. Neurocomputing 128, 296–302 (2014)
25. Wang, G.R., Zhao, Y., Wang, D.: A protein secondary structure prediction framework based on the extreme learning machine. Neurocomputing 72(1-3), 262–268 (2008)
26. Zhang, J., Martin, E., Morris, A.J.: Fault detection and classification through multivariate statistical techniques. In: Proceedings of American Control Conference, vol. 1, pp. 751–755 (1995)
27. Good, R.P., Kost, D., Cherry, G.A.: Introducing a unified PCA algorithm for model size reduction. IEEE Transactions on Semiconductor Manufacturing 23(2), 201–209 (2010)
28. Zhao, J., Wang, W., Liu, Y., Pedrycz, W.: A two-stage online prediction method for a blast furnace gas system and its application. IEEE Transactions on Control Systems Technology 19(3), 507–520 (2011)

Distributed Learning over Massive XML Documents in ELM Feature Space

Xin Bi, Xiangguo Zhao*, Guoren Wang, Zhen Zhang, and Shuang Chen

College of Information Science and Engineering,
Northeastern University, Liaoning, Shenyang, China 110819
edijasonbi@gmail.com,
zhaoxiangguo@mail.neu.edu.cn

Abstract. Since centralized learning solutions are unable to meet the requirements of mining applications with massive training samples, a solution to distributed learning over massive XML documents is proposed in this paper, which provides distributed conversion of XML documents into representation model in parallel based on MapReduce, and a distributed learning component based on Extreme Learning Machine for mining tasks of classification or clustering. Within this framework, training samples are converted from raw XML datasets with better efficiency and information representation ability and taken to distributed learning algorithms in ELM feature space. Extensive experiments are conducted on massive XML documents datasets to verify the effectiveness and efficiency for both distributed classification and clustering applications.

Keywords: XML, Extreme Learning Machine, classification, clustering, distributed computing.

1 Introduction

Extreme Learning Machine (ELM) was proposed by Huang, *et al.* in [11,12] based on generalized single-hidden layer feedforward networks (SLFNs). With its variants[10,2,13,7,6], ELM achieves extremely fast learning capacity and good generalization performance due to its *universal approximation capability* and *classification capability*. Recently, paper [8] pointed out that from the optimization method point of view, ELM for classification and SVM are equivalent. Furthermore, it is proved in [9] that ELM provides a unified learning platform with a widespread type of feature mappings.

It is generally believed that all the ELM based algorithms consist of two major stages[5]: 1) random feature mapping; 2) output weights calculation. The first stage is the key concept in ELM theory. Most existing ELM based classification algorithms can be viewed as *supervised learning in ELM feature space*. While in [3], the *unsupervised learning in ELM feature space* is studied, drawing the conclusion that the proposed ELM kMeans algorithm can get better clustering results than in original feature space.

---

* Corresponding author.

Recently, the volume of XML documents keeps explosively increasing. MapReduce[1] provides tremendous parallel computing power without concerns for the underlying implementation and technology. However, MapReduce framework requires distributed storage of the datasets and no communication among mappers or reducers, which brings challenges to: 1) converting XML datasets into representation model; 2) implementing learning algorithms in ELM feature space.

To our best knowledge, massive XML mining problems in this paper is discussed for the *first* time. The contributions can be summarized as

1. A *distributed representing algorithm* DXRC is proposed to convert massive XML documents into XML representation model in parallel;
2. Existing *distributed supervised learning algorithms in ELM feature space* are implemented to make comparison of massive XML documents classification performance, including PELM and POS-ELM;
3. A *distributed unsupervised learning algorithm* DEK is proposed based on ELM kMeans[3] to realize distributed clustering over massive XML documents in *ELM feature space*;
4. Empirical experiments are conducted on clusters to verify the performance of our solution.

The remainder of this paper is structured as follows. Section 2 proposes a distributed representation converting algorithm. ELM feature mapping is presented in Section 3. Section 4 presents classification algorithms based on distributed ELMs and proposes a distributed clustering algorithm in ELM feature space based on MapReduce. Section 6 shows the experimental results. Section 7 draws conclusions of this paper.

2   Distributed XML Representation

In this section, we propose a distributed converting algorithm, named Distributed XML Representation Converting (DXRC), to calculate TFIDF[14] for DSVM based on MapReduce.

The *map* function in Algorithm 1 accepts key-value pairs as input. The key of key-value pairs is the XML document ID and the value is the corresponding XML document content. A HashMap *mapEle* (Line 1) is used to cache the all the elements of one XML document (Lines 2-11), using element name as key and another HashMap *mapEleTF* (Line 4) as value. The *mapEleTF* caches the TF values of all the words in one element (Lines 5-10). That is, for each XML document, there are as many items in *mapEle* as there are elements; for each element, there are as many items in *mapEleTF* as there are distinct words in this element. Each item in *mapEle* and *mapEle* will be emitted as output in the form of $\langle term, \langle docID, element, times, sum \rangle \rangle$ (Lines 12-17).

After the $\langle term, \langle docID, element, times, sum \rangle \rangle$ pairs are emitted by map function, all the key-value pairs with the same key, which are also the key-value pairs of the same word in XML documents, are combined and passed to the same *reduce* function in Algorithm 2 as input. For each key-value pair processed by

---

Algorithm 1. Mapper of DXRC

Input: $\langle docID, content \rangle$
Output: $\langle term, \langle docID, element, times, sum \rangle \rangle$

1 Initiate HashMap $mapEle$;
2 **foreach** $element \in content$ **do**
3   Initiate $sum = 0$;
4   Initiate HashMap $mapEleTF$;
5   **foreach** $term \in element$ **do**
6          $sum++$;
7   **if** $mapEleTF.containsKey(term)$ **then**
8              $mapEleTF$.put(term, mapEleTF.get(term)+1);
9   **else**
10             $mapEleTF$.put($term$,1);
11     $mapEle$.put($element, mapEleTF$);
12 **foreach** $itrEle \in mapEle$ **do**
13     $element = ele$.getKey();
14   **foreach** $itrEleTF \in itrEle$ **do**
15         $term = itrEleTF$.getKey();
16         $times = itrEleTF$.getValue();
17   emit($term, \langle docID, element, times, sum \rangle$);

---

reduce function, two HashMaps $mapDocEleTF$ (Line 1) and $mapTDocs$ (Line 2) are initiated. The HashMap $mapDocEleTF$ is to cache the tf values of a word in each element in the corresponding XML document and $mapTDocs$ is to cache the number of documents containing this word. The total number of document N (Line 3) and the vector $weights$ (Line 4), which indicates the wights of all the elements in each XML document, are obtained through distributed cache defined in MapReduce job configuration. Since Reduce now have all the tf values grouped by XML elements along with their weights, weighted tf values (Line 6) and the number of documents containing each word are calculated and cached in $mapDocEleTF$ and $mapTDocs$ respectively (Lines 5-12). Then the idf value can be calculated (Line 14) and multiplied by each item in $mapDocEleTF$. The output of reduce is the $\langle position, tfidf \rangle$ pairs, of which $position$ is $\langle docID, element \rangle$ indicating the index of DSVM matrix and $tfidf$ is the value of the matrix.

3 ELM Feature Mapping

In ELM, the n input nodes correspond to the n-dimensional data space of original training samples; while L hidden nodes correspond to the L-dimensional *ELM feature space*. With the m-dimensional output space, the decision function output the class label of the training sample.

Algorithm 2. Reducer of DXRC

Input: $\langle term, list(\langle docID, element, times, sum\rangle)\rangle$

Output: training samples matrix in the form of $\langle position, tfidf\rangle$

1 Initiate HashMap $mapDocEleTF$;
2 Initiate HashMap $mapTDocs$;
3 N=DistributedCache.get("totalDocsNum");
4 $weights$ = DistributedCache.get("elementWeightsVector");
5 **foreach** $itr \in list$ **do**
6      $weightedDocEleTF = weights[docId, element] * itr.times/itr.sum$;
7      $mapDocEleTF$.put($\langle docId, element\rangle, weightedDocEleTF$);
8   **if** $mapTDocs.containsKey(docId)$ **then**
9          $newTimes = mapTF$.get(docId) + irt.getValue().$times$;
10         $mapTDocs$.put($docID, newTimes$);
11   **else**
12         $mapTDocs$.put($docID$,itr.getValue().$times$);

13 $docsNumber = mapTDocs$.size();
14 $idf = \log(mapTDocsSize/N)$;
15 **foreach** $itrDocEleTF \in mapDocEleTF$ **do**
16     $position = itrDocEleTF$.getKey();
17     $tfidf = itrDocEleTF$.getValue()$*idf$;
18   emit($position, tfidf$);

The *ELM feature mapping* denoted as $\mathbf{H}$ is calculated as

$$\mathbf{H} = \begin{bmatrix} \mathrm{h}(\mathbf{x}_1) \\ \vdots \\ \mathrm{h}(\mathbf{x}_N) \end{bmatrix} = \begin{bmatrix} G(\mathbf{w}_1, b_1, \mathbf{x}_1) & \cdots & G(\mathbf{w}_L, b_L, \mathbf{x}_1) \\ \vdots & \cdots & \vdots \\ G(\mathbf{w}_1, b_1, \mathbf{x}_N) & \cdots & G(\mathbf{w}_L, b_L, \mathbf{x}_N) \end{bmatrix}_{N \times L} \tag{1}$$

4   Distributed Classification in ELM Feature Space

In this section, we introduce the learning procedure of classification problems in ELM feature space, and distributed implementations based on two existing representative distributed ELM algorithms, which are PELM[4] and POS-ELM[15].

4.1   Supervised Learning in ELM Feature Space

In supervised learning applications, since ELM is to minimize the training error and the norm of the output weights[11,12], that is

$$\text{Minimize:} \quad \|\mathbf{H}\boldsymbol{\beta} - \mathbf{T}\|^2 \quad \text{and} \quad \|\boldsymbol{\beta}\| \tag{2}$$

where $\mathbf{T} = \left[\mathbf{t}_1^T, ..., \mathbf{t}_L^T\right]_{m \times L}^T$ is the vector of class labels.

The matrix $\boldsymbol{\beta}$ is the output weight, which is calculated as

$$\boldsymbol{\beta} = \mathbf{H}^\dagger \mathbf{T} \tag{3}$$

where $\mathbf{H}^\dagger$ is the Moore-Penrose Inverse of $\mathbf{H}$.

If the number of training samples is much larger than the dimensionality of the feature space, the output weight calculation equation can be rewritten as

$$\beta = \left(\frac{\mathbf{I}}{C} + \mathbf{H}^{\mathrm{T}}\mathbf{H} \right)^{-1} \mathbf{H}^{\mathrm{T}}\mathbf{T} \tag{4}$$

4.2 Distributed Implementations

The original ELM was parallelized by PELM in [4]; Online sequential ELM (OS-ELM) was implemented on MapReduce as POS-ELM in [15].

Parallel ELM. Since the major cost in ELM is the calculation of generalized inverse of matrix $\mathbf{H}$, The matrix multiplication $\mathbf{U} = \mathbf{H}^{\mathrm{T}}\mathbf{H}$ and $\mathbf{V} = \mathbf{H}^{\mathrm{T}}\mathbf{T}$ can be calculated by a MapReduce job. In map function, each term of $\mathbf{U}$ and $\mathbf{V}$ is calculated in parallel. In reduce function, all the intermediate results are merged and summed to the corresponding elements of the result matrix.

Parallel Online Sequential ELM. The basic idea of Parallel Online Sequential ELM (POS-ELM) is to calculate $\mathbf{H}_1, \cdots, \mathbf{H}_B$ in parallel. POS-ELM takes advantage of the calculation of partial ELM feature matrix $\mathbf{H}_i$ with a chunk of training data of OS-ELM, in each map function calculates its $\mathbf{H}_i$ with its own data chunk. The reduce function collects all the $\mathbf{H}_i$ and calculates β_{k+1} as

$$\beta_{k+1} = \beta_k + \mathbf{P}_{k+1}\mathbf{H}_{k+1}^{\mathrm{T}}(\mathbf{T}_{k+1} - \mathbf{H}_{k+1}\beta_k) \tag{5}$$

where

$$\mathbf{P}_{k+1} = \mathbf{P}_k - \mathbf{P}_k\mathbf{H}_{k+1}^{\mathrm{T}}(\mathbf{I} + \mathbf{H}_{k+1}\mathbf{P}_k\mathbf{H}_{k+1}^{\mathrm{T}})^{-1}\mathbf{H}_{k+1}\mathbf{P}_k \tag{6}$$

5 Distributed Clustering in ELM Feature Space

Generally, kMeans algorithm in ELM feature space, as ELM kMeans for short, has two major steps: 1) transform the original data into ELM feature space; 2) implement traditional clustering algorithm directly. Clustering in the ELM feature space is much more convenient than kernel based algorithms.

We present Distributed ELM k-Means (DEK) based on ELM kMeans[3]. Algorithm 3 presents the map function of DEK. For each sample stored on this mapper (Line 1), the distance between the sample and each cluster centroids is calculated (Lines 2, 3). Then each sample is assigned to the cluster whose centroid is the nearest one to this sample (Line 4). The intermediate key-value pair is emitted in the form of $\langle c_{max}, \mathbf{x}_i \rangle$ (Line 5), in which $\mathbf{x}_i$ is the specific sample and c_{max} is the assigned cluster of $\mathbf{x}_i$.

Algorithm 4 presents the reduce function of DEK. We add up the sum distance in Euclidean space of all the samples $\mathbf{x}_i$ in list($\mathbf{x}$) (Lines 1, 2), and then calculate the mean value to represent the new version of the centroid $c_j^{updated}$ of cluster c_j

---

Algorithm 3. Mapper of DEK

Input: Training samples $\mathbf{X}$, k centroids $\mathbf{C}$
Output: $\langle$ centroid c_{max}, sample $\mathbf{x}_i \rangle$
1 **foreach** $\mathbf{x}_i \in \mathbf{X}$ **do**
2   **foreach** $c_j \in \mathbf{C}$ **do**
3       $\lfloor$ Calculate distance d_{ij} between $\mathbf{x}_i$ and c_j;
4   Assign $\mathbf{x}_i$ to the cluster c_{max} with $\max_j(d_{ij})$;
5   Emit $\langle c_{max}, \mathbf{x}_i \rangle$;

(Line 3). When all the k cluster centroids are updated, if this version of centroids are the same as the older one, or if the maximum iteration number is reached, DEK holds that the clustering job is done; otherwise, DEK continues to the next round of MapReduce job until convergence.

Algorithm 4. Reducer of DEK

Input: $\langle$ centroid c_j, samples list$(\mathbf{x})\rangle$
Output: Updated set of centroids $\mathbf{C}_{updated}$
1 **foreach** $\mathbf{x}_i \in$ list$(\mathbf{x})$ **do**
2     $\lfloor$ Add $\mathbf{x}_i$ to squared sum S;
3 Calculate $c_j^{updated}$ of cluster c_j as $c_j^{updated} = S/$list$(\mathbf{x}).length$;

---

6 Performance Evaluation

6.1 Experiments Setup

Three datasets *Wikipedia XML Corpus* provided by INEX, *IBM DeveloperWorks* articles and *ABC News* are used in our paper. We choose the same numbers of XML documents out of all the three datasets, that is 6 classes and 500 documents in each class. The only parameter of learning algorithms in ELM feature space, i.e., the number of hidden nodes L, is set to 800.

To evaluate the performance, three sets of evaluation criteria are utilized: 1) for scalability evaluation, we compare the criteria of speedup, sizeup and scaleup; 2) for classification problems, accuracy, recall and F-measure are used; 3) for clustering problems, since we treat this class label as the cluster label, the same evaluation criteria are used as classification problems.

All the experiments are conducted on a Hadoop cluster of nine machines, each of which is equipped with an Intel Quad Core 2.66GHZ CPU, 4GB of memory and CentOS 5.6 as operating system. The MapReduce framework is configured with Hadoop version 0.20.2 and Java version 1.6.0_24.

6.2   Evaluation Results

Scalability of DXRC. The scalability of representation converting algorithm DXRC is first evaluated. Figure 1 demonstrates good speedup, sizeup and scaleup of DXRC. The representation ability DSVM applied in DXRC can be found in our previous work [16].

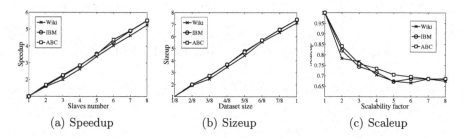

(a) Speedup   (b) Sizeup   (c) Scaleup

Fig. 1. Scalability of DXRC

Scalability of Massive XML Classification in ELM Feature Space. The speedup comparison between PELM and POS-ELM on three datasets are presented in Figure 2.

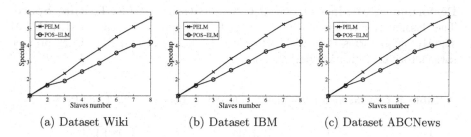

(a) Dataset Wiki   (b) Dataset IBM   (c) Dataset ABCNews

Fig. 2. Comparison of speedup between PELM and POS-ELM

Since the centralized calculation reduces the scalability of both PELM and POS-ELM to some degree, especially for POS-ELM. Thus, the speedup of PELM is better than POS-ELM.

Figure 3 demonstrates the sizeup comparison between PELM and POS-ELM, from which we find that the sizeup of PELM is better than POS-ELM.

For the scaleup comparison, Figure 4 demonstrates that both PELM and POS-ELM have good scaleup performance, and PELM outperforms POS-ELM on each of the three datasets.

In summary, PELM has better scalability than POS-ELM, but both of them have good scalability for massive XML documents classification applications.

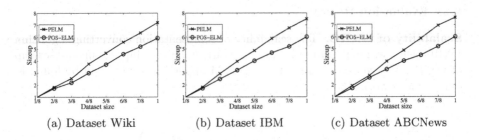

Fig. 3. Comparison of sizeup between PELM and POS-ELM

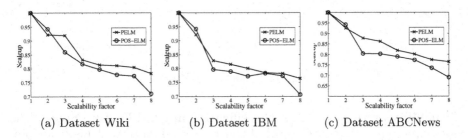

Fig. 4. Comparison of scaleup between PELM and POS-ELM

Performance of Massive XML Classification in ELM Feature Space.
The classification results are shown in Table 1, from which we can see that PELM
slightly outperforms POS-ELM. However, both PELM and POS-ELM provide
satisfactory classification performance.

Table 1. Classification performance comparison between PELM and POS-ELM

Datasets	PELM			POS-ELM		
	Accuracy	Recall	F-measure	Accuracy	Recall	F-measure
Wikipedia	0.7886	0.7563	0.7721	0.7923	0.7745	0.7833
IBM developWorks	0.7705	0.8145	0.7919	0.7711	0.7863	0.7891
ABC News	0.8681	0.8517	0.8598	0.8517	0.8335	0.8425

Scalability of Massive XML Clustering in ELM Feature Space. In
theory, the scalability of distributed k-Means in ELM feature space and in original feature space are the same, we only presents the scalability of DEK without
comparison with distributed k-Means in original feature space in Figure 5, which
demonstrates good scalability of DEK.

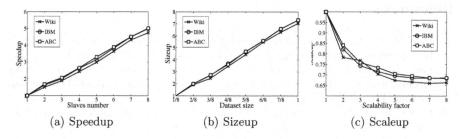

(a) Speedup   (b) Sizeup   (c) Scaleup

Fig. 5. Scalability of DEK

Performance of Massive XML Clustering in ELM Feature Space. The comparison results on three different datasets are presented in Table 2. It can be seen from the comparison results that DEK gets better clustering performance due to its ELM features mapping.

Table 2. Clustering performance of DEK compared with parallel k-Means

Dataset	Parallel k-Means			DEK		
	Accuracy	Recall	F-measure	Accuracy	Recall	F-measure
Wikipedia	0.7602	0.7426	0.7513	0.7737	0.7648	0.7692
IBM developerWorks	0.7985	0.8268	0.8126	0.8124	0.8277	0.8200
ABC News	0.8351	0.8125	0.8201	0.8529	0.8192	0.8357

7 Conclusion

This paper addresses the problem of distributed XML documents learning in ELM feature space, which has no previous work to our best knowledge. Parallel XML documents representation converting problem is discussed by proposing algorithm DXRC. Massive XML documents classification in ELM feature space is studied by implementing PELM and POS-ELM; while for massive XML documents clustering in ELM feature space, a distributed ELM k-Means algorithm DEK is proposed. Experimental results demonstrate that the distributed XML learning in ELM feature space shows good scalability and learning performance.

Acknowledgment. This research is partially supported by the National Natural Science Foundation of China under Grant Nos. 61272181, and 61173030; the National Basic Research Program of China under Grant No. 2011CB302200-G; the 863 Program under Grant No. 2012AA011004; and the Fundamental Research Funds for the Central Universities under Grant No. N120404006.

References

1. Dean, J., Ghemawat, S.: MapReduce: Simplied Data Processing on Large Clusters. In: Operating Systems Design and Implementation, pp. 137–150 (2004)
2. Feng, G., Huang, G.B., Lin, Q., Gay, R.K.L.: Error minimized extreme learning machine with growth of hidden nodes and incremental learning. IEEE Transactions on Neural Networks 20, 1352–1357 (2009)
3. He, Q., Jin, X., Du, C., Zhuang, F., Shi, Z.: Clustering in extreme learning machine feature space. Neurocomputing 128, 88–95 (2014)
4. He, Q., Shang, T., Zhuang, F., Shi, Z.: Parallel extreme learning machine for regression based on mapreduce. Neurocomputing 102, 52–58 (2013)
5. Huang, G., Song, S., Gupta, J., Wu, C.: Semi-supervised and unsupervised extreme learning machines. IEEE Transactions on Cybernetics PP(99), 1–1 (2014)
6. Huang, G.B., Chen, L.: Convex incremental extreme learning machine. Neurocomputing 70, 3056–3062 (2007)
7. Huang, G.B., Chen, L.: Enhanced random search based incremental extreme learning machine. Neurocomputing 71, 3460–3468 (2008)
8. Huang, G.B., Ding, X., Zhou, H.: Optimization method based extreme learning machine for classification. Neurocomputing 74, 155–163 (2010)
9. Huang, G.B., Zhou, H., Ding, X., Zhang, R.: Extreme Learning Machine for Regression and Multiclass Classification. IEEE Transactions on Systems, Man, and Cybernetics 42, 513–529 (2012)
10. Huang, G.B., Zhu, Q.Y., Mao, K.Z., Siew, C.K., Saratchandran, P., Sundararajan, N.: Can threshold networks be trained directly? IEEE Transactions on Circuits and Systems II: Analog and Digital Signal Processing 53, 187–191 (2006)
11. Huang, G.B., Zhu, Q.Y., Siew, C.K.: Extreme learning machine: a new learning scheme of feedforward neural networks. In: International Symposium on Neural Networks, vol. 2 (2004)
12. Huang, G.B., Zhu, Q.Y., Siew, C.K.: Extreme learning machine: Theory and applications. Neurocomputing 70, 489–501 (2006)
13. Rong, H.J., Huang, G.B., Sundararajan, N., Saratchandran, P.: Online sequential fuzzy extreme learning machine for function approximation and classification problems. IEEE Transactions on Systems, Man, and Cybernetics 39, 1067–1072 (2009)
14. Salton, G., Buckley, C.: Term-weighting approaches in automatic text retrieval. Information Processing and Management 24, 513–523 (1988)
15. Wang, B., Huang, S., Qiu, J., Liu, Y., Wang, G.: Parallel Online Sequential Extreme Learning Machine Based on MapReduce. To appear in Neurocomputing (2014)
16. Zhao, X., Bi, X., Qiao, B.: Probability based voting extreme learning machine for multiclass xml documents classification. In: World Wide Web, pp. 1–15 (2013)

Hyperspectral Image Nonlinear Unmixing by Ensemble ELM Regression

Borja Ayerdi* and Manuel Graña

Grupo de Inteligencia Computacional (GIC), Universidad del País Vasco
(UPV/EHU), San Sebastián, Spain

Abstract. In this paper we use the best properties of Extreme Leaning
Machines (ELM) for the task of obtaining a nonlinear unmixing of hyperspectral images. The best advantage of ELM is its speed of training,
provided by the random input layer weight setting process. On the other
hand, ELM has a big uncertainty in its input, which can be coped with
by aggregating several ELM into an ensemble which obtain a reduction of
the uncertainty by combining the results of the individual ELM. Unmixing consists in the estimation of the composition of the pixels obtaining
a sub-pixel segmentation resolution containing the abundance the composing materials. Linear unmixing assumes that there is a collection of
endmember spectra so that the pixels are convex combinations of them.
In the case of non-linear unmixing, we do not have a characterization
of the composing material spectra, but we try to estimate the decomposition of the pixel into predefined classes of materials characterized by
pure pixels extracted from the image on the basis of known ground truth
information.

1 Introduction

Hyperspectral image pixels are high dimensional vector produced by a high resolution sampling of the light spectrum in the visual and infrared bands [1].
Working with hyperspectral image is driven by the expectation that it will be
possible to identify the materials in the imaged scene by the interpretation or
classification of the pixel spectra. Among the challenges posed by hyperspectral
image [2] one of the most salient is the unmixing, aka subpixel resolution material identification, that provides the abundances of the composing materials in
each pixel. The early approaches [3] have been linear methods which assume a
linear mixing of constituent spectra, i.e. the endmembers. This linear mixture
is usually explained by the existence of well defined regions inside the area covered by a pixel (which can be hundred of meters in some sensors) of materials
with well differentiated spectra. In general, the problem of linear unmixing is
decomposed into two problems:

* This work has been partially supported by the "Ayudas para la Formación de Personal Investigador" fellowship from the Gobierno del País Vasco.

- Finding the library of endmembers, i.e. the spectra of the constituent materials, that will serve as the basis for the decomposition of the image pixel spectra. This library may be extracted from some public database of material reflectance measurements, (i.e. USGS spectral library[1]), or may be induced from the image by some endmember extraction algorithm. A critical problem in this induction process is the a priori determination of the number of endmembers, a case of intrinsic dimensionality of the data [4]. Endemem-ber induction relies on the assumption of the data lying in a convex region, methods to decompose non-convex data into convex partitions [5] allow to extend conventional endmember induction to this case.
- Solving a linear problem that gives the abundances of the endmembers in the pixel. Solving this problem maybe tricking if we enforce the conditions expected from the mixing process, that is, that all abundances are positive and add up to 1. Least squares solutions do not guarantee these properties, so it is required to resort to non-negative matrix factorization (NNMF) or similar linear decomposition tools.

Some approaches, such as the Independent Component Analysis (ICA) or the NNMF solve both problems simultaneously. Other approaches, are sequential, performing first the endmember identification and afterwards the unmixing. In this paper, we do not try to find the constituent material spectra, instead we seek directly to obtain the abundances of the target classes by training a regression model implemented by Extreme Learning Machines (ELM). In fact, the approach is a generalization of the unmixing process covering also non-linear unmixing, and it is not restricted by the limitations imposed by other methods, such as the assumption of statistical independence underlying ICA, or the need to find a basis of non-negative vectors imposed by NNMF. The endmember basis may be somehow embedded in the weights from the input to the hidden units of the ELM, where the hidden units activation functions introduce the non-linear aspect of the unmixing process.

2   Non Linear Unmixing

Figure 1 illustrates how non-linear mixing may happen at a given pixel. Reflections and scattering processes produce such nonlinear mixing of material reflectances, in the figure the ray that goes through the canopy to the soil and then to the sensor suffers this kind of effect. The basic non-linear model accounting for second-order interactions between spectra is the bi-linear model [6]. A more sophisticated approach, still preserving the role of the enmembers as the extrema of the convex polytope enclosing the iamge data, is the Geodesic Simplex Projection [6] which models the data as a manifold embedded in the simplex defined by the endmembers, solving the abundances as the projection distance into this manifold. The selection of landmark points on these manifolds helps to reduce the computational burden imposed by the search of the nearest point for projection

---

[1] http://speclab.cr.usgs.gov/spectral-lib.html

on the manifold [7,8]. Other approaches use the kernel trick to model the non-linear manifold of the data [9], proposed as non-linear fluctuations taking into account non-linear interactions among the endmembers. The specification of multiple kernels improve the flexibility of the approach allowing class-dependent tuning of the parameters [10].

The application of Markov Random Field (MRF) modeling provides a graceful way to define Bayesian algorithms to perform non-linear unmixing. For instance, [11] assumes a non-linear term in the mixing equation, whose spatial distribution is analyzed by MRF modeling to perform non-linear detection and unmixing. In [12], a second order polynomial on the endmembers is postulated as the non-linear mixing model, and Hamiltonian Monte Carlo algorithm is applied to estimate the parameters of the model at each pixel, profiting from spatial continuity of the MRF. Spatial correlation is also considered in [13] as a regularization of the parameter estimation in the non-linear unmixing algorithm based on kernel transformations. Other approaches consider local windows in order to refine the relevance of the endmembers for each pixel decomposition [14]. These approaches are extended to the use of local NNMF approaches [15,16].

The approach in this paper consists in formulating the problem of unmixing as a non-linear regression problem, where we want to extract the abundance parameters by applying the non-linear ELM model trained with some of the pixels of the image. The output of the ELM is a collection of output units with range [0, 1], each modeling the abundance of a given class. We associate classes with materials. As we do not know the exact mixtures of the pixels, we train the ELM with pure pixels per class, generalization would provide the abundances in non-pure pixels. For robustness against the uncertainty of the ELM training, we use ensembles of ELM whose output is combined by selecting the maximum value for each output variable.

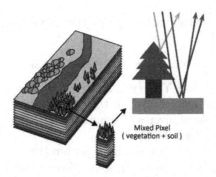

Fig. 1. Illustration of a physical process that may account for a non-linear spectral mixing at pixel

3   Materials and Methods

3.1   ELM Ensemble

The classifiers tested in this paper are ensembles of ELM, similar to the Voting ELM (V-ELM) [17], which uses majority voting for achieving the decision on the classification result. In this paper we deal with regression instead of classification. We have used the basic ELM as the ensemble building block and tested different output combination criterions, like mean, maximum, minimum and median value. Upon inspection we have decided on the maximum as the most illustrative combinator.

Basic ELM. The Extreme Learning Machine (ELM) [18] is a very fast training algorithm for single-layer feedforward neural networks (SLFN). The key idea of ELM is the random initialization of the SLFN hidden layer node weights. Consider a set of M data samples $(\mathbf{x}_i, y_i)$ with $\mathbf{x}_i \in \mathbb{R}^{d_1}$ and $y_i \in \mathbb{R}^{d_2}$. Then, a SLFN with N hidden neurons is modeled as the following expression:

$$\mathbf{y} = \Phi(\mathbf{x}) = \sum_{i=1}^{N} \beta_i f(\mathbf{w}_i \cdot \mathbf{x}_j + b_i), j \in [1, M], \tag{1}$$

where $f(x)$ is the activation function, $\mathbf{w}_i$ the input weights to the i-th neuron in the hidden layer, b_i the hidden layer unit bias and β_i are the output weights. The application of this equation to all available data samples can be written in matrix form as

$$\mathbf{H}\beta = \mathbf{Y}, \tag{2}$$

where $\mathbf{H}$ is the hidden layer output matrix defined as the output of the hidden layer for each input sample vector

$$\mathbf{H} = \begin{pmatrix} f(\mathbf{w}_1 \cdot \mathbf{x}_1 + b_1) & \cdots & f(\mathbf{w}_N \cdot \mathbf{x}_1 + b_N) \\ \vdots & \ddots & \vdots \\ f(\mathbf{w}_1 \cdot \mathbf{x}_M + b_1) & \cdots & f(\mathbf{w}_N \cdot \mathbf{x}_M + b_N) \end{pmatrix}, \tag{3}$$

and $\beta = (\beta_1, \ldots, \beta_N)^T$ and $\mathbf{Y} = (\mathbf{y}_1, \ldots, \mathbf{y}_M)^T$. The way to calculate the output weights β from the hidden-layer to the target values is computing the of a Moore–Penrose generalized inverse of the matrix $\mathbf{H}$, denoted as $\mathbf{H}^\dagger$. The mean least squares solution is $\beta = \mathbf{H}^\dagger \mathbf{Y}$.

ELM universal approximation has been proven [19], however model selection in ELM has been an active research area, where some incremental strategies [20,21,22,23] provide efficient near-optimal hidden layer unit selection. Also, ensembles of ELM have been reported [17] providing enhanced results. Robust estimation of the output weights for situations where the pseudo-inverse is singular is proposed in [24].

Algorithm 1. ELM

Input A set of training samples $(\mathbf{x}_i, y_i)$ with $\mathbf{x}_i \in \mathbb{R}^{d_1}$ and $y_i \in \mathbb{R}^{d_2}$, an activation function $f(x)$, and a number of hidden nodes N:

1. Randomly generate bias matrix $\boldsymbol{b} = (b_1, \ldots, b_N)^T$ and weight matrix $\mathbf{W} = (\mathbf{w}_1, \ldots, \mathbf{w}_N)$.
2. Calculate $\mathbf{H}$ as in equation 3.
3. Calculate the output weight matrix $\boldsymbol{\beta} = \mathbf{H}^\dagger \mathbf{Y}$.

3.2 Benchmarking Images

We introduce the benchmark images with due references. We do not show the ground truth images, which can be found elsewhere[2].

Salinas Complete. Salinas hyperspectral dataset was collected over the Valley of Salinas, Southern California, in 1998. It contains 217×512 pixels of 224 spectral bands from 0.4 to 2.5 μm, with nominal spectral resolution of 10 nm. It was taken at low altitude with a pixel size of 3.7 m. The data include vegetables, bare soils, and vineyard fields. The Salinas A sub-scene comprises 83×86 pixels and is known to represent a difficult classification scenario with highly mixed pixels.

Indian Pines. Indian Pines scene was collected over Northwestern Indiana in June of 1992[1]. This scene of 145×145 pixels was acquired over a mixed agricultural/forest area early in the growing season. The spectral dimension is 224 spectral channels in the wavelength range from 0.4 to 2.5 μm, nominal spectral resolution of 10 nm, and spatial resolution of 20 m by pixel. This scene contains 16 mutually exclusive ground-truth classes. These data, including ground-truth information, are available online, a fact which has made this scene a widely used benchmark for testing the accuracy of hyperspectral data classification and segmentation algorithms.

4 Results

Results presented in this paper are preliminary, qualitative appraisal of the distribution of the non-linear found class abundances by means of ELM. The design of the classifiers is as follows: Each ELM takes as input a pixel spectrum, and its output is a vector whose components take value in the [0,1] interval, i.e. each output unit corresponds to the abundance of a ground truth class. We do not try to discover the number of classes, we use the available ground truth information for the ELM design, as well as for the supervised training. The number

---

[2] http://www.ehu.es/ccwintco/index.php/Hyperspectral_Remote_Sensing_Scenes

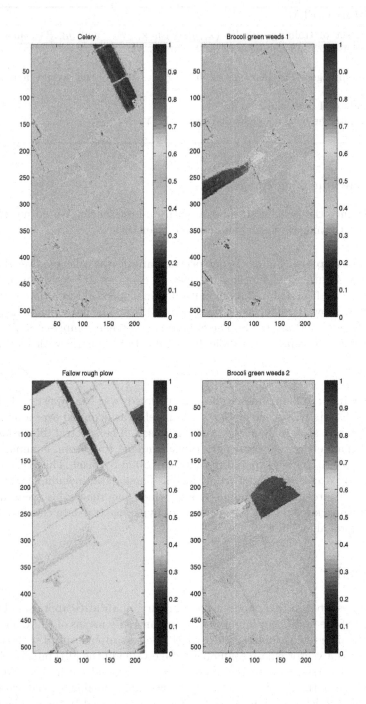

Fig. 2. Salinas unmixing of some of the classes by ELM membership

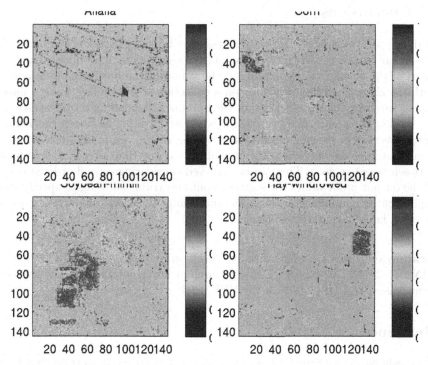

Fig. 3. Indian Pines unmixing of some of the classes by ELM membership

of hidden units was fixed at 100. The size of the ensembles was 30, and the ensemble output operator was the maximum value, i.e we return the maximum of each output unit found in all the ensemble ELM for the input spectra. Training data consists of a small number of pure pixels (10 pixels per class). We have not performed cross-validation tests yet, we compute the result of the ensemble over all the image, showing the abundance map for each class using a heat color map, so that red corresponds to highest values and blue to the lowest. Each image corresponds to the values of an output unit over all the pixels of the image.

Figures 2, and 3 show the abundance maps for some (lack of space impedes comprehensive visualization of results) of the ground truth classes of the corresponding benchmark image. Some general comments can be made from these images. Results are widely variable from image to image. The Salinas image contains big homogeneous regions, so that the approach works very well for some of the classes, such as the ones shown in figure 2, where the contrast of values of the output variable are evident. The target classes are clearly identified as red regions. However, the output generation operator (maximum) produces inordinate big values for the other regions, so that contrast is less than it would be desirable. Notice that low values (blue) do also identify meaningful regions in the image, such as roads. Indian Pines results in figure 3 also contains some good detections. However it must be said that this image is not very well labeled, so that results suffer also from mislabeling.

5   Conclusions

In this paper we apply ensembles of regression ELMs to the task of estimating the non-linear unmixing of hyperspectral images. This approach produces abundance maps of the classes, corresponding to the values of the output node associated with the class. Preliminary results on a collection of benchmark images show a strong varability of the efficiency of the approach, depending on the sensor and the kind of the scene. The best case is when we have big homogeneous regions. Future work will be addressed to fine tuning of the approach, improving the combinaiton of the outputs of the ensemble, in a way that increases the contrast between target regions. This can be achieved considering second order statistics of the output units values for a given input spectrum, allowing to specify more sophisticated combinations, such as diminishing the value of confusing pixels, i.e. with large output values variance.

Acknowledgments. This work has been supported by the MICINN grant TIN2011-28753-C02-02, university research group grant IT874-13 of the Gobierno Vasco, and UFI 11/07 of the UPV/EHU

References

1. Landgrebe, D.A.: Signal Theory Methods in Multispectral Remote Sensing. Wiley, Hoboken (2003)
2. Bioucas-Dias, J., Plaza, A., Camps-Valls, G., Scheunders, P., Nasrabadi, N., Chanussot, J.: Hyperspectral remote sensing data analysis and future challenges. IEEE Geoscience and Remote Sensing Magazine 1(2), 6–36 (2013)
3. Keshava, N., Mustard, J.: Spectral unmixing. IEEE Signal Processing Magazine 19(1), 44–57 (2002)
4. Heylen, R., Scheunders, P.: Hyperspectral intrinsic dimensionality estimation with nearest-neighbor distance ratios. IEEE Journal of Selected Topics in Applied Earth Observations and Remote Sensing 6(2), 570–579 (2013)
5. Zare, A., Gader, P., Casella, G.: Sampling piecewise convex unmixing and endmember extraction. IEEE Transactions on Geoscience and Remote Sensing 51(3), 1655–1665 (2013)
6. Burazerovic, D., Heylen, R., Geens, B., Sterckx, S., Scheunders, P.: Detecting the adjacency effect in hyperspectral imagery with spectral unmixing techniques. IEEE Journal of Selected Topics in Applied Earth Observations and Remote Sensing 6(3), 1070–1078 (2013)
7. Chi, J., Crawford, M.: Selection of landmark points on nonlinear manifolds for spectral unmixing using local homogeneity. IEEE Geoscience and Remote Sensing Letters 10(4), 711–715 (2013)
8. Chi, J., Crawford, M.: Active landmark sampling for manifold learning based spectral unmixing. IEEE Geoscience and Remote Sensing Letters 11(11), 1881–1885 (2014)
9. Chen, J., Richard, C., Honeine, P.: Nonlinear unmixing of hyperspectral data based on a linear-mixture/nonlinear-fluctuation model. IEEE Transactions on Signal Processing 61(2), 480–492 (2013)

10. Gu, Y., Wang, S., Jia, X.: Spectral unmixing in multiple-kernel hilbert space for hyperspectral imagery. IEEE Transactions on Geoscience and Remote Sensing 51(7), 3968–3981 (2013)
11. Altmann, Y., Dobigeon, N., McLaughlin, S., Tourneret, J.Y.: Residual component analysis of hyperspectral images - application to joint nonlinear unmixing and nonlinearity detection. IEEE Transactions on Image Processing 23(5), 2148–2158 (2014)
12. Altmann, Y., Dobigeon, N., Tourneret, J.: Unsupervised post-nonlinear unmixing of hyperspectral images using a hamiltonian monte carlo algorithm. IEEE Transactions on Image Processing 23(6), 2663–2675 (2014)
13. Chen, J., Richard, C., Honeine, P.: Nonlinear estimation of material abundances in hyperspectral images with ell_1-norm spatial regularization. IEEE Transactions on Geoscience and Remote Sensing 52(5), 2654–2665 (2014)
14. Cui, J., Li, X., Zhao, L.: Nonlinear spectral mixture analysis by determining per-pixel endmember sets. IEEE Geoscience and Remote Sensing Letters 11(8), 1404–1408 (2014)
15. Eches, O., Guillaume, M.: A bilinear-bilinear nonnegative matrix factorization method for hyperspectral unmixing. IEEE Geoscience and Remote Sensing Letters 11(4), 778–782 (2014)
16. Yokoya, N., Chanussot, J., Iwasaki, A.: Nonlinear unmixing of hyperspectral data using semi-nonnegative matrix factorization. IEEE Transactions on Geoscience and Remote Sensing 52(2), 1430–1437 (2014)
17. Cao, J., Lin, Z., Huang, G.B., Liu, N.: Voting based extreme learning machine. Information Sciences 185(1), 66–77 (2012)
18. Huang, G.B., Zhu, Q.Y., Siew, C.K.: Extreme learning machine: Theory and applications. Neurocomputing 70(1-3), 489–501 (2006)
19. Huang, G.B., Chen, L., Siew, C.K.: Universal approximation using incremental constructive feedforward networks with random hidden nodes. IEEE Trans. Neural Networks 17(4), 879–892 (2006)
20. Huang, G.B., Chen, L.: Convex incremental extreme learning machine. Neurocomputing 70(16-18), 3056–3062 (2007)
21. Huang, G.B., Li, M.B., Chen, L., Siew, C.K.: Incremental extreme learning machine with fully complex hidden nodes. Neurocomputing 71(4-6), 576–583 (2008)
22. Huang, G.B., Chen, L.: Enhanced random search based incremental extreme learning machine. Neurocomputing 71(16-18), 3460–3468 (2008)
23. Chen, L., Huang, G.B., Pung, H.K.: Systemical convergence rate analysis of convex incremental feedforward neural networks. Neurocomputing 72(10-12), 2627–2635 (2009)
24. Horata, P., Chiewchanwattana, S., Sunat, K.: Robust extreme learning machine. Neurocomputing (2012)

Text-Image Separation and Indexing in Historic Patent Document Image Based on Extreme Learning Machine

Lijuan Duan[1,*], Bin Yuan[1], Chunpeng Wu[2,*], Jian Li[1], and Qingshuan Guo[3]

[1] Department of Computer Science and Technology, Beijing University of Technology,
Beijing 100124, China
ljduan@bjut.edu.cn, 19890926yb@emails.bjut.edu,
lijian@bjut.edu.cn
[2] Department of Electrical & Computer Engineering, University of Pittsburgh, PA 15261, USA
chunpengwu1986@gmail.com
[3] Hitachi Beijing University of Technology, Beijing 100124, China
guo-qingshuan@hbis.com.cn

Abstract. Historic patent document retrieval is still a theoretically challenging problem, while an application of great business value. Text-image separation and indexing is a crucial step in many applications related to patent document retrieval. However, inconsistent machine-printed fonts and noises prevent text-image separation in historic patent document to be completely solved. In this paper, we apply Kernel Extreme Learning Machine (kernel-ELM) to improve the separation performance of this problem because it has mild optimization constraints and strong generalization performance with fast learning speed. The first stage of our method is to segment a document into several non-overlapping blocks by using segmentation based on 8-connectivity connected region. Then our method trains a four-class kernel-ELM classifier to distinguish paragraphs of text, caption, picture and noise. On our own historic patent document dataset, experimental results show that kernel-ELM outperforms SVM on both classification accuracy and learning speed.

Keywords: Text-image separation, Extreme Learning Machine, Historic patent document.

1 Introduction

Patent documents contain lots of knowledge and information, such as design method, design technology and so on. Fully mining and discovering this knowledge can improve the innovative capacity and efficiency of a designer [1]. Applying the same analogy to patents, it can be inferred that analyzing patents is essentially worthwhile to manage the complexities of searching and inter-relating patent information [2]. Patent document retrieval plays an important role in both business and academic works. A key challenge for patent document retrieval is the effective discovery and utilization of the contents

---

[*] Corresponding authors.

© Springer International Publishing Switzerland 2015
J. Cao et al. (eds.), *Proceedings of ELM-2014 Volume 2*,
Proceedings in Adaptation, Learning and Optimization 4, DOI: 10.1007/978-3-319-14066-7_29

stored in knowledge base [3]. In this paper, we attempt to get sematic pictures by establishing the relationship between picture and related caption based on the text-image separation and indexing method. Figure 1 shows the input document image and the result of text-image separation and indexing in our method.

The early American patent documents which may belong to the category of historic patent document are used in this paper. Historic patent documents have lots of noises, chaotic structure and inconsistent fonts for captions. Compared with the recent patent documents, we can see the noises and chaotic structure in historic patent document in Figure 2. In historic patent document, the main noises are marginal noises and stains; the structures are complex and various without unified model. Moreover, the fonts of captions are inconsistent in historic patent document. In order to display various fonts of captions, the collection of 'fig1' in our historic patent document dataset are shown in Figure 3.

Text-image classification is a pivotal problem in our method. In order to make sure the high accuracy of classification, we apply Kernel-ELM (Extreme Learning Machine with kernels) to improve the text-image classification performance. The first stage of our method is to segment a document into several non-overlapping blocks by segmentation based on 8-connectivity connected region. Then our method trains a four-class kernel-ELM to classify paragraphs of text, caption, image and noise. To my knowledge, simultaneously classifying different kinds of texts, images and noises in one classifier is rarely reported [4, 5]. Finally, according to the classification results and location information, we bind the images and its relevant caption to achieve indexing.

For the purpose of comparison, in this paper, we use ELM (ELM with random hidden nodes and random hidden neurons [6]), Kernel-ELM (ELM with kernels [7]), CELM (Constrained Extreme Learning Machine [8]) and SVM (Support Vector Machine) in text-image classification with other conditions are not change. Experimental results on our Patent document dataset show that our method outperforms SVM on classification accuracy and learning speed.

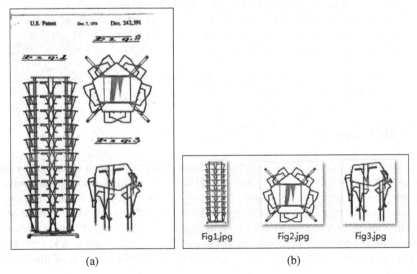

(a)   (b)

Fig. 1. (a) Input document image; (b) the result of (a) after text-image separation and indexing

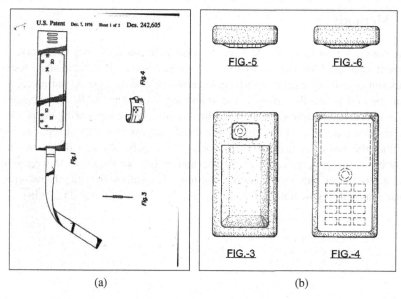

(a)   (b)

Fig. 2. (a) Historic patent document; (b) Recent patent document

Fig. 3. Various fonts of captions

The rest of the paper is organized as follows. Section 2 proposes the framework of our method, namely, the procedures of text-image separation and indexing method for historic patent document image. Then experiment and evaluates the performance of our method are illustrated in section 3. Conclusion is presented in the last section.

2   Methods

In this paper, we focus on the text-image separation and indexing in historic patent document images. Our task is to extract pictures in the historic patent document images and establish the relationship between blocks and the corresponding captions.

Our method primarily involves the following processes. a) Remove noises and other artifacts that are introduced during the scanning phase and during the production of the historic patent document. b) Divide the document into several separate blocks. c) Extract the basic features of each block. d) Classify blocks which contain four categories, namely, text, caption, picture and noise. e) Establish the relationship between blocks and the corresponding captions. We summarize the above processes as noise removal, segmentation, feature extraction, classification and indexing.

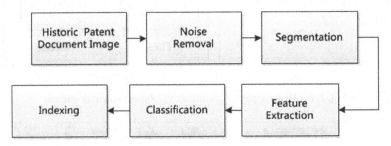

Fig. 4. Schematic diagram of our method

Figure 4 shows a schematic diagram of our method. We will discuss each of the modules in detail on the remainder of this chapter.

2.1   Document Image

A collection of scanned patent document images which are from America 1970s-1980s are used in this paper. These images are stored in JPG format with the size of 595 x 842 pixel. We have a total of 1064 historic patent document samples. Historic patent document usually has inconsistent format. In other words, font and position of captions have no strict criterion.

2.2   Noise Removal

The purpose of noise removal is to improve the quality of images. The quality of images will directly affect the performance of segmentation. Noise is a common problem in the scanned historic patent document processing. In this paper, two high frequency noises, viz., stains and marginal noises, were found by analyzing all of the samples. Figure 5 (a) shows the two noises in a historic patent document. In order to wipe off marginal noises, we first cut off the part of margin of a document (top bottom 30 pixels, left right 30 pixels). After the segmentation step, we remove the blocks which are adjacent to edge of the trimmed document image (as shown in Figure 5 (b)). And stains will be eliminated by classifier in the classification step.

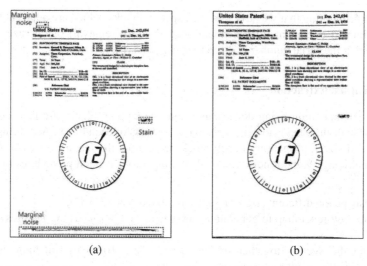

(a)   (b)

Fig. 5. (a) Original historic patent document with two high frequency noises; (b) Result of the removal operation for marginal noise

2.3   Segmentation

After cutting off the margin part, we segment the document into several non-overlapping blocks. We preprocess the document images before the segmentation. Preprocessing of document image has two tasks, colored document images (RGB) convert into gray-level image and gray-level images convert into binary image. Thresholding is utilized to accomplish the gray-level image to binary image task. Segmentation use a bounding box to encircle objects based on 8-connectivity connected region. Figure 6, shows the processing result of segmentation.

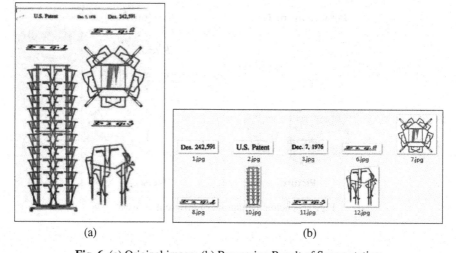

(a)   (b)

Fig. 6. (a) Original image; (b) Processing Result of Segmentation

It is worth noting that we have removed special blocks, which are adjacent to edge of the trimmed document images (mentioned in noise removal) or the size are too small, so that the serial number is discrete in Figure 6 (b).

2.4   Feature Extraction

With the segmentation being done, many non-overlapping blocks are then obtained. The purpose of this step is to extract the basic features of each block. According to the position and size information of each block, we extract the features of each block in the original image without noise removal. We have four general procedures to extract features:

1. Reshape the different size of images to a fixed-size (32x32);

2. Use grey-level map to generate a row vector (1 x 1024) as the features of each block;

3. Join the vectors together into a matrix (N x 1024) by stitching with the columns. N is the number of samples;

4. Utilize Z-score normalizing method or Max-Min value normalizing method to normalize the matrix we have obtained above.

2.5   Classification

The above matrix, each line represents the feature of a block, is used to train the classifier in this step. To utilize this trained classifier, we distinguish paragraphs of texts, captions and pictures and take out stains in historic patent document. We have four classes (shown in Figure 7) to distinguish the different blocks, namely, paragraphs of text, caption, picture and noise. According to the selection of classifier, we reserve the picture blocks and caption blocks for the final step.

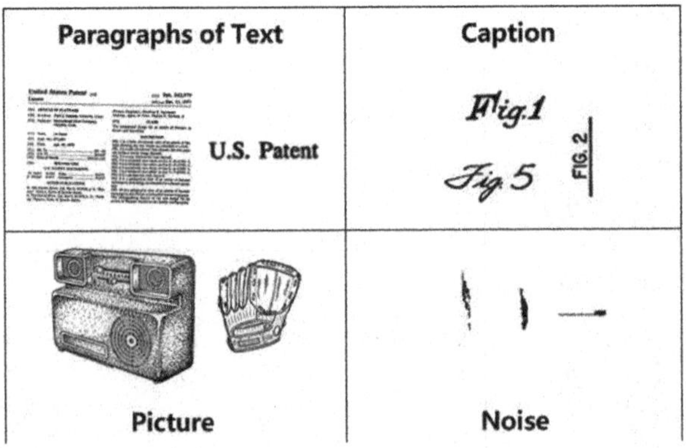

Fig. 7. Schematic diagram of different classes

2.6 Indexing

Since we have obtained labeled blocks, the finally step is to establish the relationship between picture blocks and corresponding caption blocks. In this step, first we modify the name of caption blocks according to the number it contained (we can get the number by a digital recognition method). Then we establish relationships by renaming picture blocks, that is to say, we replace name of a picture block by the name of corresponding caption block. We rename the picture block by using the name of the nearest caption block because the distance between the picture and its annotations are usually very close. At the end, as shown in Figure 1 (b), the marked pictures are stored with JPG format.

3 Experiment

In order to evaluate the performance of different classification algorithms for meaningful blocks classification in historic patent document, our patent document dataset was used in this paper. The experiment is carried out in Matlab environment. And the computer we used to run on the algorithms is Windows 7 32bit OS, Intel Core I5-2400CPU 3.10GHz with 4GB RAM memory.

Patent document dataset is consist of manually annotated blocks which we have obtained after the segmentation step in our method. Our Collection contains 582 samples, spanning four different classes, namely, caption (111 samples), picture (174 samples), paragraphs of text (217 samples) and noise (80 samples) from historic patent documents. Ten-fold cross validation is carried out on these 582 samples and the validation accuracy is produced. For the purpose of finding a best classification algorithm in our method, four different classification algorithms, namely ELM, CELM, kernel-ELM and SVM, were used on our dataset. We have carried out the simulation using the ELM and Kernel-ELM classifier form the toolbox http://www.ntu.edu.sg/home/egbhuang/elm_codes.html. And the code of CELM was given by Zhu et al. [8]. SVM classifier from the toolbox http://www.csie.ntu.edu.tw/~cjlin/libsvmtools/ is also used in this paper which is based on the LIBSVM algorithm http://www.csie.ntu.edu.tw/~cjlin/libsvm/. The parameters for each classification algorithm as following:

ELM: Number of hidden neurons 100, activation function: Sigmoid;

CELM: Number of hidden neurons 100, activation function: Sigmoid, C: 1;

Kernel-ELM: Regularization coefficient: 2^{11}, Kernel type: RBF Kernel, Kernel Para: 900;

SVM: Kernel type: RBF Kernel, Coefficient: 0, SVM type: c-SVC, others parameters are default.

Considering the randomness of ELM and CELM, we conduct 50 times iterations on the dataset and obtain their averaged results to assure the accuracy and objectivity of the experiment. Two kinds of normalization methods, namely Z-score and Max-Min value, are used to normalize data.

Table 1. Testing Rate (%) of Different Algorithms on Patent Document Dataset

	ELM	Kernel-ELM	CELM	SVM
Z-score	61.91	89.67	71.67	84.48
Max-Min	57.05	87.76	70.97	75.28

Table 2. Training Time (second) of Different Algorithms on Patent Document Dataset

	ELM	Kernel-ELM	CELM	SVM
Z-score	0.0157	0.0140	0.0176	0.7960
Max-Min	0.0176	0.0144	0.0175	0.7013

Classification Accuracy. As observed from Table 1, Kernel-ELM and SVM achieve a much higher classification accuracy than ELM and CELM. It can be noted that, for both two normalization methods, Kernel-ELM achieves the highest classification accuracy.

Training Time. As we can see in the Table 2, ELM, CELM and Kernel-ELM algorithms take much less training time than SVM. And Kernel-ELM achieves the fastest training speed, about 50 times faster than SVM, for both two normalization methods.

4   Conclusion and Discussion

In this paper, an efficient Text-image separation and classification framework for historic patent document image is presented. In order to distinguish and find pictures and captions in historic patent document images, we train a four-class classifier to classify paragraphs of text, caption, picture and noise in our method. For the purpose to find a fast and efficient classification algorithm to solve the classification problem, ELM, CELM, Kernel-ELM and SVM are used in this paper. And the experiment result on our own dataset shows that Kernel-ELM achieves a better performance than ELM, CELM and SVM both on classification accuracy and training speed.

Acknowledgement. This research is partially sponsored by NTFYPSTS No. 2013BAK02B03, the National Basic Research Program of China under contract No. 2014CB349303, Natural Science Foundation of China (No. 61175115, No. 61370113 and No. 61003105), the Importation and Development of High-Caliber Talents Project of Beijing Municipal Institutions (CIT&TCD201304035), Jing-Hua Talents Project of Beijing University of Technology (2014-JH-L06), and the International Communication Ability Development Plan for Young Teachers of Beijing University of Technology (No. 2014-16).

References

1. Xu, Y.H.: Apply text mining in analysis of patent document. In: Computer-Aided Industrial Design & Conceptual Design, CAID & CD 2009 (2009)
2. Bonino, D., Ciaramella, A., Corno, F.: Review of the state-of-the-art in patent information and forthcoming evolutions in intelligent patent informatics. World Pat. Inf. 32(1), 30–38 (2010)
3. Abbas, A., Zhang, L.M., Khan, S.: A literature review on the state-of-the-art in patent analysis. World Patent Information 37, 3–13 (2014)
4. Bukhari, S.S., Shafait, F., Breuel, T.: Improved Document Image Segmentation Algorithm using Multiresolution Morphology. In: DRR 2011, pp. 1–10 (2011)
5. Mollah, A.F., Basu, S., Nasipuri, M., Basu, D.K.: Text/Graphics Separation for Business Card Images for Mobile Devices. In: IAPR International Workshop on Graphics Recognition, pp. 263–270 (2009)
6. Huang, G.B., Zhu, Q.Y., Siew, C.: Extreme learning machine: a new learning scheme of feed forward neural networks. In: Proceedings of IEEE International Joint Conference on Neural Networks, vol. 2, pp. 985–990 (2004)
7. Huang, G.B., Zhou, H., Ding, X., Zhang, R.: Extreme Learning Machine for Regression and Multiclass Classification. IEEE Transactions on Systems, Man, and Cybernetics—Part B: Cybernetics 42(2) (April 2012)
8. Zhu, W.T., Miao, J., Qing, L.Y.: Constrained Extreme Learning Machine: a Novel Highly Discriminative Random Feedforward Neural Network. In: Proc. of IJCNN 2014 (2014)

Anomaly Detection with ELM-Based Visual Attribute and Spatio-temporal Pyramid

Tan Xiao[1,2], Chao Zhang[1], Hongbin Zha[1], and Fangyun Wei[1]

[1] Key Laboratory of Machine Perception, Peking University, Beijing, P.R. China
[2] CRSC Communication & Information Corporation, Beijing, P.R. China
pkuxiaotan@pku.edu.cn, {chzhang,zha}@cis.pku.edu.cn, wei4943005270163.com

Abstract. In this paper, a novel approach for automatic anomaly detection in surveillance video is proposed. It is highly efficient for real-time detection. It can also handle multi-scale detection and can cope with both spatial and temporal anomalies. Specifically, features capturing both appearance and motion characteristic are extracted from densely sampled spatio-temporal video volume (STV). And to bridge the semantic gap between low-level visual feature and high-level event, we use the middle-level visual attributes as the intermediary. These three-level framework is modeled as an Extreme Learning Machine (ELM) which can effectively and efficiently tell whether a STV belongs to an anomalous event. We also use the Spatio-temporal Pyramid (STP) to capture the spatial and temporal continuity of an anomalous event , enabling our approach to cope with multi-scale and complicated events. Experiments on several datasets are carried out and the superior performance compared to state-of-the-art approaches verifies the effectiveness of our approach.

Keywords: Automatic anomaly detection, Visual attributes, Extreme Learning Machine, Spatio-temporal Pyrimid.

1 Introduction

Surveillance system are widely used and detecting anomalous event plays an important role in real world. But current systems are burdensome for operators as they are required to watch screens showing the content captured by cameras. Detecting unusual individuals and events, a.k.a., anomalies, is the most important and tasks of human operators. The performance of anomaly detection is dependent on the human operators. But the task is becoming more difficult and tiring for human operators and the performance of them can degrade significantly. Fortunately, with the development of video analytics techniques, automatic anomaly detection approaches has attracted considerable attention in recent years.

Specifically, anomaly detection is defined as discovering events with a low probability of occurrence. Several approaches have been proposed and generally they can be summarized into three categories, i.e., supervised approaches [1], semi-supervised approaches [2], and unsupervised approaches [3,4,5,6]. In real-world scenario, anomalies are usually quite rare, are different between each other, and have unpredictable variations. Thus unsupervised approaches are more. And as anomalous events are indeed difficult to defined in advance, only unsupervised approaches are practical in real-world applications.

Unsupervised approaches can be divided into several subcategories. Trajectories based approaches [7] focus on the spatial location of objects and their motions are tracked. But they can only capture the abnormal track because they only consider the spatial deviations. In addition, because precise segmentation targets is almost impossible to in crowd scene, these approaches can't be applied to this scenario. And to model motion of objects, optical flow has been widely utilized [3,4]. But these approaches have very unstable performance in crowded scenes [8]. And they can only detect anomalous motion while ignoring anomalous appearance. Recently, densely sampled local spatio-temporal descriptor capturing both motion and appearance characteristic are utilized in [6], and they can possesses some degree of robustness to unimportant variations in surveillance video. Then they construct models to capture the relationship between low-level visual features and high-level semantic event. Though promising results are achieved, they ignore the semantic gap between features and event thus their performance is still unsatisfactory enough.

In this paper we propose a novel automatic anomaly detection approach with Extreme Learning Machine [9] (ELM) based visual attribute and Spatio-temporal Pyramid (STP). Specifically, spatio-temporal video volumes (STV) with pixel-by-pixel analysis which are densely sampled from surveillance video lay the foundation of our approach. Then spatio-temporal descriptor capturing both motion and appearance characteristic is extracted for each STV, and each STV is further represented by bag-of-video-words HOG feature. Then to bridge the semantic gap between low-level visual features and high-level event, we propose to use visual attributes [10] as the intermediary. This three-level (feature-attribute-event) framework can be modeled by ELM. The output of the ELM can be utilized to tell whether the STV belongs to an anomaly. And since the ELM can be constructed and updated efficiently, the model can be updated continuously, thus no offline or supervised pretraining are required. So our approach can detect anomaly which even hasn't been observed before. And the efficient update procedure also enables our approach to cope with the scene change both in motion and . Furthermore, to detect multi-scale and complicated anomalies, Spatio-temporal Pyramid, as the temporal extension of spatial pyramid [11] is proposed. Thus multi-scale event can be detected by STP. Moreover, since an event is always related to several STVs which may have different location or time or both, complicated events can be detected by discovering the relationship. STP can achieve this task, enabling our approach to detect complicated events.

2   Related Work

As has mentioned above, one of the most widely used techniques in previous is trajectory analysis. But precise tracking methods are always required. However, tracking objects is computationally expensive, especially in crowded scene. And some previous works also utilize optical flow [3,4], but their performance in crowded scenes is quite unreliable [6].

Recently, the focus of anomaly detection is on local spatio-temporal features. Several approaches have been proposed and received increasingly attention [12]. Typically, these approaches focus on pixel level and describe the local characteristic of video by low-level visual features. Then they can construct pixel-level background model and behavior template based on the local feature [13]. In addition, approaches utilizing spatio-temporal video volumes in the context of bag-of-video-words have achieved promising results [4,6]. For example, probabilistic models such as Latent Dirichlet Allocation (LDA) [14] can be straightforwardly applied to video analysis if we ignore the spatial and temporal relationship of local features [15]. But intuitively, the spatial and temporal relationship between STVs are quite essential for scene understanding and event detection [16]. Thus the efforts to incorporate either spatial or temporal compositions of STVs into the conventional probabilistic model have been made. But they can't handle online and real-time detection because they are highly time-consuming [17]. In addition, several approaches [12,18] try to construct models based on the spatio-temporal behavior and analysis the spatio-temporal pattern of each pixel as a function of time to detect low-level local anomalous events. However, they ignore the relationship between each pixel in space and time because they just process each pixel independently, which may lead to too local detection.

A multi-scale and non-parametric approach is proposed in [6]. Dense and local spatio-temporal features which can capture both motion and appearance characteristics of objects are extracted at each individual pixel. And to utilize the spatial and temporal relationship between pixels, "overlapping" features are used in their approach. They can achieve promising results but their approach indeed face the challenge of efficiency when performing accurate multi-scale anomaly detection. Actually, our Spatio-temporal Pyramid is partially motivated by by their overlapping features because they both consider the spatial and temporal relationship between features. But compared to ehum, our STP can be constructed with more efficiently and much better performance is achieved. Furthermore, our STP can cope with multi-scale detection naturally while their approach actually treats different scales independently.

Moreover, approaches mentioned above all construct models between low-level visual features and high-level events straightforwardly while ignoring the semantic gap between them which is quite important for event detection. Thus they can't achieve satisfactory results. This problem motivates us to use visual attribute as intermediary to bridge the semantic gap.

3   Detection via ELM-Based Visual Attribute

3.1   Spatio-temporal Local Features

First, we need to extract local features to capture the motion and appearance characteristics of densely sampled STVs at each pixel. Consider a pixel (x, y, t), we can construct a STV $v \in \mathbb{R}^{n_x \times n_y \times n_t}$ with the size $n_x \times n_y \times n_t$ centered at (x, y, t). Then we calculate the histogram of the spatio-temporal gradient (HOG) of the video in polar coordinates to describe the STV. We can denote the spatial gradients as $G_x(x, y, t)$, $G_y(x, y, t)$, and the temporal gradient as $G_t(x, y, t)$ respectively. To alleviate the influence of noise in videos and texture and contrast, we first normalize the spatial gradient as follows,

$$G_s(x, y, t) = \frac{\sqrt{G_x^2(x, y, t) + G_y^2(x, y, t)}}{\sum_{x', y', t' \in v} \sqrt{G_x^2(x', y', t') + G_y^2(x', y', t') + \epsilon}} \tag{1}$$

where $G_s(x, y, t)$ is the normalized spatial gradient and ϵ is a small constant to avoid numeric instabilities. Based on the normalized spatial gradient, we can construct 3D normalized gradient represented in polar coordinates as follows,

$$M_{3D}(x, y, t) = \sqrt{G_s^2(x, y, t) + G_t^2(x, y, t)} \tag{2}$$

$$\theta(x, y, t) = \tan^{-1}(\frac{G_y(x, y, t)}{G_x(x, y, t)}), \phi(x, y, t) = \tan^{-1}(\frac{G_t(x, y, t)}{G_s(x, y, t)}) \tag{3}$$

where $M_{3D}(x, y, t)$ is the magnitude of 3D normalized gradient, and $\phi(x, y, t) \in [-\frac{\pi}{2}, \frac{\pi}{2}]$ and $\theta(x, y, t) \in [-\pi, \pi]$ are the orientations of the gradient respectively. Then we can construct the histogram of oriented gradients (HOG features) for a given STV v. First, for each pixel in the give STV v, we can extract 3D normalized gradient features. Then the feature is quantized into $n_\phi + n_\theta$ bins based on their gradient orientations. If we look back to the feature extraction and construction procedure, we can observe that the local characteristics of both motion and appearance can be captured by the HOG features. Consequently, both anomalous actions and objects can be detected based on the HOG features. Moreover, because of the normalization step, it shows robustness to data noise and unimportant variations in the video such as texture and contrast. Actually, this HOG feature can be used as the input of the Extreme Learning Machine.

3.2   Constructing Normal Events

As mentioned in Section 1, our approach is unsupervised. Thus we need to define normal and anomalous events automatically. Generally, STVs belonging to normal events may form clusters in the feature space while STVs of anomalous events are outliers in the space. Thus we can construct normal events by using this property. Consequently, we can regard the clusters as the normal events and the judgement of anomaly can be given by analyzing how close a STV is related to a cluster.

In this paper, we propose to construct normal events automatically from video data. Given a set of spatio-temporal features $\mathbf{H} = [h_1, ..., h_n] \in \mathbb{R}^{d \times n}$, where n is the size of feature set. Actually we don't need a training set because the initial feature set $\mathbf{H}$ can be constructed by using the first few seconds of the video. In addition, to guarantee that our selection algorithm is computationally

Algorithm 1. Normal Events Construction

Input: H,$\lambda = 1$, $\mathbf{S}_0$ K, c
Output: S;
 1: Initialize $\mathbf{Z}_0 = \mathbf{S}_0, a_0 = 1$
 2: **for** $k = 0, 1, 2, ..., K$ **do**
 3:    $\mathbf{S}_{k+1} = \arg\min_{\mathbf{S}} : p_{\mathbf{Z}_k, L}(\mathbf{S}) = D_{\frac{\lambda}{L}}(\mathbf{Z}_k - \frac{1}{L}\nabla f(\mathbf{Z}_k))$
 4:    **while** $f_0(\mathbf{S}_{k+1}) > p_{\mathbf{Z}_k, L}(\mathbf{S}_{k+1})$ **do**
 5:      $L = L/c$
 6:      $\mathbf{S}_{k+1} = \arg\min_{\mathbf{S}} : p_{\mathbf{Z}_k, L}(\mathbf{S}) = D_{\frac{\lambda}{L}}(\mathbf{Z}_k - \frac{1}{L}\nabla f(\mathbf{Z}_k))$
 7:    **end while**
 8:    $a_{k+1} = (1 + \sqrt{1 + 4a_k^2})/2$
 9:    $\mathbf{Z}_{k+1} = (\frac{a_{k+1} + a_k - 1}{a_{k+1}})\mathbf{S}_{k+1} - (\frac{a_k - 1}{a_{k+1}})\mathbf{S}_k$
10: **end for**

---

feasible, we can also randomly select some features to reduce n. Then we need to select some features from **H** as the representatives of clusters, i.e., the normal events. In our method, the number of clusters is determined automatically by the algorithm, which is self-adaptive to the test data. Following the idea in [5], we'd like to select an optimal subset of **H** , such that we can well reconstructed the rest of features from them. This criterion can be formulated as follows,

$$min_{\mathbf{S}} = \frac{1}{2}\|\mathbf{H} - \mathbf{HS}\|_F^2 + \lambda\|\mathbf{S}\|_{2,1} \tag{4}$$

where $\mathbf{S} \in \mathbb{R}^{n \times n}$ is the selection matrix, $\|\mathbf{S}\|_F$ is the Frobenius norm of matrix $\mathbf{S}$, $\|\mathbf{S}\|_{2,1} = \sum_{i=1}^n \|\mathbf{S}_{i.}\|_2$ is the $L_{2,1}$-norm of matrix, and λ is the model parameter. Finally, the selection can be done by selecting index i which satisfies $\|\mathbf{S}_{i.}\| > 0$. Consequently we can obtain clusters representing the normal events.

To solve this problem, first we can construct a function $p_{\mathbf{Z}, L}(\mathbf{S})$ as

$$p_{\mathbf{Z}, L}(\mathbf{S}) = f(\mathbf{Z}) + \langle \nabla f(\mathbf{Z}), \mathbf{S} - \mathbf{Z}\rangle + \frac{L}{2}\|\mathbf{S} - \mathbf{Z}\|_F^2 + g(\mathbf{Z}) \tag{5}$$

And we can define another function $D_\tau(.) : \mathbf{M} \in \mathbb{R}^{n \times n} \mapsto \mathbf{N} \in \mathbb{R}^{n \times n}$

$$\mathbf{N}_{i.} = \begin{cases} 0, & \|\mathbf{M}_{i.}\| \leq \tau \\ (1 - \tau/\|\mathbf{M}_{i.}\|)\mathbf{M}_{i.}, & \text{otherwise} \end{cases} \tag{6}$$

Based on the two functions above, we can obtain the selection matrix. We don't show the detail of derivation here because of the limit of space. The whole algorithm is summarized in Algorithm 1.

3.3   Detection via ELM-Based Visual Attributes

Intuitively, one simple way to tell whether a STV represented by low-level visual feature belongs to a anomaly is to consider the relationship between the low-level feature and all high-level normal events. Thus we just need to define a function or model to measure the relationship. However, directly constructing model between features and events may suffer from the semantic gap between them. Recent research on visual attribute [10] has pointed out this problem and proposed to use visual attribute to bridge the semantic gap.

Theoretically, we can apply any models to both steps. And we find that linear model can always lead to satisfactory performance while guaranteing the efficiency. Given a set of STVs and the corresponding low-level visual features $\mathbf{H} = [h_1, ..., h_n] \in \mathbb{R}^{d \times n}$, and the corresponding event labels $\mathbf{E} = [e_1, ..., e_n] \in \{-1, 1\}^{m \times n}$, where $e_{ij} = 1$ if the j-th STVs belongs to the i-th event and $e_{ij} = -1$ otherwise. Here we utilize a middle-level $\mathbf{A} = [a_1, ..., a_n] \in \mathbb{R}^{k \times n}$ to bridge the semantic gap, where k is the number of visual attributes which we set $k = 256$ in this. Actually, a_j can be regarded as the visual attribute for the i-th STV. Thus we can construct two linear models as follows,

$$a_{ij} = g(w_i \cdot h_j + b_i), \ i = 1, ..., k \tag{7}$$

$$e_{ij} = \sum_{t=1}^{k} \beta_{it} \cdot a_{tj}, \ , i = 1, ..., m \tag{8}$$

where a_{ij} is the i-th attribute for the j-th STV, e_{ij} is the relation degree between the i-th event and the j-th STV which is used to tell the STV belongs to an anomaly or not. Now we just need to specify two model parameters, i.e., $\mathbf{W} = [w_1, ..., w_k] \in \mathbb{R}^{d \times k}$ and $\mathbf{B} = [b_1, ..., b_k]^T \in \mathbb{R}^k$ for the model between low-level visual feature and middle-level visual attribute, and $\beta = [\beta_1, ..., \beta_k] \in \mathbb{R}^{m \times k}$ for the model between middle-level visual attribute and high-level event. And $g(\cdot)$ is the activation function which is infinitely differentiable. In this paper we use sigmoidal function as the activation function.

In fact, learning two models simultaneously is quite difficult and inefficient because we need to adjust parameters in two models iteratively, such that it can't be applied to online and real-time applications. But fortunately, this three-level (feature-attribute-event) framework can be formulated as an Extreme Learning Machine (ELM) [9] which is a variant of artificial neural network (ANN), where feature is the input layer, attribute is the hidden layer and event is the output layer. In the theory of ELM, the parameters between input layer and hidden layer can be totally random, i.e., we actually don't need to learn these parameters from the training data. Thus we just need to compute the parameters between the hidden layer and the output layer, which is extremely.

Now we can randomize the parameters $\mathbf{W}$ and $\mathbf{B}$. Because we will not change $\mathbf{W}$ and $\mathbf{B}$, we can compute the visual attributes for the training data as follows,

$$\mathbf{A} = g(\mathbf{W}^T \mathbf{H} + \mathbf{B} \mathbf{1}_k^T) \tag{9}$$

where $\mathbf{1}_k = [1, ..., 1]^T$. Then we just need to compute β by minimizing

$$O = \|\beta \mathbf{A} - \mathbf{E}\|_F^2 \tag{10}$$

and we can get the solution for β as

$$\hat{\beta} = \mathbf{E}\mathbf{A}^\dagger \tag{11}$$

where $\mathbf{A}^\dagger$ is the Moore-Penros generalized inverse of matrix $\mathbf{A}$ [19]. Then given any STV represented by $h^* \in \mathbb{R}^d$, we can compute its relationship with any event as $e^* \in \mathbb{R}^m$ via the middle-level visual attribute by the ELM as follows,

$$e^* = \hat{\beta}g(\mathbf{W}^T h + \mathbf{B}) \tag{12}$$

As discussed above, a normal STV is always close to a cluster while an abnormal STV is always outliers. Thus if it belongs to an normal event, it may have strong relationship with an event, leading to $e_i^* \approx 1$ for some j, while no the elements in e^* may have large value if it's an outlier and has quite weak relationship with all events. So we can define the degree of anomaly as

$$d_{anomaly} = 1 - max_i(e_i^*) \tag{13}$$

Now we can select an threshold δ such that a STV who satisfies $d_{anomaly} > \delta$ is judged as abnormal. We can define the anomaly probability p_a, empirically from $10^{-2}, 10^{-3}, 10^{-4}$ and 10^{-5}. Then we can compute $d_{anomaly}$ for all STVs in the first one or two seconds in a test video, and set δ such that the ratio of STVs whose $d_{anomaly}$ are smaller than δ is about p_a. So about p_a STVs will be treated as anomalies. We have a postprocessing step on the initial judgement to obtain better results, which will be introduced in detail in Section 4.

4   Spatio-temporal Pyramid

Fortunately, We can observe that an anomalous event is continuous in space and time, i.e., it's always related to different parts in the camera and may last for a period of time. Thus, taking the spatial and temporal relationship of STVs into consideration can significantly promote the detection performance. In this paper, we propose to use Spatio-temporal Pyramid (STP) to capture the relationship. We can use any levels based on the specific situation. Here we just use two-level STP but we find that satisfactory result can be achieved under this setting.

There is an interesting and important phenomenon, which is also essential for achieving satisfactory performance. The judgement on upper level STV tends to have high precision but low recall, implying that our approach can highly confidently claim that a upper level STV is anomalous but some anomalous STV may be missed. This is reasonable because the upper level STV can capture the global information of an event where the spatial and temporal continuity of an event can be adequately taken into consideration while some important local details will be ignored. On the other hand, the judgement on the lower level STV usually has low precision but high recall because it's too sensitive to local details and noise and ignores the spatial and temporal relationship between STVs, but it can capture more local information than upper level STV. Thus we propose the

Spatio-temporal Pyramid to combine these two levels's STVs to take advantage of both local details and global information simultaneously as follows.

On one hand, if it's judged to be an normal STV for an upper one, it still has marked probabilistic to be anomalous. Thus the following results should be take into consideration too. 1) its six neighbors and 2) its lower STVs. In this paper, an upper STV is finally judged to be anomalous if all of the following three criteria are satisfied, 1) it's judged to be anomalous, 2) at least three of its neighbors are anomalous, and 3) at least five of its lower level STVs are anomalous. Actually, the high-precision result for upper level STV leads to the first criterion. The second criterion is based on the spatial and temporal continuity of events. The third criterion is based on a voting scheme, because it's reasonable to assume that though one lower STV may be influenced by noise or local details, it's difficult for most STVs to generate wrong judgement.

On the other hand, the spatial and temporal continuity of events can't be captured by lower level STVs, consequently the judgement of a STV should be incorporated with its upper level STV and neighbors. So a lower level STV is considered to be anomalous if it's judged to be anomalous and 1) two of more of its neighbors are anomalous or 2) its upper level STV is anomalous.

Based on the Spatio-temporal Pyramid and criteria above, we take into consideration the spatial and temporal continuity of events, the relationship between STVs in space and time, and the local details simultaneously which can significantly promote the performance. Furthermore, the Spatio-temporal Pyramid allow us to perform multi-scale detection.

5 Experiment

To verify the effectiveness of the proposed approach, we test it in the following two publicly available datasets for anomaly detection: anomaly behavior detection dataset [20][1] and UCSD pedestrian dataset [21][2]. The evaluation and comparison of different approaches are presented in two kinds of performance curves, i.e., precision-recall, ROC curves, and Equal Error Rate (EER) at both frame level and pixel level is also reported. We use a two-level pyramid, and the size of lower level STV is $10 \times 10 \times 10$. To extract HOG features, we set $n_\phi = 8$ and $n_\theta = 16$. We set the number of visual attributes, i.e., the number of hidden units in Extreme Learning Machine, to 256, and the anomaly probabilistic $p_a = 10^{-3}$. In fact, our method doesn't require any training data because it's totally unsupervised. It just use the first one or two seconds of a test video to construct the initial normal events set. Furthermore, we set that the Extreme Learning Machine are updated every one second. Furthermore, the following several state-of-the-art approaches for anomaly detection are compared to our approach: Optical Flow [3], MDT [21], Sparse Reconstruction (Cong et al.) [5], spatio-temporal oriented energies [20], Reddy et al. [22] and Bertini et al. [6].

---

[1] http://www.cse.yorku.ca/vision/research/

[2] http://www.svcl.ucsd.edu/projects/anomaly

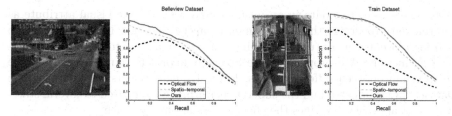

Fig. 1. Experiments on Belleview and Train Dataset

Table 1. Comparison of the proposed approach and the state-of-the-art for anomaly detection using Ped datasets. Approaches with * can perform real-time detection.

	Ped1		Ped2	
	EER (frame)	EER (pixel)	EER (frame)	EER (pixel)
Optical Flow* [3]	38%	76%	42%	80%
MDT [21]	25%	58%	25%	55%
Cong *et al.* [5]	19%	-	20%	-
Reddy *et al.** [22]	22.5%	32%	21%	31%
Bertini *et al.** [6]	31%	70%	30%	68%
Ours*	**17%**	**28%**	**16%**	**26%**

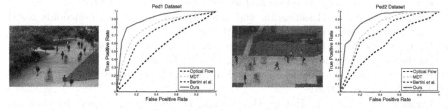

Fig. 2. Experiments on Ped Datasets

The first dataset is *Belleview*. Cars running from top to bottom is normal event, while cars entering or exiting from the intersection from left or right is the anomalous event. The second is *Train* where anomalies are moving people. The results on two datasets above are shown in Fig. 1.

In the UCSD datasets (Ped1 and Ped2),non-pedestrian entities (e.g., cyclist, skaters, small carts) and pedestrians moving in anomalous motion are considered as anomalies. The anomalous regions detected by our approach and the ROC curves of other approaches for Ped1 and Ped2 datasets are shown in Fig. 2. And the Equal Error Rate (EER) for both frame level and pixel level detection of different approaches is shown in Table 1.

6 Conclusion

In this paper, a novel automatic anomaly detection approach is proposed. Densely sampled Spatio-temporal video volumes represented by spatio-temporal local features are the fundamental of our approach. Normal event set is efficiently constructed from test data in an unsupervised way. We use visual attribute as intermediary to bridge the large semantic gap between low-level visual feature and high-level event. Extreme Learning Machine is utilized to model this three-level framework. We propose to use Spatio-temporal Pyramid to capture the relationship between different STVs. Extensive experiments on several public datasets are conducted and the superior performance compared to several state-of-the-art approaches verifies the effectiveness of our approach.

References

1. Liu, C., Wang, G., Ning, W., Lin, X., Li, L., Liu, Z.: Anomaly detection in surveillance video using motion direction statistics. In: ICIP (2010)
2. Sillito, R., Fisher, R.: Semi-supervised learning for anomalous trajectory detection. In: BMVC (2008)
3. Adam, A., Rivlin, E., Shimshoni, I., Reinitz, D.: Robust real-time unusual event detection using multiple fixed-location monitors. TPAMI (2008)
4. Kim, J., Grauman, K.: Observe locally, infer globally: a space-time mrf for detecting abnormal activities with incremental updates. In: CVPR (2009)
5. Cong, Y., Yuan, J., Liu, J.: Sparse reconstruction cost for abnormal event detection. In: CVPR (2011)
6. Bertini, M., Bimbo, A.D., Seidenari, L.: Multi-scale and realtime non-parametric approach for anomaly detection and localization. In: CVIU (2012)
7. Khalid, S.: Activity classification and anomaly detection using m-medoids based modelling of motion patterns. Pattern Recogn. (2010)
8. Kratz, L., Nishino, K.: Anomaly detection in extremely crowded scenes using spatio-temporal motion pattern models. In: CVPR (2009)
9. Huang, G., Zhu, Q.: C.Siew: Extreme learning machine: theory and applications. Neurocomputing (2006)
10. Farhadi, A., Endres, I., Hoiem, D., Forsyth, D.: Describing objects by their attributes. In: CVPR (2009)
11. Lazebnik, S., Schmid, C., Ponce, J.: Beyond bags of features: Spatial pyramid matching for recognizing natural scene categories. In: CVPR (2006)
12. Benezeth, Y., Jodoin, P.M., Saligrama, V.: Abnormality detection using low-level co-occurring events. Pattern Recogn. Lett. (2011)
13. Benezeth, Y., Jodoin, P.M., Saligrama, V., Rosenberger, C.: Abnormal events detection based on spatio-temporal co-occurences. In: CVPR (2009)
14. Blei, D.: NG, A., Jordan, M.: Latent dirichlet allocation. Journal of Machine Learning Research (2003)
15. Hospedales, T.M., Jian, L., Shaogang, G., Tao, X.: Identifying rare and subtle behaviors: A weakly supervised joint topic model. IEEE Trans. Pattern Anal. Mach. Intell. (2011)
16. Roshtkhari, M.J., Levine, M.D.: A multi-scale hierarchical codebook method for human action recognition in videos using a single example. In: Conf. Computer and Robot Vision (2012)
17. Hospedales, T.M., Jian, L., Shaogang, G., Tao, X.: Identifying rare and subtle behaviors: A weakly supervised joint topic model. IEEE Trans. Pattern Anal. Mach. Intell. (2012)
18. Jodoin, P., Saligrama, V., Konrad, J.: Behavior subtraction. IEEE Trans. Image Process., TIP (2012)

19. Rao, C.R., Mitra, S.K.: Generalized inverse of matrices and its applications. Wiley (1971)
20. Zaharescu, A., Wildes, R.: Anomalous behaviour detection using spatiotemporal oriented energies, subset inclusion histogram comparison and event-driven processing. In: Daniilidis, K., Maragos, P., Paragios, N. (eds.) ECCV 2010, Part I. LNCS, vol. 6311, pp. 563–576. Springer, Heidelberg (2010)
21. Mahadevan, V., Li, W., Bhalodia, V., Vasconcelos, N.: Anomaly detection in crowded scenes. In: CVPR (2010)
22. Reddy, V., Sanderson, C., Lovell, B.C.: Improved anomaly detection in crowded scenes via cell-based analysis of foreground speed, size and texture. In: CVPR Workshops (2011)

Modelling and Prediction of Surface Roughness and Power Consumption Using Parallel Extreme Learning Machine Based Particle Swarm Optimization

Nooraziah Ahmad[1] and Tiagrajah V. Janahiraman[2]

[1] Department of Creative Technology, Faculty of Creative Technology and Heritage, Universiti Malaysia Kelantan, Bachok, Kelantan, Malaysia
nooraziah@umk.edu.my
[2] Centre for Signal Processing and Control Systems, Dept of Electronics and Communication Engineering, College of Engineering, Universiti Tenaga Nasional, Jalan IKRAM-UNITEN,43000 Kajang, Selangor, Malaysia
tiagrajah@uniten.edu.my

Abstract. Prediction model allows the machinist to determine the values of the cutting performance before machining. Modelling using improved extreme learning machine based particle swarm optimization, IPSO-ELM has less parameters to adjust and also takes real number as particles while decreasing the norm of output weights and constraining the input weight and hidden biases within a reasonable range to improve the ELM performance. In order to solve the multi objectives modelling problem, we have proposed a parallel IPSO-ELM. In this research work, the best input weights and hidden biases for different performance were identified. The proposed method was able to model the training and the testing set with minimal error. The predicted result from the designed model was able to match the experimental data very closely.

Keywords: Modelling, Extreme Learning Machine, Particle Swarm Optimization, power consumption, surface roughness.

1   Introduction

For the last few decades, modelling cutting performances in manufacturing has become the most popular issue either to investigate or to improve the current issue. Most machinists, traditionally, used either their knowledge on the operation or manual handbook to estimate the optimal parameters of a turning operation. However, those steps are only trial and error process which needs to be checked and clarified.

Prediction model allows the machinists to determine the values of the cutting performance before machining. Thus, this allows the machining process to become more productive, competitive, machine error reduction and technical specification satisfaction [1]. The artificial intelligent (AI) has been introduced and developed in modelling the parameters to ease and simplify the manufacturing process. AI is applied in the modelling of the manufacturing operation problem through the development of Artificial Neural Network (ANN), Genetic Algorithm (GA), Fuzzy Logic and Expert System. ANN [1] is claimed to have the capability of overcoming incomplete data and produce accurate results while Support Vector Machine (SVM) is then introduced as an alternative to conventional ANN and produces less error in the prediction model [2, 3].

There are a few common problem that involves in conventional modelling techniques which include time consuming, the generalization capabilities and also an overfitting problem. Generalization is defined as the ability to train using one set of data and then, successfully classify the independent test set [5]. For ANN model, it needs to train the model to continuously obtain the best generalization capability. However, this will decrease the accuracy of the test after a certain point which will contribute to overfitting since only individual training samples are available rather than the true underlying relationship. Some researchers have even suggested some additional techniques such as stopped training/early stopping, in order to deal with the overfitting [6].

The researchers have come up with several new methods/hybrid methods in modelling as a solution to these problems. Extreme learning machine (ELM), introduced by Huang et al., proposed an algorithm for the hidden nodes determination and weight selection [4]. Unlike conventional modelling techniques, ELM simplified the ANN methods which tend to cut short of the train-test time while given the best in generalization and it automatically deals with the overfitting. However, ELM tends to require more hidden neurons than traditional gradient-based learning algorithms which results in ill-condition problem due to random selection of input weights and hidden biases [7]. Han et al. suggested an improved particle swarm optimization called IPSO-ELM which basically for single objective problem. Particle swarm optimization (PSO) is introduced as a stochastic search through an n-dimensional problem space aiming to minimize or maximize the objective function of the problem. The advantage of IPSO-ELM is that it has less parameter to adjust and also takes real number as particles while decreasing the norm of output weights and constraining the input weight and hidden biases within a reasonable range to improve the ELM performance.

Therefore, in this paper, IPSO-ELM is adopted and used in modelling the multi objectives performance function. The rest of the paper is structured as follows: in Section 2, the preliminaries of the proposed method is introduced. Next, in Section 3, the experimental design is presented while, in Section 4, the prediction results were analyzed. The confirmation test is explained and analyzed in Section 5. Finally, the concluding remarks are offered in the last section, Section 6.

2   Parallel Improved Particle Swarm Optimization-Extreme Learning Machine

In the real industry experience, most of the processes compromise multiple objectives [8]. In turning operation, surface roughness is the most important performance, which refers to the quality of part machined. The power consumption also has been a vital performance because it is directly proportional in measuring the energy in the manufacturing operation. Energy generation which is driven by the consumption demand is a key contributor to climate change and carbon dioxide emissions [9]. Therefore, in order to reduce energy generated, the manufacturer must minimize the power consumed during machining.

In this study, the objective of this paper is to model the surface roughness together with power consumption using proposed extreme learning machine based on particle swarm optimization. We propose a parallel implementation of IPSO-ELM [7] to find the best input weights and hidden biases for multi objective modeling. Fig. 1 shows the structure of parallel IPSO-ELM. The proposed method combines two performance models using basic ELM with selected input weight and hidden biases from PSO in one algorithm.

A brief explanation on the improved extreme learning machine based on particle swarm optimization proposed by Han et al. is represented in Subsection 2.1. Our proposed method is then explained in Subsection 2.2.

2.1 Improved Extreme Learning Machine (IPSO-ELM)

In IPSO-ELM, the particles are the input weights and biases which are randomly initialized. The fitness value is calculated based on Extreme learning machine (ELM). ELM is based on single hidden layer feed forward neural network (SLFN). It contains an input layer, a hidden layer and an output layer. Consider N data samples $\{(x_i, t_i)\},...,(x_N, t_N)\}$ where x_i is an element of R^n and t_i is an element of R^m. The output function for generalized SLFN with L hidden nodes is given by

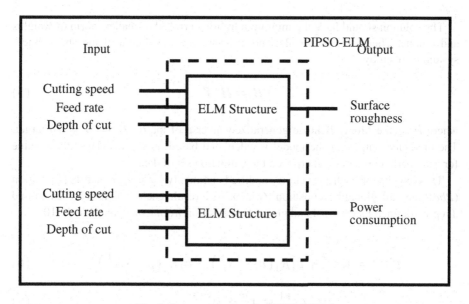

Fig. 1. Proposed structure of PIPSO-ELM

$$f(x) = \sum_{i=1}^{L} \beta_i H = \sum_{i=1}^{L} \beta_i g(w_i x_j + b_i), j = 1,.., N \qquad (1)$$

By considering that there is no error in solving the output function of the SLFN, the Eq. (1) can be written compactly as

$$f(x) = H\beta = T \qquad (2)$$

where

$$H = \begin{bmatrix} g(w_1 x_1 + b_1) & \cdots & g(w_L x_1 + b_L) \\ \vdots & \ddots & \vdots \\ g(w_1 x_N + b_1) & \cdots & g(w_L x_N + b_L) \end{bmatrix} \qquad (3)$$

and $\beta=(\beta_1, \ldots, \beta_L)^T$ and $T=(T_1, \ldots, T_N)^T$. Unlike traditional SLFN, the weight in ELM does not need to be tuned and it can perform well even with less data [4]. The weight between hidden layer and output layer can be the same as to obtain the least-squares solution of the linear system below:

$$\min_{\beta} \| H\beta - T \| \qquad (4)$$

The solution should have the minimum training error, the smallest norm of weights and unique. Therefore, the smallest norm least-squares solution of the above linear system is given by

$$\hat{\beta} = H^\dagger T \qquad (5)$$

where $H^\dagger$ is the Moore-Penrose generalized inverse of matrix H and T is the target. The predicted output for the model is calculated based on Eq. (2). The fitness value for each particle refers to an error on the validation set of data.

The velocity of each particle is modified iteratively by its personal best position (*Pbest*) and the global best position (*Gbest*). As a result, each particle searches around a region defined its personal and global best position from its neighborhood [10].

$$v_i^{k+1} = w v_i^k + c_1 R_1 \left(p_i - x_i^k \right) + c_2 R_2 \left(p_g - x_i^k \right) \qquad (6)$$

$$x_i^{k+1} = x_i^k + v_i^{k+1} \qquad (7)$$

where v_i^k is the velocity of ith particle at k^{th} iteration in the swarm while x_i^k is the current position of the particle. c_1 and c_2 are the acceleration coefficients, R_1 and R_2 are two different random number between 0 and 1, p_i is the *Pbest* and p_g is the *Gbest* achieved by the particle in its neighborhood.

The process is repeated until the stop criterion is met. The result for this algorithm will be the particle that has the best performance among its population. Figure 1 shows the pseudocode for IPSO-ELM.

2.2 Proposed Parallel Improved Extreme Learning Machine Based Particle Swarm Optimization (PIPSO-ELM)

In order to model surface roughness and power consumption effectively, a parallel model based IPSO-ELM model which has been described by Han et al [7] is proposed. We called it parallel IPSO-ELM (PIPSO-ELM). As shown in Fig. 1, the PIPSO-ELM is divided into two blocks of algorithm that works separately. Each block represents the surface roughness and the power consumption, respectively. Therefore, the input weights and the biases for each of the performance will be different. This will reflect the accuracy of the model and will be detailed in the results and discussion section. The proposed model is the one having 100 populations and a maximum number of 50 iterations with varying the number of hidden node(s). Fig. 2 shows the pseudo code of PIPSO-ELM.

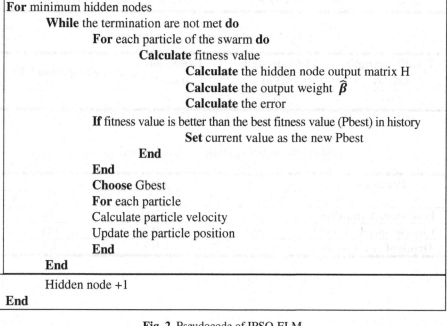

Initialization: All weights, w_i and biases, b_i are randomly assigned;
For minimum hidden nodes
      While the termination are not met **do**
            For each particle of the swarm **do**
                  Calculate fitness value
                        Calculate the hidden node output matrix H
                        Calculate the output weight $\widehat{\beta}$
                        Calculate the error
                If fitness value is better than the best fitness value (Pbest) in history
                        Set current value as the new Pbest
                End
            End
            Choose Gbest
            For each particle
            Calculate particle velocity
            Update the particle position
            End
      End
      Hidden node +1
End

Fig. 2. Pseudocode of IPSO-ELM

3   Experiment

An experimental design consisting of 15 experiments was utilized to collect the surface roughness and power consumption measurement data. The experiments were conducted on CNC lathe machine using carbide insert TNMG 160408-M3 TP2500 with the tool holder MTJNR 2525M16. The workpiece used in this experiment was carbon steel AISI 1045. The chemical and physical properties of the workpiece material are shown in Table 1 and Table 2, respectively.

This design of experiment contains three levels, three input factors with full replication. Table 3 shows the parameters and their levels employed in the experimental run. Each experiment was stopped after 100mm cutting length. New cutting edge was used in each experiment to ensure the accuracy of the reading. All of the experiments were done in a dry cutting condition. The measurement of surface roughness was recorded using surface roughness tester model Mitutoyo SJ-301 and the power consumption was recorded using clamp meter FLUKE 353.

The samples were 150mm length and 20mm in diameter. There were 15 experimental results used in this study. The model randomly selected 60% from the results as a training data while the rest was used for validation. In order to verify the model, three sets of test data were used as the inputs for the selected model. This data set was totally different from the training and validation sets.

Table 1. Chemical properties AISI 1045

C (%)	Mn (%)	P (%)	S (%)
0.42-0.50	0.60-0.90	0.04 (max)	0.05 (max)

Table 2. Physical properties AISI 1045

Tensile strength (MPa)	Yield strength (MPa)	Reduction of area (%)	Elongation (%)
565	310	40	16

Table 3. Parameters and their levels of experiments

Parameters	Levels		
	1	2	3
Feed rate, A (mm/rev)	100	200	300
Cutting speed, B (m/s)	0.1	0.2	0.3
Depth of cut, C (mm)	0.1	0.8	1.5

4   Prediction Results and Analysis

The experimental results based on response surface methodology are represented as measured data in Table 4. The PIPSO-ELM described in the previous section was used in training the models using sigmoidal activation function. The results from the predicted data, then, were analyzed using the mean absolute percentage error (MAPE) for the each model. The MAPE is calculated as follows:

$$MAPE = \frac{T_i - y_i}{y_i} \qquad (8)$$

where T_i is the measured output and y_i is the predicted output.

The numbers of hidden nodes obtained from the algorithm are different. The hidden node is varied from five to ten nodes for different performance function. Therefore, from the proposed model, seven and eight number of hidden nodes was gained in order to model for the surface roughness and the power consumption prediction model, respectively.

Table 4. Results from the measured experiments, the predicted value and the MAPE

Exp	Surface Roughness, μm			Power Consumption, Watt		
	Measured	Predicted	MAPE	Measured	Predicted	MAPE
1	0.59	0.60894	0.032	3373.42	3384.04	0.003
2	0.77	0.78067	0.009	3468.61	3481.76	0.004
3	3.42	3.37191	0.014	3927.25	3940.99	0.003
4	3.20	3.15064	0.015	6935.78	6943.73	0.001
5	1.82	1.79657	0.011	2934.98	2952.16	0.006
6	1.47	1.45307	0.009	4557.51	4572.39	0.003
7	1.62	1.62455	0.001	4212.81	4222.75	0.002
8	1.46	1.45796	0.004	7791.03	7797.21	0.001
9	0.42	0.43513	0.028	3543.61	3558.99	0.004
10	3.54	3.47959	0.017	3852.25	3868.91	0.004
11	0.76	0.77369	0.022	5024.80	5033.96	0.002
12	3.35	3.30769	0.013	6935.78	6944.20	0.001
13	1.27	1.39688	0.103	5070.95	4614.89	0.090
14	1.37	1.39688	0.017	4136.37	4614.89	0.116
15	1.58	1.39688	0.118	4853.17	4614.89	0.049

The results after the training phase are tabulated in Table 4. The average MAPE for surface roughness is 0.028 while for power consumption is 0.019. Both have lower MAPE which shows that the accuracy of learning stage for the proposed PIPSO-ELM was excellent.

5   Confirmation Test

Once the best models were selected, the model was used to predict and verify the performance functions for the testing set. The MAPE, again, will be used as the key evaluator for this confirmation test. The results can be seen in the Table 5. Based on the result of confirmation test, the average MAPE for surface roughness is 0.178 while for power consumption is 0.026. Even though the results values are higher than the training stage, but it is still an acceptable lower error.

Table 5. Results from the confirmation test

Exp	Surface Roughness, μm			Power Consumption, Watt		
	Measured	Predicted	MAPE	Measured	Predicted	MAPE
1	2.79	3.04250	0.089	2877	2991.41	0.040
2	0.64	0.43513	0.317	3606	3558.99	0.013
3	1.58	1.38014	0.128	6238	6074.91	0.026

6   Conclusion

The main emphasis of this paper is to consider more than one objective function using proposed technique. The following conclusions can be drawn based on the study done by the author.

1. For multi objectives problem, PIPSO-ELM can be used to predict the performance functions.
2. The proposed method was able to model the training and the testing set with minimal error. The predicted result from the designed model was able to match the experimental data very closely.

For future work, the proposed model can be used with the optimization technique for optimal cutting parameters.

References

1. Benardos, P., Vosniakos, G.-C.: Predicting surface roughness in machining: a review. International Journal of Machine Tools and Manufacture 43, 833–844 (2003)
2. Bouacha, K., Yallese, M.A., Mabrouki, T., Rigal, J.: Statistical analysis of surface roughness and cutting forces using response surface methodology in hard turning of AISI 52100 bearing steel with CBN tool. Int. J. Refract. Met. H. 28, 349–361 (2010)
3. Karayel, D.: Prediction and control of surface roughness in CNC lathe using artificial neural network. Journal of Materials Processing Technology 209, 3125–3137 (2009)
4. Huang, G.B., Zhu, Q.-Y., Siew, C.-K.: Extreme learning machine: theory and applications. Neurocomputing 70, 489–501 (2006)

5. Rosin, P.L., Fierens, F.: Improving neural network generalisation. In: International Geoscience and Remote Sensing Symposium, IGARSS 1995. Quantitative Remote Sensing for Science and Applications, July 10-14, vol. 2, pp. 1255–1257 (1995)

6. Sarle, W.S.: Stopped training and other remedies for overfitting. In: Proceedings of the 27th Symposium on the Interface (1995)

7. Han, F., Yao, H., Ling, Q.: An improved evolutionary extreme learning machine based on particle swarm optimization. Neurocomputing 116, 87–93 (2013)

8. Aggarwal, A., Singh, H., Kumar, P., Singh, M.: Optimization of multiple quality characteristics for CNC turning under cryogenic cutting environment using desirability function. Journal of Materials Processing Technology 205 (2008)

9. Rajemi, M.F., Mativeng, P.T., Aramcharoen, A.: Sustainable machining: selection of optimuturning conitions based on minimum energy considerations. Journal of Cleaner Production 8, 1059–1065 (2010)

10. Thepsonthi, T., Ozel, T.: Multi objective process optimization for micro end milling of Ti-6Al-4V titanium alloy. International Journal of Advance Manufacturing Technology 63, 903–914 (2012)

OS-ELM Based Emotion Recognition
for Empathetic Elderly Companion

Zixiaofan Yang*, Qiong Wu, Cyril Leung, and Chunyan Miao

Joint NTU-UBC Research Centre of Excellence in Active Living
for the Elderly, School of Computer Engineering,
Nanyang Technological University,
Singapore

Abstract. The empty nest syndrome has become a significant problem
in many countries. Caregivers can help the elderly to cope with some
of the issues related to this syndrome, but a tremendous shortage of
caregivers is expected due to the low income and high workload. The
growing popularity of mobile device use by the elderly has created a
great potential for mobile virtual companions. Studies have shown that
virtual companions that can show empathy towards its users are perceived as more caring, likeable, trustworthy and supportive. Therefore,
we believe that an empathetic companion can help to improve the elderly's emotional wellness and promote an active independent lifestyle.
In this work, we propose an empathetic elderly companion based on the
psychological theory of empathy in the form of a smart chat bot. The
empathetic companion records the elderly's speech during their conversations, extracts phonetic features, and employs a fast online learning
algorithm, namely Online Sequential Extreme Learning Machine (OS-
ELM), to recognize emotion. Based on the elderly's current emotion, the
companion will provide context-aware empathetic responses. The experimental results show that OS-ELM is superior in classifying emotions
from phonetic information.

Keywords: Elderly companion, Empathy, OS-ELM, Emotion recognition.

1   Introduction

The empty nest syndrome refers to feelings of depression and grief experienced
by parents after their children leave the family [12]. It has become a pervasive
modern social problem especially among the elderly [23]. The feeling of social iso-
lation among empty nesters leads to various kinds of negative emotions, such as
loneliness, anxiety, and depression. These may lead to many physical problems,
such as dysfunction of endocrine and central nervous system, and suppressed im-
mune system [20]. Hence, the elderly are in great need of caregivers to provide
companionship and help them relieve the symptoms. However, the low income

---

* Peking University, China.

© Springer International Publishing Switzerland 2015   331
J. Cao et al. (eds.), *Proceedings of ELM-2014 Volume 2,*
Proceedings in Adaptation, Learning and Optimization 4, DOI: 10.1007/978-3-319-14066-7_32

and high workload has resulted in an increasing shortage of caregivers, which has become an imminent problem in eldercare.

A study by Plaza et al.[15] shows a rapid increase in the usage and acceptance of mobile phones among the elderly. This represents a great opportunity for mobile companions to reduce the need for human care-givers. To this end, many mobile agents have been proposed. For example, Martín et al.[13] proposed a multi-agent system for the elderly, which is capable of detecting falls through sensors in a mobile device. Xiao et al.[24] developed a mobile system named "Canderoid" to remotely monitor the outdoor travel status of the elderly in case help is needed. Wilks et al. [21] proposed a companion which can elicit personal information from the elderly through conversations about their photographs for purpose of better 'knowing' its owners preferences and wishes. McCarthy et al.[3] proposed an application entitled "MemoryLane", which is able to weave excerpts selected from the elderlys memories into a meaningful story. Wu et al. [22] proposed a curious companion to enhance an elderly's game playing experiences.

The existing mobile elderly companions, however, give care to the elderly mainly through security guarantee, improving healthcare, or engaging them in conversations. Few agents focus on the emotional needs of the elderly and how to improve their mood through empathy. Studies have shown that virtual companions that can show empathy towards human users are perceived as more caring, likeable, trustworthy and supportive [9]. Hence, we aim to develop an empathetic companion that caters to the emotional needs of the elderly.

In this work, we propose an empathetic elderly companion based on the theory of parallel and reactive empathy [16] and the Socioemotional selectivity theory (SST) [2]. This empathetic elderly companion resides in a mobile device and is built on top of a chat bot. The empathetic companion monitors an elderly's emotion state through recording the phonetic information during chatting, and displays context-sensitive parallel or reactive empathy to improve the elderly's emotional wellness. To recognize the elderly's emotional states with phonetic information in real-time, we employ a fast online learning method referred to as the Online Sequential Extreme Learning Machine(OS-ELM) [11]. OS-ELM has been shown to provide excellent learning and generalization performance at a very fast speed. Besides, OS-ELM can perform online learning for personalized emotion recognition and handle data with varying chunk lengths which is suitable for handling the phonetic information collected by the elderly companion.

To evaluate the performance of the empathetic elderly companion on emotion recognition, experimental studies are carried out on phonetic data sets collected from five volunteers. Feature extraction is done using the OpenSMILE toolkit [4]. The performance of OS-ELM with four different activation functions are compared. The experimental results show that the sigmoid function is the best activation function to train our model with the average accuracy of 85%. The performances using three different feature sets are also compared. It is found that there is no significant difference on the performance between the different feature sets.

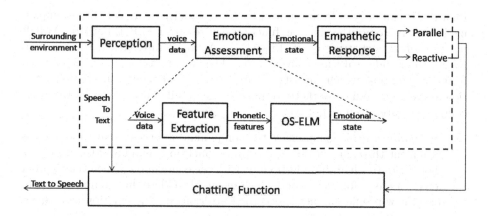

Fig. 1. The architecture of the empathetic elderly companion

The rest of the paper is organized as follows: Section 2 reviews the psychological theories on empathy and presents a general model of the elderly companion. Section 3 focuses on the "emotion assessment" component of the proposed model, which consists of two steps: feature extraction and emotion recognition. Section 4 presents the experimental details. The main findings are summarized in Section 5.

2   The Empathetic Companion

In human psychology, the concept of empathy has many different definitions. According to Strayer [17], these definitions can be divided into two major categories: affective empathy and cognitive empathy. Affective empathy is the capacity to respond with an appropriate emotion to another's mental states, while cognitive empathy is the capacity to understand another's perspective or mental state. However, the two theoretical categories overlap to some extent. Hence, Hoffman [6] proposed an integrated definition, in which empathy is considered as having an affective response that results from empathic arousal cognitive processes, such as motor mimicry, association of cues from others, and perspective-taking.

Further more, Stephan and Finlay [16] have distinguished between two kinds of empathy regarding the different affective outcomes: parallel empathy and reactive empathy. In parallel empathy, the responses are similar to those that the other person is experiencing. In reactive empathy, the empathier exhibits a higher cognitive awareness of the situation to react with empathetic behaviors that do not necessarily match with those of the target's affective states. For example, when someone is insulted, a parallel empathier will empathize the pain and discomfort, while reactive empathier may feel indignation and resentment toward the attacker.

Motivated by the findings mentioned above, we apply and integrate the concept of empathy into the model of empathetic elderly companion, as shown in

Figure 1. It can be seen from Figure 1 that the empathetic model of the companion is built on top of the chatting function realized by a ALICE bot [19]. ALICE bot is a software chat bot created in AIML (Artificial Intelligence Markup Language) to provide real time conversations. The empathetic model is highlighted in the dashed box, which consists of three major components: perception, emotion assessment, and empathetic response. The detailed discussion on the three components are given as follows:

- *Perception*: In this component, the companion observes its surrounding environment through sensors. A recording function is employed to collect the phonetic data when the elderly chats with the companion. When the recorder detects a pause for one second which means the elderly has completed a sentence, it will send the newly recorded utterance to the emotion assessment component.
- *Emotion Assessment*: The emotion assessment component gets the voice data from the perception component, analyzes the data to classify the user's emotion state. The assessment of emotion consists of two steps: feature extraction and emotion recognition. Feature extraction will extract the most informable features for the classification of emotions. Emotion recognition classifies the emotions based on the features produced by the previous step.
- *Empathetic Response*: Based on the elderly's emotion recognized by the emotion assessment component, the empathetic response component generates context-sensitive and personalized empathetic behaviors (parallel or reactive) based on a set of rules. The rules can be abstracted from psychological theories as well as empirical knowledge. The empathetic responses are customized for the elderly based on the socioemotional selectivity theory(SST) [2], which states that towards the end of the life span, the elderly become more preferred to communicate with their family and old friends. The theory indicates a desirable effect when involving their friends and relatives into the empathetic responses. For example, when the companion finds that the elderly is angry, it will express indignation towards the related object (detected by keywords) in the context (parallel empathy). When the companion recognizes sadness in the speech, it will ask for consolation from his son to help the elderly cheer up (reactive empathy). The emotion-arousing object can be extracted from the speech context stored in the database and the intimate social partner is chosen according to the times that they have been mentioned recently.

In this work, we focus on the implementation of the "emotion assessment" component. In the following section, we will present the detailed implementation of the "emotion assessment" component.

3   Emotion Recognition Based on OS-ELM

In this section, we present the "emotion assessment" component which consists of two steps: feature extraction and emotion recognition.

3.1 Feature Extraction

We employ the Opensmile toolkit [4] to extract features from the recorded voice data (.wav file). We choose features and functions covering prosodic, spectral, and voice quality information. Specifically, the low-level descriptors (LLDs) extracted directly from the data include zero-crossing rate of time signal, root-mean-square signal frame energy, pitch frequency, harmonics-to-noise ratio, and mel-frequency cepstral coefficients. The first order regression coefficients (delta coefficients) of each of the LLDs are computed. Additionally, various types of functions are applied on the contours of LLDs and coefficients: minimum and maximum value, the range of the contour, the absolute position of the extremum value, the arithmetic mean, the standard deviation, the slope and offset of a linear approximation, the quadratic error computed as the difference of the linear approximation and the actual contour, the skewness and the kurtosis. All of the extracted features are normalized for the training of OS-ELM.

3.2 Emotion Recognition

The emotion recognition problem can be viewed as a classification problem. To produce a real-time and efficient classification of recorded phonetic information, the companion agent employs a fast neural learning algorithm referred to as the Online Sequential Extreme Learning Machine(OS-ELM) [11]. OS-ELM originates from the batch learning extreme learning machine (ELM) [8] and is developed for SLFNs with both additive and radial basis function (RBF) hidden nodes. OS-ELM has the following properties:

– OS-ELM can handle training data sequentially, not only one-by-one but also chunk-by-chunk with varying chunk length, which makes it superior to the other algorithms, such as all the BP-based and RAN-based sequential learning algorithms.
– Only the newly arrived chunk of observations are learned. As soon as training procedure of the new data is completed, we can discard the data instantly.
– No prior knowledge of the amount of training data is needed, which means as long as the user can provide newly labeled data, the OS-ELM algorithm will never stop learning.

In OS-ELM, the input weights and biases of additive nodes or centers and impact factors of RBF nodes are randomly generated. Based on this, the output weights are analytically determined. Unlike other sequential learning algorithms which have many control parameters to be tuned, OS-ELM only requires the number of hidden nodes to be specified. It has been shown in literature that OS-ELM provides better generalization performance at a very fast learning speed, comparing with other sequential learning algorithms such as Stochastic gradient descent BP (SGBP)[10], resource allocation network (RAN)[14], and growing and pruning RBF (GAP-RBF)[7]. Next, we will review the OS-ELM algorithm.

For N samples $(\mathbf{x}_i, \mathbf{t}_i)$, where $\mathbf{x}_i = [x_{i1}, x_{i2}, ..., x_{in}]^T \in \mathbf{R}^n$, $\mathbf{t}_i = [t_{i1}, t_{i2}, ..., t_{im}]^T$ $\in \mathbf{R}^m$, the output of an SLFN with $\widetilde{N}$ hidden nodes (additive or RBF) can be represented by

$$f_{\widetilde{N}}(\mathbf{x}) = \sum_{i=1}^{\widetilde{N}} \beta_i G(\mathbf{a}_i, b_i, \mathbf{x}), \mathbf{x} \in \mathbf{R}^n, \mathbf{a}_i \in \mathbf{R}^n \tag{1}$$

where $\mathbf{a}_i$ and b_i are the learning parameters of hidden nodes, β_i is the weight of the ith hidden node, and $G(\mathbf{a}_i, b_i, \mathbf{x})$ is the output of the ith hidden node with respect to the input x. For additive hidden node, $G(\mathbf{a}_i, b_i, \mathbf{x}) = g(\mathbf{a}_i \cdot \mathbf{x} + b_i), b_i \in R$, where $\mathbf{a}_i$ is the input weight and b_i is the bias of the ith hidden node and $\mathbf{a}_i \cdot \mathbf{x}$ denotes the inner product. For RBF hidden node, $G(\mathbf{a}_i, b_i, \mathbf{x}) = g(b_i \parallel \mathbf{x} - \mathbf{a}_i \parallel), b_i \in R^+$, where $\mathbf{a}_i$ and b_i are the center and impact factor of ith RBF node, R^+ is the set of all positive real values.

Once the type of node, the activation function, and the hidden node number $\widetilde{N}$ are chosen and data $\aleph = \{(\mathbf{x}_i, \mathbf{t}_i) | \mathbf{x}_i \in \mathbf{R}^n, \mathbf{t}_i \in \mathbf{R}^m, i = 1, ...\}$ arrive one-by-one or chunk-by-chunk, the OS-ELM algorithm can be represented as follows:

Initialization Phase: A small chunk of the training set $\aleph_0 = \{(\mathbf{x}_i, \mathbf{t}_i)\}_{i=1}^{N_0}, N_0 \geqslant N$ is used to initialize the network, where $N_0 \geq \widetilde{N}$

1. Assign random input weights $\mathbf{a}_i$ and bias b_i or center $\mathbf{a}_i$ and impact factor b_i to hidden nodes, where $i = 1, ..., \widetilde{N}$.
2. Calculate the initial hidden layer output matrix bfH_0

$$\mathbf{H}_0 = \begin{bmatrix} G(a_1, b_1, x_1) & \cdots & G(\mathbf{a}_{\widetilde{N}}, b_{\widetilde{N}}, \mathbf{x}_1) \\ \vdots & \ddots & \vdots \\ G(a_1, b_1, \mathbf{x}_{N_0}) & \cdots & G(\mathbf{a}_{\widetilde{N}}, b_{\widetilde{N}}, \mathbf{x}_{N_0}) \end{bmatrix}_{N_0 \times \widetilde{N}} \tag{2}$$

3. Estimate the initial output weight $\beta^{(0)}$: $\beta^{(0)} = \mathbf{P}_0 \mathbf{H}_0^T \mathbf{T}_0$, where $\mathbf{P}_0 = (\mathbf{H}_0^T \mathbf{H}_0)^{-1} = \mathbf{K}_0^{-1}$ and $\mathbf{T}_0 = [\mathbf{t}_1, \cdots, \mathbf{t}_{N_0}]^T$.
4. Set $k = 0$, where k is the number of chunks that is trained currently.

Sequential Learning Phase: Given the $(k + 1)$th chunk of new observations $\aleph_{k+1} = \{(\mathbf{x}_i, \mathbf{t}_i)\}_{i=\sum_{j=0}^{k} N_j + 1}^{\sum_{j=0}^{k+1} N_j}$, where N_{k+1} denotes the number of observations in the kth chunk.

1. Calculate the partial hidden layer output matrix $\mathbf{H}_{k+1}$

$$\mathbf{H}_0 = \begin{bmatrix} G(a_1, b_1, x_{\sum_{j=0}^{k} N_j + 1}) & \cdots & G(\mathbf{a}_{\widetilde{N}}, b_{\widetilde{N}}, \mathbf{x}_{\sum_{j=0}^{k} N_j + 1}) \\ \vdots & \ddots & \vdots \\ G(a_1, b_1, \mathbf{x}_{\sum_{j=0}^{k+1} N_j}) & \cdots & G(\mathbf{a}_{\widetilde{N}}, b_{\widetilde{N}}, \mathbf{x}_{\sum_{j=0}^{k+1} N_j}) \end{bmatrix}_{N_{k+1} \times \widetilde{N}} \tag{3}$$

2. Set $\mathbf{T}_{k+1} = [\mathbf{t}_{\sum_{j=0}^{k} N_j + 1}, \cdots, \mathbf{t}_{\sum_{j=0}^{k} N_j + 1}]_{N_{k+1} \times m}^T$.
3. Calculate the output weight $\beta^{(k + 1)}$:

$$\mathbf{K}_{k+1} = \mathbf{K}_k + \mathbf{H}_{k+1}^T \mathbf{H}_{k+1} \tag{4}$$

$$\beta(k+1) = \beta(k) + \mathbf{K}_{k+1}^{-1}\mathbf{H}_{k+1}^{T}(\mathbf{T}_{k+1} - \mathbf{H}_{k+1}\beta(k)) \tag{5}$$

It can be observed from Eq. 5 that $\mathbf{K}_{k+1}^{-1}$ is used for computing $\beta(k+1)$ from $\beta(k)$. Hence, an updated formula for $\mathbf{K}_{k+1}^{-1}$ can be derived as follows:

$$\begin{aligned}\mathbf{K}_{k+1}^{-1} &= (\mathbf{K}_k + \mathbf{H}_{k+1}^{T}\mathbf{H}_{k+1})^{-1}\\ &= \mathbf{K}_k^{-1} - \mathbf{K}_k^{-1}\mathbf{H}_{k+1}^{T}(\mathbf{I} + \mathbf{H}_{k+1}\mathbf{K}_k^{-1}\mathbf{H}_{k+1}^{T})^{-1} \times \mathbf{H}_{k+1}\mathbf{K}_k^{-1}\end{aligned} \tag{6}$$

Hence, $\beta(k+1)$ can be rewritten by $\mathbf{P}_{k+1}$ as follows:

$$\beta(k+1) = \beta(k) + \mathbf{P}_{k+1}\mathbf{H}_{k+1}^{T}(\mathbf{T}_{k+1}\beta(k)) \tag{7}$$

where $\mathbf{P}_{k+1} = \mathbf{P}_k - \mathbf{P}_k\mathbf{H}_{k+1}^{T}(\mathbf{I} + \mathbf{H}_{k+1}\mathbf{P}_k\mathbf{H}_{k+1}^{T})^{-1}\mathbf{H}_{k+1}\mathbf{P}_k$ and $\mathbf{P}_{k+1} = \mathbf{K}_{k+1}^{-1}$.

4. Set $k = k + 1$. Go to Sequential learning phase 1.

As shown in the algorithm, the training of OS-ELM only consists of a series of matrix calculations and no loop over the samples is needed. The fast learning speed indicates a superiority on real-time emotion recognition. Next, various experiments regarding the performance of OS-ELM on emotion recognition is presented.

4   Experiment

To perform emotion recognition, a fundamental task is to determine the emotion categories for classification. Based on the research of Gaurav et el.[18], recognising harmful emotions to change them as well as perceiving healthy emotions to maintain them, is crucial to the well-being of elderly. The most representative detrimental emotions are anger, sadness, and anxiety [5]. As an elderly companion, recognizing boredom is also a basic need to get the condition for entertaining them. Therefore, we select anger, sadness, anxiety/fear, happiness, boring, and neutrality as the basic emotion categories for classifying recorded speech.

For data collection, we obtained experimental data from 5 volunteers. Each volunteer recorded in total 480 utterances for the 6 emotional states listed above. The volunteers expressed the emotions in their habitual way, and the voice data were labeled explicitly by themselves during the recording. To train a personalized companion, OS-ELM is employed to learn the data set of each individual separately. For each individual, 360 utterances are randomly chosen for training and the rest 120 utterances for testing.

In this experiment, we compare the performance of OS-ELM with four different types of activation functions: sigmoid function (Sig), sine curve function (Sin), sub-function (Hardlim) for additive nodes, and radial basis function (RBF) nodes. The experimental results are shown in Table 1. It can be observed from Table 1 that sigmoid function (Sig) presents a significant superiority over the

Table 1. Performance Comparison Between Four Activation Functions

Data set	Activation function	Testing accuracy	
		Mean	STD
Volunteer 1	Sig	0.9142	0.0294
	Hardlim	0.8262	0.0381
	RBF	0.5404	0.0835
	Sin	0.2021	0.0443
Volunteer 2	Sig	0.7863	0.0345
	Hardlim	0.6925	0.0501
	RBF	0.4825	0.0830
	Sin	0.1875	0.0448
Volunteer 3	Sig	0.8717	0.0334
	Hardlim	0.8500	0.0236
	RBF	0.5083	0.0746
	Sin	0.1667	0.0509
Volunteer 4	Sig	0.8650	0.0364
	Hardlim	0.8350	0.0468
	RBF	0.6350	0.0894
	Sin	0.1517	0.0419
Volunteer 5	Sig	0.8333	0.0314
	Hardlim	0.7683	0.0493
	RBF	0.5133	0.0706
	Sin	0.1883	0.0305

Table 2. Feature Comparison of 3 Feature Sets

Feature set	Number of LLDs	Delta coefficients	Number of functions
IS09	16	16	12
Medium	26	26	19
Large	56	112	39

others, while sine curve function (Sin) generates the worst accuracy among the given functions.

To check for the best feature sets, we compare the performance of OS-ELM with sigmoid activation functions on three feature sets. The details of the three feature sets are shown in Figure 2. It can be shown that "IS09" which stands for the feature set of "The INTERSPEECH 2009 Emotion Challenge" [1] contains $(16 + 16) \times 12 = 384$ features, including 16 LLDs, their delta coefficients and 12 functions of the contour. "Medium" is a set of $(26 + 26) \times 19 = 988$ features, including 26 LLDs, their delta coefficients and 19 functions of the contour. "Large" is a set containing $(56 + 112) \times 39 = 6552$ features with 39 functions on 56 LLDs, their 56 delta coefficients, and 56 additional delta coefficients computed on the delta coefficients.

The performance comparison between the three feature sets are shown in Table 3. It can be observed that though the "large" set covers more phonetic information and uses storage space 20 times larger than the "IS09" set uses,

Table 3. Performance Comparison between Different Feature Sets

Data set	Feature Sets	Testing accuracy	
		Mean	STD
	IS09	0.9142	0.0294
Volunteer 1	Medium	0.9092	0.0226
	Large	0.9233	0.0360
	IS09	0.7863	0.0345
Volunteer 2	Medium	0.7883	0.0447
	Large	0.8017	0.0365
	IS09	0.8717	0.0334
Volunteer 3	Medium	0.8933	0.0335
	Large	0.9000	0.0261
	IS09	0.8650	0.0364
Volunteer 4	Medium	0.8933	0.0196
	Large	0.8800	0.0414
	IS09	0.8333	0.0314
Volunteer 5	Medium	0.8200	0.0383
	Large	0.8367	0.0378

there is not a significant improvement of testing accuracy. Taking into account the feature extraction time and the consumption of storage space in mobile device, features of "IS09" is sufficient for basic emotion recognition.

5 Conclusion

In this paper, we proposed an empathetic elderly companion to address the empty nest syndrome. This empathetic companion is built on top of a chat bot in mobile devices. The companion assesses an elderly's emotion and provide appropriate context-sensitive empathetic responses to help improve positive emotions. To the best of our knowledge, this is the first elderly companion which focuses on using empathetic methods to improve elderly's emotional wellness. The companion consists of three major components: perception, emotion assessment, and empathetic response. In this work, we mainly focused on implementing the emotion assessment component with two steps: feature extraction and emotion recognition. The result of experiments indicates that our elderly companion based on OS-ELM is superior for the emotion recognition task and is not only a promising design but also practical for implementation.

In the future, we will explore the knowledge for the production of parallel and reactive empathetic responses. Such knowledge can be generated and abstracted from psychological theory as well as empirical knowledge, including when the agent should use parallel or reactive responses, what the proper emotion state and intensity of the response should be, and how to make the empathetic expressions more convincing to the elderly.

References

1. 10th Annual Conference of the International Speech Communication Association, INTERSPEECH 2009, Brighton, United Kingdom (2009)
2. Carstensen, L.L., Isaacowitz, D.M., Charles, S.T.: Taking time seriously: A theory of socioemotional selectivity. American Psychologist 54(3), 165–181 (1999)
3. Carthy, S.M., Kevitt, P.M., McTear, M., Sayers, H.: MemoryLane: An intelligent mobile companion for elderly users. In: The 7th Information Technology and Telecommunication Conference: Digital Convergence in a Knowledge Society, pp. 72–82 (2007)
4. Eyben, F., Wöllmer, M., Schuller, B.: Opensmile: The munich versatile and fast open-source audio feature extractor, pp. 1459–1462 (2010)
5. Gallo, L.C., Matthews, K.A.: Understanding the association between socioeconomic status and physical health: Do negative emotions play a role? Psychological Bulletin 129(1), 10–51 (2003)
6. Hoffman, M.L.: Empathy and moral development: Implications for caring and justice. Cambridge University Press (2001)
7. Huang, G.-B., Saratchandran, P., Sundararajan, N.: An efficient sequential learning algorithm for growing and pruning RBF (GAP-RBF) networks. IEEE Transactions on Systems, Man, and Cybernetics, Part B: Cybernetics 34(6), 2284–2292 (2004)
8. Huang, G.-B., Zhu, Q.-Y., Siew, C.-K.: Extreme learning machine: Theory and applications. Neurocomputing 70(1-3), 489–501 (2006)
9. Koster, T.: The persuasive qualities of an empathetic agent. In: 5th Twente Student Conference on IT (2006)
10. LeCun, Y.A., Bottou, L., Orr, G.B., Müller, K.-R.: Efficient backprop. In: Montavon, G., Orr, G.B., Müller, K.-R. (eds.) Neural Networks: Tricks of the Trade, 2nd edn. LNCS, vol. 7700, pp. 9–48. Springer, Heidelberg (2012)
11. Liang, N.-Y., Huang, G.-B., Saratchandran, P., Sundararajan, N.: A fast and accurate online sequential learning algorithm for feedforward networks. IEEE Transactions on Neural Networks 17(6), 1411–1423 (2006)
12. Lowenthal, M., Chiriboga, D.: Transition to the empty nest: Crisis, challenge, or relief? Archives of General Psychiatry 26(1), 8–14 (1972)
13. Martín, P., Sánchez, M., Álvarez, L., Alonso, V., Bajo, J.: Multi-agent system for detecting elderly people falls through mobile devices. In: Novais, P., Preuveneers, D., Corchado, J.M. (eds.) ISAmI 2011. AISC, vol. 92, pp. 93–99. Springer, Heidelberg (2011)
14. Platt, J.: A resource-allocating network for function interpolation. Neural Computation 3(2), 213–225 (1991)
15. Plaza, I., Martín, L., Martin, S., Medrano, C.: Mobile applications in an aging society: Status and trends. Journal of Systems and Software 84(11), 1977 (2011)
16. Stephan, W.G., Finlay, K.: The role of empathy in improving intergroup relations. Journal of Social Issues 55(4), 729–743 (1999)
17. Strayer, J.: Affective and cognitive perspectives on empathy. Cambridge University Press (1987)
18. Suri, G., Gross, J.J.: Emotion regulation and successful aging. Trends in Cognitive Sciences 16(8), 409–410 (2012)
19. Wallace, R.S.: The anatomy of ALICE. Springer (2009)
20. Wang, J., Zhao, X.: Empty nest syndrome in China. International Journal of Social Psychiatry 58(1), 110 (2012)

21. Wilks, Y.: Artificial companions as dialogue agents. In: Proceedings of the SIG-DIAL 2009 Conference: The 10th Annual Meeting of the Special Interest Group on Discourse and Dialogue, pp. 216–216 (2009)
22. Wu, Q., Miao, C., Tao, X., Helander, M.: A curious companion for elderly gamers. In: Southeast Asian Network of Ergonomics Societies Conference (SEANES), pp. 1–5 (2012)
23. Wu, Q., Shen, Z., Leung, C., Zhang, H., Cai, Y., Miao, C.: Internet of things based data driven storytelling for supporting social connections. In: IEEE and Internet of Things (iThings/CPSCom), pp. 383–390 (2013)
24. Xiao, B., Asghar, M., Jamsa, T., Pulii, P.: "Canderoid": A mobile system to remotely monitor travelling status of the elderly with dementia. In: 2013 International Joint Conference on Awareness Science and Technology and Ubi-Media Computing (iCAST-UMEDIA), pp. 648–654 (2013)

Access Behavior Prediction in Distributed Storage System Using Regularized Extreme Learning Machine

Wan-Yu Deng[1,2], Chit Lin Su[1], and Yew-Soon Ong[1,*]

[1] School of Computer Engineering, Nanyang Technological University, Singapore
[2] School of Computer, Xi'an University of Posts&Telecommunications, China
{wydeng,asysong,lschit}@ntu.edu.sg

Abstract. In this paper, we propose a fast and accurate block-level operation (writing or reading) and *transferred size* prediction method based on Regularized Extreme Learning Machine, which represents a key component towards sustainable, green data center. The proposed RELM-based method can produce competitive performance at fast learning speed. Benefitting from the random weights of RELM, these two prediction tasks can be unified as one, thus reducing the training time to half. Experiments on SNIA shows that block-level *operation type* prediction can reach an accuracy of 99.04%, while the *transferred size* prediction is at 0.0234 NRMSE.

Keywords: Machine Learning, Big data, Access behavior prediction, Distributed storage system.

1 Introduction

Data centers require an enormous amount of power to perform various data operations efficiently and effectively. Power consumption in data centers has increased gradually over the years and this attracted much attention from the public, private and government sectors. According to the DCD Intelligence Industry Census[1], the global electricity consumption of data centers increased by 19% between 2011 and 2012. And, the growth in power consumption remains unwavering at a rate of over 7% from 2012 to 2013. In Singapore, there has been a 9.7% increase in power from 930MW in 2012 to 1,020MW in the following year. Based on the report of Data Center Energy Efficiency Benchmarking from the National Environment Agency [2], 3.6% of total electrical energy consumption came from data centers in 2010. Furthermore, it has been estimated that the demand in energy consumption by data centers will continue to increase to 2,260GWh by 2015. Note that this represents an estimated growth of 51% in energy consumption from 2010. With the rapid growth in power consumption, energy-related expenditures is becoming a core financial factor of many organizations. The ballooning power consumption made by data centers worldwide, in

---

* Corresponding author.

particular, has been one of the key highlights of many countries and companies recently. From the 2013 Data Center Industry Survey of the Uptime Institute [3], it has been reported that 36% of data centers have increased their amount of year-over-year budget compared to 32% in 2012 and 27% in 2011. More than 10% increase in data center budgets year-over-year has been reported [3]. The cost of power consumption in data centers thus contributes a significant percentage to the total expenses spent in data center organizations. Taking this cue, many new implementations of power management and energy efficiency measures have been sought after for the purpose of lowering the growth rate in power consumption and expenditures of data centers. Power management and energy-efficient measures can be applied in different areas, extending from Information Technology (IT) systems, environmental conditions, air management of data center, cooling and electrical systems, on-site generation to the reduce, reuse and recycle of waste heat [4]. In data center, energy usage of Information Technology (IT) equipment occupies more than half of the total usage of the entire facility, hence there is a need for more efficient usage of IT equipment. Usage of energy-efficient servers, network equipment and power supplies, consolidated storage devices and implementation of virtualization are some effective means to help reduce energy consumption in data centers [4]. Among the various areas, data storage devices used in data center have garnered increasing attention, since the power consumption is proportional to the number of data storage devices in a linear manner. In order to support power management and energy-efficient measures, different types of data storage technologies have been applied in data centers. For example, switching from 3.5-inch disks to 2.5-inch disks have reported a reduction in power consumption of 45% to 50% and makes disk spinning easier [5]. And, spinning down disks either when they are not in use can also help save power in storage systems [5]. Furthermore, solid state disks have been greatly recommended and used in lieu of hard disk drives due to its higher performance, as well as power and space saving properties [6]. Other types of disks such as SAS and SATA, which are energy-efficient, environmentally friendly and less expensive have also been considered [7]. Nowadays, data center organizations have conducted various researches in the development of ideal data storage system architecture in order to reduce cost and power consumption. For instance, the usage of next generation Non-Volatile Memory (NVM) technologies has received increasing attentions from large scale data centers due to its characteristics of high performance and capacity, as well as low power con-sumption and space footprint [8]. To date, numerous research studies in computational intelligence have contributed to the discovery and development of novel energy-efficient power saving strategies in data storage devices. Among them, machine learning plays an important role. In the quest towards reducing power consumption by data storage devices, various forms of machine learning problem formulations have been considered on the different types of data collected, including knowledge extraction, classification, prediction, pattern recognition, optimization and others. Google considered the use of artificial neural networks to perform prediction on the Power Usage Effectiveness (PUE) of data centers and optimization on energy efficiency [9].

Various other researches have also used a variety of other machine learning and data management techniques on storage related data to assist in storage system architecture and provide efficient energy aware operations and valuable feedback to enhance decision making. For instance, application of machine learning techniques on data storage workload information can provide performance related statistical and prediction results which are valuable to decision making process in data storage management such as data placement and data migration. Data access prediction and optimization is among the popular use of machine learning adopted in data storage systems. With the large amount of data produced and managed by data-intensive applications, the need to optimize the performance and storage demands of data storage systems has been paramount, since data access prediction to support the operations of workload monitoring and analysis can lead to improved performance with capacity scaling in storage systems. In this paper, Regularized Extreme Learning Machine (RELM), one of the recent paradigms of machine learning, is investigated for predicting the data access pattern in data center storage systems. Particularly, RELM is considered for the predictions of block-level read and write operations, and *transferred size*. In the era of Big data, there is common consensus that the data involved possesses the characteristics of big instance size, thus making it a challenge to design reliable prediction approach for improved data storage optimization. To simulate this, the data access pattern of time series data will be of interest in the present study.

2   Related Work

Several prior studies have investigated various statistical techniques for data analysis and developed probabilistic models to predict the characteristic and behaviors of application workloads in distributed environments. Devarakonda and Iyer [10] proposed a statistical approach for the prediction of CPU and memory consumptions, as well as the I/O behaviors of file system. For the purpose of file transfer estimation in distributed systems, Faerman et al. [11] proposed the Adaptive Regression Modeling (AdRM). Wang et al. [12] employed machine learning techniques to model the data request of storage devices using Classification and Regression Trees (CART). Senger et al. [13] applied an online approach in classification of process behaviors. Moreover, in order to provide scheduling decision support in large-scale environments, Senger et al. [14] proposed prediction model which can determine the execution times in parallel applications. Oldfield and Kotz [15] proposed the Armada framework for applications management. Kim et al. [16] described a heuristic approach for data access improvement in data-intensive applications. Chervenak et al. AL-Mistarihi and Yong [17] also considered an approach for the best replica selection. However, the main drawback of the above mentioned approaches is that they do not consider the dynamic behavior of applications, as well as predictions and estimations for data access optimization. Renato Porfirio Ishii [18]thus introduced an adaptive window approach to overcome this problem, it however did not take the dependencies of data access behaviors into account. These issues shall be taken into consideration in the current work.

3   The Proposed Approach

This paper proposes an effective approach to support the fast prediction of access behavior for data access optimization in distributed systems. The prediction tasks include operation type (writing and reading) and *transferred size* (data size read or written to disk). This approach is divided into the following three steps (Fig.1): feature extraction, model building and prediction of observations.

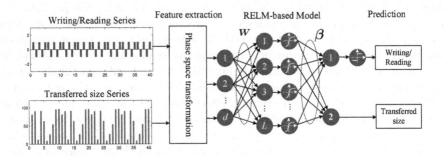

Fig. 1. Structure of access behavior prediction

3.1   Multi-dimensional Time Series and Feature Extraction

Application knowledge acquisition is responsible for monitoring access behavior by using event interception. Different libraries and tools such as Unix DL-Sym [19] and Unix Ptrace [20] provide event interception. Upon extracting the access behavior, the sequence of *read-and-write* events can be represented using a multi-dimensional time series. Every event is defined by a quintuple $\mathbf{tr} = \{pid, inode, trz, time, op\}$ [18] in which *pid* is the process identifier, *inode* is the file selection of techniques to model times series identifier, *trz* is the block-level *transferred size* of data read or written to disk, *time* represents the interval between consecutive operations, and *op* is the *operation type*, i.e., read or write. A series containing n samples can then be represented as $\mathbf{TR} = \{tr_0, tr_1, ..., tr_n\}$. With such organization, the time series is stored in a trace file. In what follows, our focus is placed on *operation type* and *transferred size*. Denote this series as $\mathbf{ts} = [\mathbf{s}_o, \mathbf{s}_z]$ where $\mathbf{ts}(i) = (\mathbf{s}_o(i), \mathbf{s}_z(i)), i = 1, 2, ..., n$. By multi-dimensional phase transformation we have,

$$\mathbf{x}_k = \{\mathbf{s}_o(k), \mathbf{s}_o(k - \tau_o), ..., \mathbf{s}_o(k - (d_o - 1))\tau_1, \mathbf{s}_z(k), ..., \mathbf{s}_z(k - (d_z - 1))\tau_z\} \quad (1)$$

where τ_o and d_o are time delay and embedding dimensions for *op* series while τ_z and d_z are time delay and embedding dimensions for *transferred size* series. τ and d can be determined by Empirical validation, Average displacement [21], Mutual information [22], Autocorrelation methods [23] and so on. According to the embedding delay theorem [24], if embedding dimensions $d = d_o + d_z$ is sufficiently large, a smooth function $f : \mathbf{R}^d \to \mathbf{R}$ exists to satisfy $\mathbf{ts}(k+1) = f(\mathbf{x}_k)$. Hence it is possible to construct a model from dataset $\aleph = \{(\mathbf{x}_k, \mathbf{t}_k)\}_{i=k}^N$ that simulate this function using machine learning approaches. $\mathbf{x}_k$ is described in (1), and $\mathbf{t}_k$ is the value in $(k+1)$-th moment, i.e. $\mathbf{t}_k = (\mathbf{s}_o(k+1), \mathbf{s}_z(k+1))$. Noticed that, $\mathbf{s}_o(k+1)$ is Boolean value $\{-1, 1\}$ where we let -1 denote reading and 1 denote writing.

3.2   Data Access Prediction Model Using RELM

The Regularized Extreme Learning Machine (RELM) is an efficient learning algo-rithm of the Single-hidden-Layer Feedforward Neural networks (SLFN) in which the input weight and bias are randomly assigned, while the output weight can be analytically obtained by means of the simple least square method. Regularized parameter is applied to obtain the minimization of structural risk so as to improve the generalized performance of the model.

Given generated data $\aleph = \{(\mathbf{x}_k, \mathbf{t}_k)\}_{k=1}^N$ and the number of hidden nodes L, the mathematic model of RELM can be described as

$$\underset{\beta}{\text{minimize}} \quad \frac{1}{2}||\boldsymbol{\beta}||^2 + \frac{1}{2}\gamma||\boldsymbol{\xi}||^2$$

$$\text{subject to} \quad \sum_{i=1}^{L} \beta_i g(\mathbf{w}_i \cdot \mathbf{x}_k + b_i) - \mathbf{t}_k = \xi_k, k = 1, ..., N, \tag{2}$$

where $\boldsymbol{\xi} = [\xi_1, ..., \xi_N]$ is error vector, $\mathbf{w}_i$ is input weights, b_i is bias, $\boldsymbol{\beta} = [\beta_1, ..., \beta_L]$ is output weights, and γ is regularized parameter. The Lagrangian for (2) can be represented as follows

$$\ell(\boldsymbol{\beta}, \boldsymbol{\xi}, \boldsymbol{\alpha}) = \frac{1}{2}\gamma||\boldsymbol{\xi}||^2 + \frac{1}{2}||\boldsymbol{\beta}||^2 - \boldsymbol{\alpha}(\mathbf{H}\boldsymbol{\beta} - \mathbf{T} - \boldsymbol{\xi}) \tag{3}$$

where $\alpha = [\alpha_1, ..., \alpha_n]$ is the Lagrangian multiplier,

$$\mathbf{H} = \begin{bmatrix} g(\mathbf{w}_1 \cdot \mathbf{x}_1 + b_1) & \cdots & g(\mathbf{w}_L \cdot \mathbf{x}_1 + b_L) \\ \vdots & \cdots & \vdots \\ g(\mathbf{w}_1 \cdot \mathbf{x}_N + b_1) & \cdots & g(\mathbf{w}_L \cdot \mathbf{x}_N + b_L) \end{bmatrix}_{N \times L} \tag{4}$$

is hidden layer output matrix, and $\mathbf{T} = [\mathbf{t}_1, ..., \mathbf{t}_N]^T$ is expected outputs. Setting the gradients of Lagrangian with respect to $(\boldsymbol{\beta}, \boldsymbol{\xi}, \boldsymbol{\alpha})$ equal to zero and we have

$$\beta = \left(\frac{\mathbf{I}}{\gamma} + \mathbf{H}^T\mathbf{H}\right)^{\dagger} \mathbf{H}^T\mathbf{T} \tag{5}$$

For the regularized parameter γ, we will use fast cross validation method to determine its value. It is unnecessary to re-compute $\left(\frac{\mathbf{I}}{\gamma} + \mathbf{H}^T\mathbf{H}\right)^{\dagger}$ for different γ, we just only need to pre-compute the eigenvalue decomposition of $\mathbf{H}^T\mathbf{H}$,

$$\mathbf{H}^T\mathbf{H} \overset{\text{svd}}{\leftarrow} \mathbf{U}\boldsymbol{\Sigma}\mathbf{U}^T = \sum_{i=1}^{L} \sigma_i \mathbf{u}_i \mathbf{u}_i^T \tag{6}$$

where $\mathbf{U} = [\mathbf{u}_1, ..., \mathbf{u}_L]^T$ is a singular matrix and $\boldsymbol{\Sigma} = \text{diag}\{\sigma_1, ..., \sigma_L\}$ is a singular values matrix. Then for each $\gamma \in \{\gamma_1, \gamma_2, ..., \gamma_q\}$, $\left(\frac{\mathbf{I}}{\gamma} + \mathbf{H}^T\mathbf{H}\right)^{\dagger}$ can be computed by

$$(\frac{\mathbf{I}}{\gamma} + \mathbf{H}^T\mathbf{H})^{\dagger} = \mathbf{U}\left(\boldsymbol{\Sigma} + 1/\gamma\right)\mathbf{U}^T \tag{7}$$

Given new instance $\mathbf{x}$, the prediction value of $\mathbf{y} = h(\mathbf{x})\boldsymbol{\beta} \in \mathbf{R}_{[-1,1]} \times \mathbf{R}$, the first output $y_1 \in \mathbf{R}_{[-1,1]}$ is the corresponding *operation type* and the second output $y_2 \in \mathbf{R}$ is the corresponding *transferred size*. We transform y_1 to binary class label with $\text{Sign}(y_1 > 0) = 1$ (Writting) and $\text{Sign}(y_1 < 0) = -1$ (Reading). The proposed method can then be summarized as:

Algorithm 1: RELM-Based Access Prediction

---

1 Collect the trace of distributed storage system: $\mathbf{TR} = \{\mathbf{tr}_0, \mathbf{tr}_1, ..., \mathbf{tr}_n\}$ where $\mathbf{tr} = \{\text{pid}, \text{inode}, \text{amt}, \text{time}, \text{op}\}$;
2 Filter series op and trz as two-dimensional series $\mathbf{ts} = [\mathbf{s}_o, \mathbf{s}_z]$;
3 Construct data set $\aleph = \{(\mathbf{x}_k, \mathbf{t}_k)\}_{i=k}^N$ by multi-dimensional phase space transformation, where
   $\mathbf{x}_k = \{s_o(k), s_o(k-\tau_o), ..., s_o(k-(d_o-1))\tau_1, s_z(k), s_z(k-\tau_z), ..., s_z(k-(d_z-1))\tau_z\}$ and
   $\mathbf{t}_k = (s_o(k+1), s_z(k+1))$;
4 Given candidate $\gamma = \{\gamma_1, \gamma_2, ..., \gamma_q\}$, validation set $\aleph_v$ (splitting one part from $\aleph$) , the number of hidden neurons L , and activation function $g(x)$;
5 Randomly assign input weights $\mathbf{w}$ and bias $\mathbf{b}$;
6 Calculate $\mathbf{H}$ as (4) ;
7 Compute the eigenvalue decomposition $\mathbf{H}^T\mathbf{H} \overset{\text{svd}}{\leftarrow} \mathbf{U}\boldsymbol{\Sigma}\mathbf{U}^T$;
8 **foreach** $\gamma_i \in \{\gamma_1, \gamma_2, ..., \gamma_s\}$ **do**
9 | a. Calculate the output weights $\boldsymbol{\beta} = \mathbf{U}(\boldsymbol{\Sigma} + 1/\gamma_i)\mathbf{U}^T\mathbf{H}^T\mathbf{T}$;
10 | b. Get validation error $e_{\gamma_i}^k = ||\mathbf{T}_v - \mathbf{H}_v\boldsymbol{\beta}||$;
11 | c. Get the optimized γ_i with the least e_{γ_i} and corresponding $\boldsymbol{\beta}$ as the final result.;
12 **end**
13 For new instance $\mathbf{x}$, compute the output by $\mathbf{y} = h(\mathbf{x})\boldsymbol{\beta}$;
14 Get *operation type* by $\text{Sign}(y_1 > 0) = 1$ (Writting) and $\text{Sign}(y_1 < 0) = -1$ (Reading);
15 Get block-level *transferred size* by y_2;

---

4   Experimental Study

In this section, experiments on predictions of block-level operation and *transferred size* have been performed in order to support data access optimization in intelligent workload monitoring and analysis. Since the data storage workload of large scale storage system directly influences the performance and capacity of data storage system architecture, it is essential to analyze the behavior of real-world storage workload for support in operational decision making. Moreover, it is fundamental to achieve the accurate prediction results which can provide the feedback mechanism to relevant components in data storage systems. In this experiment, RELM-based block-level operation and *transferred size* predictions have been performed on block-level I/O traces of enterprise storage volumes in order to predict future workload operations and transferred sizes.

4.1   The Datasets

Real world traces provided by the Storage Networking Industry Association (SNIA) are used in the present experimental study. 1-week block I/O traces were collected from enterprise servers at Microsoft Research Cambridge. Block I/O traces come from 13 servers, 36 logical volumes, and 179 hard disks as shown in Table 1. Event Tracing for Windows (ETW) was utilized for collection of data traces. Each trace event defines the I/O request which includes the 1) timestamp; 2) disk number; 3) initial logical number; 4) block-level *transferred size*; and 5) type of operation (read or write). We filtered only 3 attributes for our present study: 1) timestamp, 2) block-level *transferred size*, and 3) type of operation. In order to proceed with the experiments, we focused on two of those traces (denoted by the boldface rows in Table 1): server **proj** and **mds**. The dataset generated from these two traces by means of phase transformation (with an embedding dimensionality of $d=7$) is then given in Table 2.

Table 1. Description of Data Center Servers

No.	Server	Function	#Volumes
1	usr	User home directories	3
2	**proj**	**Project directories**	**5**
3	prn	Print Server	2
4	hm	Hardware monitoring	2
5	rsrch	Research projects	3
6	prxy	Firewall/Web proxy	2
7	src1	Source control	3
8	src2	Source control	3
9	stg	Web stagin	2
10	ts	Terminal Server	1
11	web	Web/SQL Server	4
12	**mds**	**Media Server**	**2**
13	wdev	Test web server	4

The measurement considered to measure the prediction performance of *transferred size* is the Normalized Root Mean Squared Error (NRMSE), which is defined as:

$$NRMSE = \frac{\sqrt{\sum_{i=1}^{n}(\hat{x}_i - x_i)^2}}{n(x_{max} - x_{min})} \qquad (8)$$

Table 2. Description of Data Center Servers

	# Training data	# Testing data	Extracted Features
mds	2800000	248738	14
proj	1000000	48569	14

where x_{max} is the maximum observed value and x_{min} is the minimum observed value. x_i defines the expected value at instant i, $\hat{x}_i$ defines the obtained value at time instant i, and n defines the number of observations predicted. For *operation type* prediction, we evaluate prediction performance based on the classification testing accuracy.

$$Testing accuracy = \frac{\text{Number of samples with right prediction}}{\text{Total number of samples}} \times 100\% \qquad (9)$$

4.2 Results

We first classify the **proj** and **mds** generated time series data according to the properties of stochasticity and stationarity in order to select the most adequate learning models to include as counterpart algorithms for comparison in the present study.

The ACF (Auto-correlation Function) [25] measures the correlation of a time series with itself shifted by some time delay and has been typically employed to

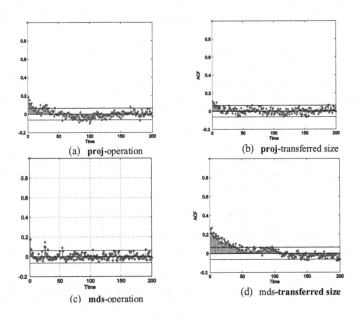

Fig. 2. Auto-correlation Function Measures of **proj** and **mds** time series

measure the stationarity characteristics of the series. Specifically, a time series with a high rate of decay in ACF indicates strong stationary property in the data. For proj, the ACF measurements are presented in Fig.2 (a) and (b); it delay very fast initially and then exhibit a constant auto-covariance structure that is stable over time. The results obtained indicated that **proj** is more likely a stationary time series. For **mds**, the ACF charts are presented in Fig.2 (c) and (d). The trend in the figures indicates that the ACF value decays slowly at beginning, thus indicating that **mds** is likely to be a non-stationary time series data.

The RP (Recurrence plot) [26], on the other hand, provides a way to visualize the periodic nature of a trajectory through a phase space, and can be employed to estimate the stochasticity characteristics of a time series. The more lines parallel to the diagonal the more deterministic the time series is. The RP charts of **proj** are presented in Fig.3 (a) and (b), where a low frequency of diagonals is observed. This implies that **proj** is likely to be stochastic. Likewise, the RP charts of **mds** in Fig.3 (c) and (d) also displayed a low frequency of diagonals. Taking the cue from [27], **proj** can be better modelled using statistical tools, such as the AR(Auto-Regressive) and ARMA(Auto-Regressive and Moving Average Model) models [25]; **mds** on the other hand should thus be better modelled using RBF (Radial Basis Function approximation) [27], BP(Back-propagation neural network) [28], Polynomial approaches [27] and SVM(Support Vector Machine)[29] To showcase the prediction accuracy of our proposed approach, we randomly selected 60 samples in **mds** to visualize the predicted results obtained along with the available target ground truth, which is depicted Fig.4. We can observed that in most of cases the prediction on *transferred size* is close to the expected value, however, in some moments, the predictions have been erroneous. On the *operation type* prediction problem, the results obtained have been considerably positive in most cases. Performance of the RELM against several state-of-the-art algorithms including BP, AR, ARMA, Polynomial approach and RBF are then reported in Tables 3 and 4.

Table 3. Performances of the different machine learning methods on the mds dataset. Results of AR, ARMA, Polynom and RBF have been taken directly from [27] for comparison study. - in table indicates that the information is not available in [27]

Methods	Transferred size			Block-level operation		
	NRMSE	Train. time(s)	Nodes	Test. accuracy(%)	Train. time(s)	Nodes
MV-ELM	0.0234	15.65s	200	99.04%	15.65s	200
BP	0.0311	1463.2s	50	99.02%	1653.5s	50
AR(1)	0.51	–	–	–	–	–
ARMA(5,5)	0.5	–	–	–	–	–
Polynom	0.88	–	–	–	–	–
RBF	0.25	–	–	–	–	–

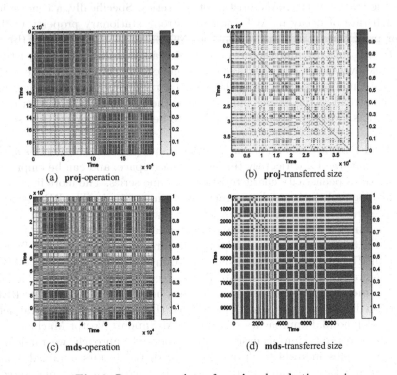

(a) **proj**-operation

(b) **proj**-transferred size

(c) **mds**-operation

(d) **mds**-transferred size

Fig. 3. Recurrence plots of **proj** and **mds** time series

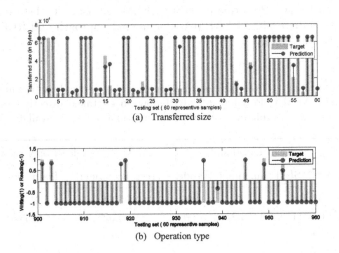

(a) Transferred size

(b) Operation type

Fig. 4. RELM prediction and target ground truth of **mds**

Table 4. Performances of the different machine learning methods on the proj dataset. Results of AR, ARMA, Polynom and RBF have been taken directly from [27]for comparison study. - in table indicates that the information is not available in [27]

Methods	Block-level Operation			Transferred size		
	NRMSE	Train. time(s)	Nodes	Test. accuracy(%)	Train. time(s)	Nodes
ELM	0.2279	8.94s	400	97.78%	9.52s	400
BP	0.2360	251.7s	50	97.72%	314.7s	50
Polynom	0.4	—	–	–	–	–
RBF	0.4	—	–	–	–	–
AR(1)	0.33	—	–	–	–	–
ARMA(1,2)	0.38	—	–	–	–	–

From the results, RELM is noted to showcase improved performance over all the other counterparts considered in both *operation type* and block-level *transferred size* predictions and having done so with a much shorter learning time incurred. On **mds**, RELM reported an accuracy that is above 99.04% in *operation type* prediction while generating a NRMSE of 0.0234 on block-level *transferred size* prediction.

4.3 Effects of Embedding Dimensions

In this subsection, we conducted further study on the effects of the embedding dimensions d on performance accuracy. To do so, we vary the embedding dimensions d incrementally from 1 to 9, i.e., $d = 1, 2, ..., 9$ to observe the changes in performance accuracy of the proposed RELM. The empirical results obtained for both multi-dimensional and single-dimensional phase transformations are then summarized in Fig.5. From the results, it can be concluded that the choice of embedding dimensions used greatly impacts the resultant performance accuracy. With reference to proj, the *transferred size* prediction at $d = 2$, is 0.235 (NRMSE), and decreases to 0.228 at $d = 7$. In the region of an embedding dimension d of 7 the performance accuracy is noted to remain relatively stable. This implies that although the embedding theorem [24] states that a smooth function exist to model the series if d is sufficiently large, it may not be necessary to use an overly large value of d in practice. Further, between the multi-dimensional and single-dimensional method of phase transformations, the former is found to exhibit higher accuracies on both the *operation type* and block-level *transferred size* prediction problems.

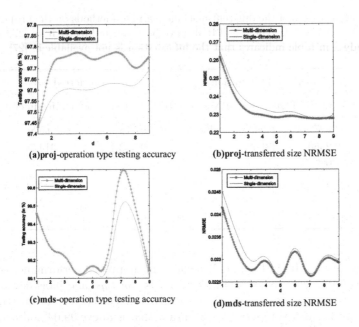

(a)proj-operation type testing accuracy   **(b)proj**-transferred size NRMSE

(c)mds-operation type testing accuracy   **(d)mds**-transferred size NRMSE

Fig. 5. RELM performance w.r.t embedding dimension d

5   Conclusion

In this paper, Regularized Extreme Learning Machine (RELM) has been presented and investigated for the prediction of block-level operation and *transferred size* in distributed storage system. Real-world block-level I/O traces of enterprise storage volumes at Microsoft Research at Cambridge provided by Storage Networking Industry Association (SNIA) have been used for experimental study and performance evaluations. Experimental results show that the proposed method provides improved prediction accuracy and learning time on the time series I/O traces compared to other state-of-art approaches including BP, AR and ARMA methods. The approach has been shown to apply well on both non-stationary and stationary time series data. The higher prediction accuracy on of future workload behaviors, i.e., block-level operation and *transferred size*, is expected to contribute in the optimization of distributed storage workload monitoring and analysis operations in data centers. In the near future, we hope to incorporate RELM into live distributed storage systems so as to better measure and study the true contributions brought about on lowering the power consumption and expenditures of data centers.

Acknowledgment. This work is partially supported by the ASTAR Thematic Strategic Research Programme (TSRP) Grant No. 1121720013 and the Computational Intelligence Research Laboratory at NTU.

References

1. DCD Intelligence, DCD Industry Census 2013: Data Center Power - Is the data center industry getting better at using power? (2014)
2. National Environment Agency (NEA), Data Center Energy Efficiency Benchmarking (2012)
3. Uptime Institute - The Global Data Center Authority, 2013 Data Center Industry Survey (2013)
4. National Renewable Energy Laboratory, Best Practices Guide for Energy-Efficient Data Center Design (2011)
5. Data Storage Technology brings Data Center Power Consumption benefits
6. Understanding the Behaviour of Solid State Disk
7. SAS and SATA, Solid-State Storage Lower Data Center Power Consumption
8. Leong, Y.K.: Future of Data Centers: Next Generation NVM and Hybrid Integration, Data Center Technologies Division (2012)
9. Google is improving its data centers with the power of machine learning
10. Devarakonda, M.V., Iyer, R.K.: Predictability of process resource usage: a measurement-based study on UNIX. IEEE Trans. Softw. Eng. 15(12), 1579–1586 (1989)
11. Faerman, M., Su, A., Wolski, R., Berman, F.: Adaptive Performance Prediction for Distributed Data-intensive Applications. In: Proceedings of the 1999 ACM/IEEE Conference on Supercomputing, New York, NY, USA (1999)
12. Wang, M., Au, K., Ailamaki, A., Brockwell, A., Faloutsos, C., Ganger, G.R.: Storage device performance prediction with CART models. In: Proceedings of the IEEE Computer Society's 12th Annual International Symposium on Modeling, Analysis, and Simulation of Computer and Telecommunications Systems (MASCOTS 2004), pp. 588–595 (2004)
13. Senger, L.J., Fernandes de Mello, R., Santana, M.J., Helena, R., Santana, C., Yang, L.T.: An On-Line Approach for Classifying and Extracting Application Behavior on Linux. In: Yang, L.T., Guo, M. (eds.) High-Performance Computing, pp. 381–401. John Wiley & Sons, Inc. (2005)
14. Senger, L.J., et al.: An Instance-based Learning Approach for Predicting Execution Times of . . . (2004)
15. Oldfield, R., Kotz, D.: Improving Data Access for Computational Grid Applications. Clust. Comput. 9(1), 79–99 (2006)
16. Kim, J., Chandra, A., Weissman, J.B.: Using Data Accessibility for Resource Selection in Large-Scale Distributed Systems. IEEE Trans. Parallel Distrib Syst. 20(6), 788–801 (2009)
17. AL-Mistarihi, H.H.E., Yong, C.H.: On Fairness, Optimizing Replica Selection in Data Grids. IEEE Trans. Parallel Distrib. Syst. 20(8), 1102–1111 (2009)
18. Ishii, R.P., Fernandes de Mello, R.: An Online Data Access Prediction and Optimization Approach for Distributed Systems. IEEE Trans. Parallel Distrib. Syst. 23(6), 1017–1029 (2012)
19. Jung, C., Woo, D.-K., Kim, K., Lim, S.-S.: Performance Characterization of Prelinking and Preloadingfor Embedded Systems. In: Proceedings of the 7th ACM & IEEE International Conference on Embedded Software, New York, NY, USA, pp. 213–220 (2007)
20. Spillane, R.P., Wright, C.P., Sivathanu, G., Zadok, E.: Rapid File System Development Using Ptrace. In: Proceedings of the 2007 Workshop on Experimental Computer Science, New York, NY, USA (2007)

21. Rosenstein, M.T., Collins, J.J., De Luca, C.J.: Reconstruction Expansion As a Geometry-based Framework for Choosing Proper Delay Times. Phys. D 73(1-2), 82–98 (1994)
22. Fraser, A.: Information and entropy in strange attractors. IEEE Trans. Inf. Theory 35(2), 245–262 (1989)
23. Abarbanel, H., Brown, R., Sidorowich, J., Tsimring, L.: The analysis of observed chaotic data in physical systems. Rev. Mod. Phys. 65(4), 1331–1392 (1993)
24. Robinson, J.C.: A topological delay embedding theorem for infinite-dimensional dynamical systems. Nonlinearity 18(5), 2135 (2005)
25. Box, G.E.P., Jenkins, G.: Time Series Analysis, Forecasting and Control. Holden-Day, Incorporated (1990)
26. Marwan, N., Carmen Romano, M., Thiel, M., Kurths, J.: Recurrence plots for the analysis of complex systems. Phys. Rep. 438(5-6), 237–329 (2007)
27. Ishii, R.P., Rios, R.A., Mello, R.F.: Classification of time series generation processes using experimental tools: a survey and proposal of an automatic and systematic approach. Int. J. Comput. Sci. Eng. 6(4), 217 (2011)
28. Rumelhart, D.E., Hinton, G.E., Williams, R.J.: Learning representations by back-propagating errors. Nature 323(6088), 533–536 (1986)
29. Chang, C.-C., Lin, C.-J.: LIBSVM: A Library for Support Vector Machines. ACM Trans. Intell. Syst. Technol. 2(3), 1–27 (2011)

ELM Based Fast CFD Model with Sensor Adjustment*

Hongming Zhou, Yeng Chai Soh, Chaoyang Jiang, and Wenjian Cai

School of Electrical and Electronic Engineering,
Nanyang Technological University, Nanyang Avenue, Singapore 639798
{hmzhou,eycsoh,CJIANG003,ewjcai}@ntu.edu.sg

Abstract. Computational fluid dynamics (CFD) simulation is a useful tool to provide temperature and air flow patterns of an indoor environment. However, one big drawback of the CFD simulation is that it usually requires high computational power and takes long time for one simulation. Thus it is hard to apply the CFD results in the HVAC real-time dynamic visualization system. In this paper, we proposed a fast machine learning approach to solve the CFD problem. The extreme learning machine (ELM) is used to learn the CFD model and it can estimate the CFD results approximately 100 times faster than the conventional CFD simulator. The estimated results could be further calibrated based on the indoor sensor information. ELM is also helpful to generate the error surface to correct the previous estimated CFD surface. The whole system only takes a few seconds to complete the CFD task in our case and thus it is possible to achieve the real-time visualization purpose.

Keywords: HVAC, Computational Fluid Dynamics, K-means, Extreme Learning Machine.

1 Introduction

Computational fluid dynamics (CFD) [1] is the science of using numerical methods and algorithms to solve fluid-related problems. In an indoor environment, the CFD model is a useful tool to provide heat transfer patterns and air flow dynamics. It has been used in a wide range of HVAC applications [2]. However, due to its complex internal mathematics, typically the Navier-Stokes equations [3] that defines the continuous fluid flow, CFD simulations cost heavily on computational power. A typical 3-D CFD simulation requires high-performance computers and takes long time for obtaining the fluid flow results. The CFD simulation also requires a lot of engineering expertise to obtain a validated solution.

In many real-time HVAC applications, there are needs of fast updating of the indoor conditions. However, the standard CFD simulation may take too long time to be implemented in a real-time manner. In our project that requires real-time monitoring of room conditions, there is an urgent need to come out a solution to increase the speed of CFD drastically. At the same time, maintaining or even increasing the CFD accuracy is desirable. To solve a CFD problem, there are many existing numerical methods.

---

* This work is supported by Singapore's National Research Foundation under NRF2011NRF-CRP001-090. It is also partially supported by the Energy Research Institute at NTU (ERIAN).

Two main approaches are the finite volume method(FVM) [4] which is a relatively fast and has low memory requirement, and the finite element method (FEM) [5] which is much more stable than FVM but is slower and requires large more memory [6]. There are certainly many other approaches as well, including finite difference method, spectral element method, boundary element method and etc. However, none of the above solutions can give satisfactory computational speed for the real-time updating purpose.

Extreme learning machine [7, 8] is an emergent machine learning technique that can approximate any continuous target functions. Due to its proven universal approximation ability [9] and extremely fast learning speed, ELM can potentially be used as a learning algorithm to solve the CFD problems to achieve real-time updating purpose. In this paper, the ELM regression model is used to predict the detailed CFD results based on the boundary conditions (obtained from sensors located at boundary) and location information. The predicted CFD result follows closely to the actual CFD result. Based on the sensors located within the room, further calibration to the predicted CFD can be carried out to make the predicted results as accurate as possible.

The remaining part of the paper is organized as follows: section 2 reviews some related works such as extreme learning machine and K-means clustering algorithm; section 3 describes the proposed approaches of how to implement ELM to solve CFD related problems and the adjustment based on sensor information; section 4 concludes the paper.

2   Related Works

2.1   Extreme Learning Machine

Extreme learning machine was originally proposed by Huang et al. [7] for the single-hidden-layer feedforward neural networks and then extended to the generalized single-hidden-layer feedforward networks where the hidden layer need not be neuron alike [10]. The hidden layer parameters are randomly generated without tuning and are independent of the training data. The input data is mapped from the input space to the L-dimensional hidden layer feature space (ELM feature space). The output of ELM can be written as:

$$f_L(\mathbf{x}) = \sum_{i=1}^{L} \beta_i h_i(\mathbf{x}) = \mathbf{h}(\mathbf{x})\beta \tag{1}$$

where $\beta = [\beta_1, \cdots, \beta_L]^T$ is the matrix of the output weights from the hidden nodes to the output nodes, and $\mathbf{h}(\mathbf{x}) = [h_1(\mathbf{x}), \cdots, h_L(\mathbf{x})]$ is the row vector representing the outputs of L hidden nodes with respect to the input $\mathbf{x}$. $\mathbf{h}(\mathbf{x})$ actually maps the data from the d-dimensional input space to the L-dimensional hidden layer feature space (*ELM feature space*) $\mathbf{H}$, and thus, $\mathbf{h}(\mathbf{x})$ is indeed a feature mapping.

The above linear equations can be written in the matrix form:

$$\mathbf{H}\beta = \mathbf{T} \tag{2}$$

where $\mathbf{H}$ is the hidden layer output matrix:

$$\mathbf{H} = \begin{bmatrix} \mathbf{h}(\mathbf{x}_1) \\ \vdots \\ \mathbf{h}(\mathbf{x}_N) \end{bmatrix} = \begin{bmatrix} h_1(\mathbf{x}_1) & \cdots & h_L(\mathbf{x}_1) \\ \vdots & \vdots & \vdots \\ h_1(\mathbf{x}_N) & \vdots & h_L(\mathbf{x}_N) \end{bmatrix} \qquad (3)$$

and $\mathbf{T} = [t_1, \cdots, t_N]^T$ is a vector of target labels. The solution of above equation is given as: $\beta = \mathbf{H}^\dagger \mathbf{T}$, where $\mathbf{H}^\dagger$ is the Moore-Penrose generalized inverse [11, 12] of matrix $\mathbf{H}$.

2.2 K-means Clustering Algorithm

K-means is one of the simplest unsupervised learning algorithms that solve the well known clustering problems. It classifies a given dataset through a certain number of clusters.

Let $X = \{x_1, x_2, x_3, \cdots, x_n\}$ be the set of data points and $V = \{v_1, v_2, v_3, \cdots, v_c\}$ be the set of centers, the K-means algorithm aims at minimizing an objective function known as squared error function given by:

$$J(V) = \sum_{i=1}^{c} \sum_{j=1}^{c_i} (\| x_i - v_i \|)^2 \qquad (4)$$

where, $\| x_i - v_i \|$ is the Euclidean distance between x_i and v_i, c_i is the number of data points in i^{th} cluster, c is the number of cluster centers.

The algorithmic steps for K-means clustering algorithm are given as follows:

1. Randomly select c cluster centers;
2. Calculate the distance between each data point and cluster centers;
3. Assign the data point to the cluster center whose distance from the cluster center is minimum of all the cluster centers;
4. Recalculate the new cluster center using:

$$v_i = (1/c_i) \sum_{j=1}^{c_i} x_i \qquad (5)$$

5. Recalculate the distance between each data point and new obtained cluster centers;
6. If no data point was reassigned then stop, otherwise repeat from step 3.

The CFD simulation results are dense data that represent the complete thermal status of the room. One slice of CFD data at a specific height of the room can contains up to millions of data. Thus, the K-means clustering algorithm could be extremely useful to group the data for the ease of further analysis. The K-means clustering results could also be very useful to extract common features between different CFD results.

3   Fast CFD Estimation Using ELM

In our approach, firstly we did 27 different 2D CFD simulation results based on different boundary conditions as shown in Table 1. The CFD results were used for training the ELM network to obtain the ELM estimation model to approximate the CFD solution. Secondly based on the estimated CFD results and existing CFD results, we did Kmeans clustering analysis. The clustering results were used to help us better understand the similarities of temperature distribution between estimated and actual CFD result. Such similarity were used to reproduce the limited indoor sensor information. We aimed to use ELM to learn the error surface as well. once the error surface between estimated and actual CFD results are learnt, the estimated CFD results can be enhanced by adding this error surface. The proposed approach are shown and explained by Fig. 1.

Table 1. Specification of boundary conditions

Wall Temp (° C)	Inlet Air Velocity (m/s)	Inlet Temp (° C)
24.5	0.9	20
25.0	1.0	21
25.5	1.1	22

3.1   Fast Estimation of the CFD Results Based on ELM

3.1.1   Problem Formulation

Give boundary conditions $\mathbf{B} = \{\mathbf{b_1}, \mathbf{b_2}, \cdots, \mathbf{b_n}\}, \mathbf{b_i} \in \mathbb{R}^N$ and the CFD result $\mathbf{w}(\mathbf{x}, t), \mathbf{x} = (x, y, z) \in \mathbb{X}, t \in \mathbb{T}$, we could find a $f(.)$ such that:

$$f(\mathbf{B}, \mathbf{x}) = \mathbf{w}(\mathbf{x}, t) \qquad (6)$$

that is, at each specific location $\mathbf{x}$, there exists a model that can learn $f(.)$ if given $\mathbf{B}$ and $\mathbf{w}$. Instead of solving the complex partial differential equations, which may take long time and require high computational cost, we try to solve the problem using machine learning approach. As an emergent machine learning technique, extreme learning machine (ELM) is proven to be an universal approximator [9] and is very fast with minimal computational cost. Thus ELM could definitely be a suitable model to learn the CFD equations. As shown in Fig. 2, at each specific location of the CFD result, one ELM is used to learn the relationship between boundary conditions, location and the CFD result at the corresponding location.

3.1.2   Simulated CFD Results Using ELM

In this paper, we simulated 27 CFD results based on the boundary conditions defined in Table. 1. Out of the 27 CFD results, 26 of them are used for training and 1 CFD result is left for validating the accuracy of the trained model. For better visualized view, the 2D CFD image is transformed to a 3D surface, where it is easier to identify the differences of absolute temperature values at different locations. Fig. 3 shows the 3D representation of the 2D CFD image.

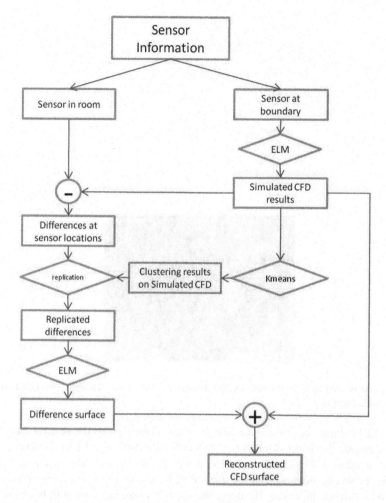

Fig. 1. The thermal sensation scale of ELM predicted, PMV calculated and actual experimental results

When using ELM to train the 26 CFD results, the boundary condition and the location information (x, y) are used as the training features and the temperature results that obtained by CFD are used as the label information. Each CFD result contains 79×79 sample points. So there are totally $26 \times 79 \times 79 = 162266$ training data points and $79 \times 79 = 6241$ testing data points. The equality constrained optimization method based ELM [8] is used to learn the model. During parameter selection process, we aim to select the best pair of paramters that can result in the most close result to the actual case. In our simulation, the hidden nodes chosen is $10,000$ and the regularization parameter is chosen as 2^{16}. Fig. 4(a) shows the 3D surface of ELM simulated results. For better analysis of the errors and do further calibration based on sensor data, we also plot the difference between actual surface (Fig. 3) and ELM simulated surface (Fig. 4(a)) as shown in Fig. 4(b).

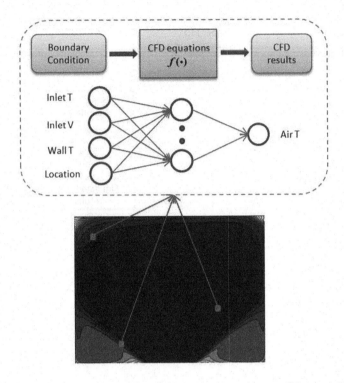

Fig. 2. ELM is used to learn the relationship between the boundary condition and CFD result at each specific location

The ELM simulated surface has similar shape compared to the actual CFD shapes. It can describe the general temperature distribution well. For ELM to generate such surface, it takes 2.36 seconds, whereas if we do CFD simulation, it takes around 3 minutes. From Fig. 4(b), we can see that the ELM results are still not very accurate and need to be further improved. By using the indoor sensor data, we aim to improve the accuracy of the ELM results.

3.2 ELM Estimation of the Error Surface

3.2.1 Problem Formulation
The task of improving the accuracy of ELM simulated surface is the same as estimation of an accurate error surface as shown in Fig. 4(b). In our simulation setup, we placed 8 sensors around the boundary as shown in Fig. 5. That is given 8 sensor readings $\{x_i, y_i, \Delta z_i\}_{i=1}^{8}$, where (x_i, y_i) are the sensor locations, $\Delta z_i = z_{Sensor} - z_{ELM}$ represent the "true" error. z_{Sensor} are sampled from the testing CFD data, which represents the actual temperature distribution. Based on the limited (8 in our case) sensor information, we need to reconstruct the error surface.

3.2.2 Kmeans Clustering
From the observation on the actual condition (Fig. 3), ELM simulated results (Fig. 4(a)) and error surface (Fig. 4(b)), it is found that they follows similar distribution pattern.

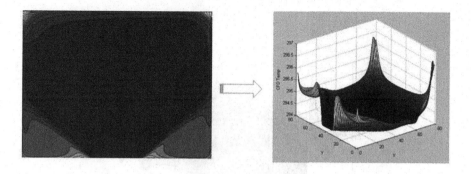

Fig. 3. The 2D CFD image is transformed to 3D surface for better visulized analysis

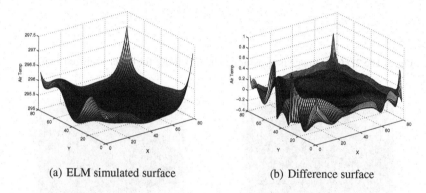

(a) ELM simulated surface   (b) Difference surface

Fig. 4. The ELM simulated surface and the difference surface between ELM and CFD

Thus it is interesting to extract such similarities and use them as the guideline to reproduce sensor information. Fig. 6 shows the Kmeans clustering results on both ELM simulated results and error surface. Both images follow similar pattern. Thus the clustering results on ELM simulated surface could be used as a guideline to reproduce the indoor sensor data, which would be helpful for us to reconstruct the error surface more accurate.

3.2.3 Sensor Data Replication

Based on the similarity found from ELM simulated surface and error surface, the 8 sensor points are extended to 49 data points as shown in Fig. 7. If one of the sensor is addressed within a certain cluster, all the replicated data points within this cluster will have the same value as the sensor point. If two or more sensors are addressed within the same cluster, the replicated data will be assigned the mean value of the sensor data.

The 8 sensors' location are: $z_{(1,1)}$, $z_{(1,40)}$, $z_{(1,79)}$, $z_{(40,1)}$, $z_{(40,79)}$, $z_{(79,1)}$, $z_{(79,40)}$, $z_{(79,79)}$. According to Fig. 7, it is found that $z_{(1,1)}$ belongs to one cluster, $z_{(1,40)}$, $z_{(40,79)}$, $z_{(79,40)}$ belong to one cluster, $z_{(40,1)}$ belongs to one cluster and $z_{(79,1)}$, $z_{(1,79)}$, $z_{(79,79)}$ belong to one cluster. All the other duplicated points are assigned with the corresponding mean value of the sensor points in the same cluster.

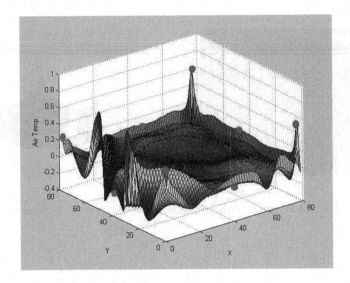

Fig. 5. The 2D CFD image is transformed to 3D surface for better visulized analysis

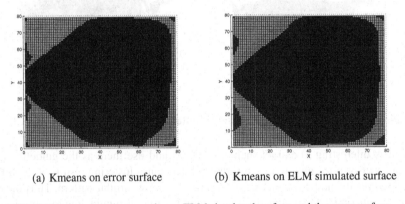

(a) Kmeans on error surface   (b) Kmeans on ELM simulated surface

Fig. 6. Kmeans clusteirng results on ELM simulated surface and the error surface

3.2.4 Reconstruction of the Error Surface Using ELM

Given the 8 sensor data and 41 extra data points, we aim to reconstruct the error surface. The 49 known data points are used as the training data. The locations are the training features and the temperature differences are the label. When reconstructing the surface, all the pixel locations 79×79 are the testing features. In the training phase, the 49 training points are randomly split into 36 training and 13 validation points. The best parameters are chosen based on 100 trails. The reconstructed error surface (Fig. 8(a)) is based on the testing results of the 79×79 location information. The testing results are used to correct the error surface to make it smooth and as close to "0" as possible. It takes ELM 0.038 seconds to generate such surface in our case. The visualized ELM corrected error surface is shown in Fig. 8(b). As we can observe from Fig. 8(a), the ELM error surface tends to

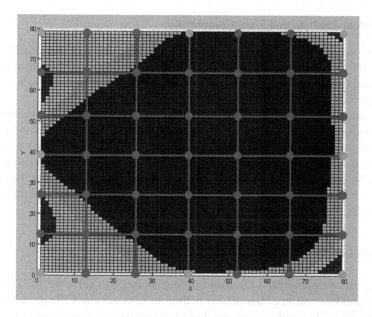

Fig. 7. Baed on the kmeans clustering results on ELM simulated results, extra 41 data points (red dots) are created on top of the original 8 sensors (green dots)

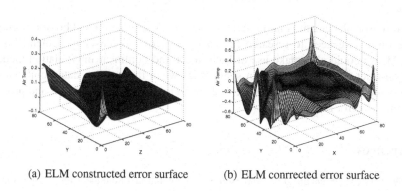

(a) ELM constructed error surface   (b) ELM conrrected error surface

Fig. 8. The visualized error surface of ELM, and the corrected error surface after subtracting ELM error surface

compromise the sharp change region of original error surface. Thus when subtracting the ELM predicted error surface from original error surface, the edge sharp fluctuation will be reduced. To quantize the effect of ELM, we can look at the mean errors of the different results. The root mean square error (RMSE) of the original error surface is 0.0845. The ELM corrected error surface has a RMSE of 0.0708, which shows an improvement of 16.21%. Indeed we can expect a much higher improvement if the indoor sensors are optimally placed.

3.3 System Analysis and Time Comparison

The previous two sub-sections introduce how ELM is used to generate CFD results and modify the results afterwards. It takes ELM 2.36s to generate the estimated CFD surface and 0.038 seconds to generate the error surface. If we use the CFD simulator, it takes approximately 3 mins to complete the simulation.

The performance of ELM on correcting the error surface depends heavily on the sensor placement. From Fig. 5, we can see that the sensors are not placed at the highly fluctuated points, thus many important information could be lost. Though the Kmeans clustering results can bring in the distribution information so that the sensor information could be replicated according to the distribution, the high fluctuation information of the surface is still incomplete. Thus in the future, the optimal sensor placement is really desirable for better accuracy.

4   Conclusion

In this paper, we presented a machine learning approach to solve the CFD problems. The extreme learning machine (ELM) is used to estimate the CFD model based on many existing CFD simulation results. The ELM approach is an economic fast solution that can generate the estimated CFD results quickly (in a few seconds). It does not take heavy computational cost as well. Based on the indoor sensor information, the ELM estimated results can further be calibrated to make it as close to the actual situation as possible.

The sensor placement is an important factor to determine the calibration accuracy. In the future, we need to research further on the optimal sensor placement. Instead of ELM, we could also make use of both linear and nonlinear interpolation methods to reconstruct the error surface. In some cases when the training data points are not enough, the ELM simulated surface could vary a lot during different runs and provides meaningless error estimation.

References

1. Pletcher, R.H., Tannehill, J.C., Anderson, D.: Computational fluid mechanics and heat transfer. CRC Press (2012)
2. Ladeinde, F., Nearon, M.D.: CFD applications in the HVAC&R industry. ASHRAE Journal 39, 44–48 (1997)
3. Aris, R.: Vectors, Tensors, and the basic Equations of Fluid Mechanics. Courier Dover Publications, New York (1990)
4. Patankar, S.V.: Numerical Heat Transfer and Fluid Flow. CRC Press, BR (1980)
5. Zienkiewicz, O.C., Taylor, R.L., Zhu, J.Z.: The Finite Element Method: Its Basis and Fundamentals, 7th edn. Butterworth-Heinemann, Oxford (2013)
6. Surana, K.S., Allu, S., Tenpas, P.W., Reddy, J.N.: K-version of finite element method in gas dynamics: higher-order global differentiability numerical solutions. International Journal for Numerical Methods in Engineering 69(6), 1109–1157 (2007)
7. Huang, G.-B., Zhu, Q.-Y., Siew, C.-K.: Extreme learning machine: Theory and applications. Neurocomputing 70, 489–501 (2006)

8. Huang, G.-B., Zhou, H., Ding, X., Zhang, R.: Extreme Learning Machine for Regression and Multiclass Classification. IEEE Transactions on Systems, Man, and Cybernetics - Part B: Cybernetics 42(2), 513–529 (2012)
9. Huang, G.-B., Chen, L., Siew, C.-K.: Universal approximation using incremental constructive feedforward networks with random hidden nodes. IEEE Transactions on Neural Networks 17(4), 879–892 (2006)
10. Huang, G.-B., Chen, L.: Convex Incremental Extreme Learning Machine. Neurocomputing 70, 3056–3062 (2007)
11. Serre, D.: Matrices: Theory and Applications. Springer-Verlag New York, Inc. (2002)
12. Rao, C.R., Mitra, S.K.: Generalized Inverse of Matrices and its Applications. John Wiley & Sons, Inc., New York (1971)

Melasma Image Segmentation
Using Extreme Learning Machine

Yunfeng Liang[1], Zhiping Lin[1], Jun Gu[1], Wee Ser[1],
Feng Lin[2], Evelyn Yuxin Tay[3], Emily Yiping Gan[3],
Virlynn Wei Ding Tan[3], and Tien Guan Thng[3]

[1] School of Electrical and Electronic Engineering,
Nanyang Technological University, Singapore
[2] School of Computer Engineering, Nanyang Technological University, Singapore
[3] National Skin Center, Singapore
lian0102@e.ntu.edu.sg

Abstract. This paper introduces an image segmentation method based
on the extreme learning machine (ELM) to detect melasma in human
face images. In this work, skin texture features are extracted and fed
into ELM classifier for segmentation. The results obtained by the proposed method show better segmentation performance visually than that
obtained by a conventional threshold based image segmentation method.
The proposed method could potentially lead to the development of a
more reliable computerized melasma severity assessment system.

Keywords: Extreme Learning Machine, Image Segmentation, Melasma,
Skin.

1 Introduction

Melasma is a commonly acquired disorder of hyperpigmentation that affects millions of people with Fitzpatrick skin type IV to VI [1][2][3][4]. It has been shown
in research that melasma has a significant negative impact on life quality [5]. Assessment of melasma severity is an important process in clinical treatment as it
reveals the workability of treatment to doctors and patients. Currently, melasma
area and severity index (MASI) and modified MASI (mMASI) are the most fre-
quently used instruments to measure disease severity and treatment response
in clinical trials. Although MASI and mMASI are reliable, they have certain
limitations. Firstly, they are subjective measurements which result in intra- and
inter-assessor variability. Secondly, the score is digitized and small change will
not be reflected if it is within one band or overstated if it crosses bands. Thirdly,
they are time consuming and complex in calculations. These are the factors that
motivates the authors to develop a computerized system to assess the severity
of melasma. The work in this paper is the first step towards the development
of the computerized assessment system: melasma image segmentation using extreme learning machine (ELM).

Many researches have been conducted in the area of digital image analysis of skin pigmentation, especially melanoma [6]. In melanoma diagnosis, dermoscopic images are often used to provide a better view of the subsurface structures for more accurate diagnosis. However, for the case of melasma, the main concern is to have an objective assessment of the severity of melasma and the involved area is much larger than the detection region in dermoscopy. In this work, the Visia complexion analysis system is utilized to acquire standardized digital images.

A typical image obtained from the Visia system is firstly divided into regions which contain different parts of a face according to the clinical scoring system MASI. Non-face area is subsequently detected and eliminated from the image. Various texture features of each pixel in the face area are extracted for classification and segmentation. We manually select melasma and normal regions in face skin area to generate the training data set. ELM is then applied to classify each face pixel based on the features for image segmentation.

Recently, a melasma segmentation method using a conventional threshold technique was proposed and is applied to giving computerized mMASI scores [7]. Comparing with the threshold based method, the ELM based segmentation method presented in this paper shows better segmentation performance visually and is easier to implement as there are fewer parameters to tune. As accurate melasma area detection is the most crucial part for the assessment of melasma, the proposed ELM method gives a promising way to build a reliable computerized mMASI calculation system.

2   MASI and mMASI

Melasma Area and Severity Index (MASI) is a subjective assessment to quantify the severity of melasma at various time points during treatment. MASI score is calculated using three parameters: area of involvement (A), darkness (D) and homogeniety (H). Forehead (f), right malar (rm), left malar (lm) and chin (c) region occupies 30%, 30%, 30% and 10%, of the total face area respectively. The area of involvement in each of these 4 regions is given a numerical value of 0 to 6 (0= no involvement, 1 = lees than 10%, 2= 10-29%, 3= 30-49%, 4= 50-69%, 5= 70-89%, 6= 90-100%). Darkness (D) and homogeneity (H) are numbered from 0 to 4 based on ratings (0= absent, 1= slight, 2= mild, 3=marked, 4=maximum). MASI is calculated as the sum of the severity ratings for the weighted value of the area involved multiplied by the sum of darkness and homogeneity for each of the 4 areas (see Fig. 1), the formula is as following:

$$MASI = 0.3A(f)(D(f) + H(f)) + 0.3A(lm)(D(lm) + H(lm))$$
$$+ 0.3A(rm)(D(rm) + H(rm)) + 0.1A(c)(D(c) + H(c)) \quad (1)$$

where f, lm, rm, c stands for forehead, left malar, right malar and chin regions respectively.

A modified version of MASI (mMASI) was proposed [8], which proved that the homogeneity (H) was the most difficult component to be reliably assessed

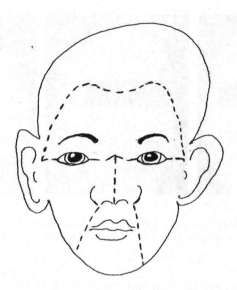

Fig. 1. A typical face image showing forehead(f), right malar(rm), left malar (lm) and chin(c) for MASI calculation.

and was removed from MASI calculation. The mMASI score is given as:

$$mMASI = 0.3A(f)D(f) + 0.3A(lm)D(lm) + 0.3A(rm)D(rm) + 0.1A(c)D(c)$$

$$(2)$$

In creating the computerized melasma assessment method, the mMASI system is adopted. The A and D values for each of the four different parts of face should be determined. Segmentation is the first step towards the development of computerized method of assessing melasma.

3 Methods

To have a complete view of the whole face, three images are taken for each patient. A full frontal image shows the forehead and the chin, two images including the left and right side of the face illustrates the malar regions (Fig. 2).

3.1 Preprocessing

Before applying ELM to detect the melasma area, the images were preprocessed to eliminate the non-face areas. Many studies have been conducted in the area of skin segmentation [9] [10] and the illumination variation condition remains a challenging problem. However, standardized images are taken with consistent light illumination in this work for the ease of skin region segmentation. We adopt the color difference method (CIE76) which is accurate and time efficient.

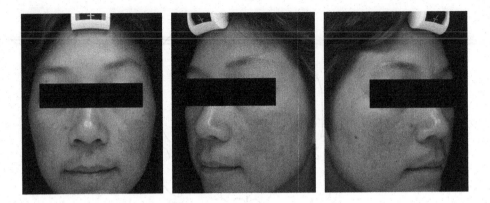

Fig. 2. Three images of a typical patient face obtained from different angles using Visia system. Frontal view(left),left side view (middle) and right side view(right).

Color difference between the skin and the origin $(L, a, b = 0, 0, 0)$ is bigger than difference between other regions and the origin. Therefore, a suitable threshold could be selected to segment the skin regions:

$$\Delta E = \sqrt{(L_1 - L_2)^2 + (a_1 - a_2)^2 + (b_1 - b_2)^2} \tag{3}$$

$$The\ pixel\ is \begin{cases} skin & , \text{if } \Delta E > t \\ non\text{-}skin & , \text{else} \end{cases} \tag{4}$$

where $(L_1\ a_1\ b_1)$ are the image pixel values form $[L, a, b]$ color space, $(L_2\ a_2\ b_2) = (0\ 0\ 0)$ is the origin, t is the threshold.

For some non-face area in the image such as the neck and ear, it is difficult to differentiate with threshold method. Geometric information is utilized to make eliminations.

To align the proposed computerized method with MASI, four regions defined by MASI are identified and divided. The frontal view images are used to obtain the forehead and chin regions. Local thresholding method is performed to locate eyes, nose and mouth. The forehead region is divided by cropping the image along the horizontal line that cross eyes. For the chin part, as the boundary is a curve that crossing the bottom of nose, left most and right most points of lips, we propose to cut along a polynomial curve that pass through these points. The same method is used to eliminate the forehead and chin regions in the side view images so that only malar regions remain in the left part.

3.2 Feature Extraction

Image texture is a set of metrics which could be acquired for image analysis. It is extensively used in image segmentation and classification. It reveals information of the visual patterns and spatial arrangement of colors or intensities in an image.

In our proposed method, 20 texture features are used to characterize each pixel which include co-occurrence matrix and the Laws Texture Energy Measures [11]. The features are stored in an $n \times 20$ matrix $\boldsymbol{x}$, where n is the number of training pixels.

Gray Level Co-occurrence Matrix. A gray level co-occurrence matrix (GLCM) is a 2D matrix that unveils the information about the positions of pixels which shares similar gray level values with each other. A GLCM $\mathbf{P}_d[i,j]$ is defined by first specifying a displacement vector $d = (d_x, d_y)$ and counting all pairs of pixels separated by d having gray levels i and j. The GLCM characterize the texture of an image by calculating how often a pixel with specific grey level values i occurs horizontally adjacent to a pixel with the value j. Each element (i,j) in GLCM specifies the number of times that the pixel with value i occurrs horizontally adjacent to a pixel with value j.

Gray level co-occurrence matrices capture texture properties of an image. Numeric features computed from the GLCM can be used to represent the texture more compactly. There are five major texture attributes for an image and they are all derived from the GLCM. They are entropy, contrast, correlation, energy and homogeneity.

Laws Texture. Laws texture measure describes texture via the output of properly designed linear filters. In his work, Laws defines the following five one dimensional convolutional filters as building blocks for higher order texture descriptors:

$$\begin{aligned}
L5 &= \begin{bmatrix} 1 & 4 & 6 & 4 & 1 \end{bmatrix} \ (level) \\
E5 &= \begin{bmatrix} -1 & -2 & 0 & 2 & 1 \end{bmatrix} \ (edge) \\
S5 &= \begin{bmatrix} -1 & 0 & 2 & 0 & -1 \end{bmatrix} \ (spot) \\
W5 &= \begin{bmatrix} -1 & 2 & 0 & -2 & 1 \end{bmatrix} \ (wave) \\
R5 &= \begin{bmatrix} 1 & -4 & 6 & -4 & 1 \end{bmatrix} \ (ripple)
\end{aligned} \tag{5}$$

From these simple 1D filters, more complicated 2D texture filters can be created by performing multiplication of two 1D filters (e.g. $L5^T E5, W5^T R5$ etc.). To get the Laws textures, the input image and the texture filters are convloved.

3.3 Image Segmentation Using ELM

Extreme Learning Machine (ELM) is an efficient neural network learning algorithm [12][13]. In ELM, there is only one layer of hidden nodes, the output of a single layer of L hidden nodes and input $\boldsymbol{x}$, which is the feature matrix obtained from previous step, can be expressed as:

$$f_L(\boldsymbol{x}) = \sum_{i=1}^{L} \beta_i h_i(\boldsymbol{x}) = \boldsymbol{h}(\boldsymbol{x})\boldsymbol{\beta} \tag{6}$$

where $\boldsymbol{\beta}$ is the output weight vector, $\boldsymbol{h}(\boldsymbol{x})$ is the output vector of the hidden layer.

$$h_i(\boldsymbol{x}, \boldsymbol{a}_i, b_i) = exp(-b_i||\boldsymbol{x} - \boldsymbol{a}_i||^2) \tag{7}$$

where $\boldsymbol{a}_i, b_i$ are the parameters of the Gaussian activation function. Then the output weight β_i is determined analytically by using the following formula:

$$\mathbf{H} \cdot \beta = \mathbf{T} \tag{8}$$

where

$$\mathbf{H} = \begin{bmatrix} G\left(\boldsymbol{x}_1; b_1, \boldsymbol{a}_1\right) & \cdots & G\left(\boldsymbol{x}_1; b_L, \boldsymbol{a}_L\right) \\ \vdots & \ddots & \vdots \\ G\left(\boldsymbol{x}_N; b_1, \boldsymbol{a}_1\right) & \cdots & G\left(\boldsymbol{x}_N; b_L, \boldsymbol{a}_L\right) \end{bmatrix}_{N \times L} \tag{9}$$

$$\beta = \left(\beta_1^T \ \beta_2^T \ \cdots \ \beta_L^T\right)_{d \times L}^T \tag{10}$$

and the target output

$$\mathbf{T} = \left(t_1^T \ t_2^T \ \cdots \ t_L^T\right)_{d \times N}^T . \tag{11}$$

$\mathbf{H}$ is the hidden layer output matrix with the ith row representing the output vector of hidden layer with respect to input $\boldsymbol{x}_i$ and the ith column representing the output of the ith hidden node with regard to all the inputs $(\boldsymbol{x}_1, \cdots, \boldsymbol{x}_N)$. The least square solution $\widehat{\beta}$ is calculated in ELM so that the training error is minimum.

$$\left\|\mathbf{H}\widehat{\beta} - \mathbf{T}\right\|_2 = \min_{\beta} \left\|\mathbf{H}\widehat{\beta} - \mathbf{T}\right\|_2 . \tag{12}$$

$\widehat{\beta}$ is calculated using the following formula:

$$\widehat{\beta} = \mathbf{H}^{\Theta}\mathbf{T} \tag{13}$$

where $\mathbf{H}^{\Theta}$ is the Moore-Penrose generalized inverse of $\mathbf{H}$ which achieves the smallest norm of β and minimizes the training error at the same time. In theory, the least square solution of minimum norm β is unique [14].

Accordingly, the training process of ELM involves three steps once a training data $(\boldsymbol{x}_i, t_i), i = 1, \cdots, N$ is given:

- Generate L hidden nodes with parameters $(\mathbf{a}, b)$ of each node randomly assigned;
- Calculate the hidden layer output matrix $\mathbf{H}$;
- Calculate the output weight $\widehat{\beta}$.

The output weight $\widehat{\beta}$ is then used to determine the output for the samples in the testing set.

4   Results

In our experiments on melasma images, ELM segmentation method shows better performance in comparison with the threshold method. One example will be illustrated in the following. Figure 3(a) shows the left malar region obtained by performing the preprocessing in section 3.1 on the original image of size 209×204. To get the training data set, the melasma and normal skin areas are

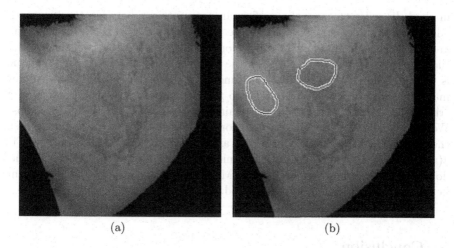

(a)   (b)

Fig. 3. Training data set selection. (a) Original image. (b) Training data set selected image.

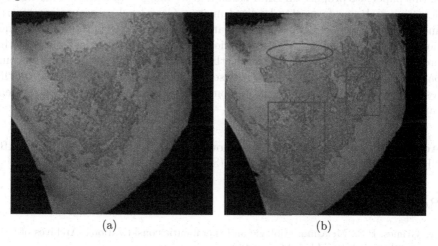

(a)   (b)

Fig. 4. Segmentation results of two different methods. (a) Proposed method. (b) Threshold method with main wrong classification areas highlighted.

manually selected. As shown in Fig.3(b), the pixels within the two circles drawn by the user are the selected training pixels. In this example, there are 880 and 811 pixels in the selected melasma and normal skin area correspondingly. 20 features introduced in section 3.2 are extracted for each image pixel and the result feature matrix x is substituted in the ELM training process in equation (6). The output weight $\hat{\beta}$ is calculated. The features of each face pixel in the image will be fed into the ELM system to do the classification. The segmentation result is shown in Fig. 4(a), where the melasma area is bounded by the black line. Fig. 4(b) shows the result obtained from applying threshold based segmentation method. In order to show the performance comparison between the two methods, two rectangles and one ellipse are added to Fig. 4(b) to locate the main areas that contain different classification results between the two methods. In the areas bounded by the rectangles, the false positive rate of detecting melasma is very high if observe visually, which means many pixels belonging to the normal skin set are wrongly classified into the melasma set. For the other kind of error, false negative detections, are mainly in the area bounded by the ellipse. We can see nearly half of the brown color pixels in the ellipse which should be in the melasma set are wrongly classified as normal skin pixels. If we observe Fig. 4(a) and compare with Fig. 3(a), the misclassification problems just mentioned do not show up, showing that the ELM based segmentation method performs better than the threshold based method.

5   Conclusion

In this paper, we propose an ELM based texture image segmentation method for detecting melasma from standardized images. Gray level co-occurrence matrix and Laws textures are utilized to generate features. ELM with radial basis function kernel is applied in the classification. By assessing the result visually, the proposed ELM based segmentation method shows a better performance in compare with a conventional threshold based segmentation method. Based on the result, a more reliable and accurate computerized melasma assessment system can be developed in the future.

Acknowledgement. We wish to acknowledge the funding support by ASTAR-NHG-NTU Skin Research Grant 2014 (SRG\14011).

References

1. Grimes, P.E.: Melasma: etiologic and therapeutic considerations. Archives of Dermatology 131(12), 1453–1457 (1995)
2. Sheth, V.M., Pandya, A.G.: Melasma: a comprehensive update. Journal of the American Academy of Dermatology 65(4), 689–714 (2011)
3. Tamega Ade, A., Miot, L., Bonfietti, C., Gige, T., Marques, M., Miot, H., et al.: Clinical patterns and epidemiological characteristics of facial melasma in brazilian women. Journal of the European Academy of Dermatology and Venereology 27(2), 151–156 (2013)
4. Handel, A., Lima, P., Tonolli, V., Miot, L., Miot, H.: Risk factors for facial melasma in women: a case-control study. British Journal of Dermatology (2014)
5. Balkrishnan, R., McMichael, A., Camacho, F., Saltzberg, F., Housman, T., Grummer, S., Feldman, S., Chren, M.-M.: Development and validation of a health-related quality of life instrument for women with melasma. British Journal of Dermatology 149(3), 572–577 (2003)
6. Korotkov, K., Garcia, R.: Computerized analysis of pigmented skin lesions: a review. Artificial Intelligence in Medicine 56(2), 69–90 (2012)

7. Liang, Y., Lin, Z., Ser, W., Lin, F., Gan, E., Tay, E., Thng, S.T.G.: Automated scoring of melasma using computerized digital image analysis of clinical photographs - a pilot study. Presented in International Pigment Cell Conference (IPCC 2014), Singapore, September 4-7 (2014)
8. Pandya, A.G., Hynan, L.S., Bhore, R., Riley, F.C., Guevara, I.L., Grimes, P., Nordlund, J.J., Rendon, M., Taylor, S., Gottschalk, R.W., et al.: Reliability assessment and validation of the melasma area and severity index (MASI) and a new modified MASI scoring method. Journal of the American Academy of Dermatology 64(1), 78–83 (2011)
9. Vezhnevets, V., Sazonov, V., Andreeva, A.: A survey on pixel-based skin color detection techniques. In: Proc. Graphicon, Moscow, Russia, vol. 3, pp. 85–92 (2003)
10. Phung, S.L., Bouzerdoum, A., Chai Sr, D.: Skin segmentation using color pixel classification: analysis and comparison. IEEE Transactions on Pattern Analysis and Machine Intelligence 27(1), 148–154 (2005)
11. Laws, K.I.: Textured image segmentation. Tech. rep., DTIC Document (1980)
12. Huang, G.-B., Zhu, Q.-Y., Siew, C.-K.: Extreme learning machine: theory and applications. Neurocomputing 70(1), 489–501 (2006)
13. Huang, G.-B., Zhou, H., Ding, X., Zhang, R.: Extreme learning machine for regression and multiclass classification. IEEE Transactions on Systems, Man, and Cybernetics, Part B: Cybernetics 42(2), 513–529 (2012)
14. Rao, C.R., Mitra, S.K.: Generalized inverse of matrices and its applications, vol. 7. Wiley, New York (1971)

Detection of Drivers' Distraction Using Semi-Supervised Extreme Learning Machine*

Tianchi Liu[1,**], Yan Yang[2,**,***], Guang-Bin Huang[1], and Zhiping Lin[1]

[1] School of Electrical and Electronic Engineering,
Nanyang Technological University, Singapore
{tcliu,egbhuang,ezplin}@ntu.edu.sg
[2] Energy Research Institute @ NTU (ERI@N), Interdisciplinary Graduate School,
Nanyang Technological University, Singapore
y.yang@ntu.edu.sg

Abstract. Monitoring drivers' visual behavior using machine learning techniques has been identified as an effective approach to detect and mitigate driver distraction to enhance road safety. In our previous work, detection system based on supervised Extreme Learning Machine (ELM) was developed and tested with satisfactory performance. However, supervised ELM requires all training data to be labeled, which can be costly and time-consuming. This paper proposed and evaluated a semi-supervised distraction detection system based on Semi-Supervised Extreme Learning Machine (SS-ELM). The experimental results show that SS-ELM outperformed supervised ELM in both accuracy (95.5% for SS-ELM vs. 93.0% for ELM) and model sensitivity (97.6% for SS-ELM and 95.5% for ELM), suggesting that the proposed semi-supervised detection system can extract information from unlabeled data effectively to improve the performance. SS-ELM based detection system has the potential of improving accuracy and alleviating the cost of adapting distraction detection systems to new drivers, and thus is more promising for real world applications.

Keywords: Semi-supervised Extreme Learning Machine, Eye Tracking, Driver Cognitive Distraction.

1   Introduction

Distracted driving has emerged as one of the key risk factors for road traffic injuries, due to the advancement and prevalence of personal communication devices [1]. For example, the use of cell phones during driving causes thousands

---

* Tianchi Liu's work was supported by a grant from Singapore Academic Research Fund (AcRF) Tier 1 under Project RG 80/12 (M4011092).

** These authors contribute equally to this work.

*** The permanent address is with State Key Laboratory of Millimeter Waves, School of Information Science and Engineering, Southeast University, Nanjing 210096, China. Yan Yangs work was supported in part by the National Science Foundation of China under Grant No. 61101216, and in part by the 111 Project under Grant No. 111-2-05.

of additional fatalities every year in the United States [2]. Efficient driver distraction detection systems are required to monitor drivers' states in real time. Machine learning methods have drawn much attention because of their ability to discover hidden patterns revealing the complicated relationship between observable behavior changes and drivers' inner states [3,4,5,6,7,8].

As distraction affects drivers' behavioral, psychological and physiological measures, they are considered as potential features for machine learning algorithms. Several studies have been focus on developing feature combinations, including mainly 4 types [9]: 1) subjective ratings from drivers, 2) drivers' driving performance, such as lane position and steering control, 3) physiological measures, such as ECG and EEG signals, and 4) behavior measures, such as eye movement pattern and head position. Non-intrusive measures like eye movement, head position and vehicle dynamics are more promising for real-time distraction detection system [10], compared to subjective ratings and intrusive measures like EEG and ECG signals.

Previous work used various machine learning techniques for driver state detection. For example, Liang's group developed one Support Vector Machines (SVM) based [3] and one Bayesian Network based [11] detection systems. Studies on their performance indicated that both methods can provide satisfactory detection accuracy (81.1% and 80.1% on average respectively). One novel technique, namely Long Short-Term Memory (LSTM) recurrent neural networks [12], was developed specifically for distraction detection and proved to outperform SVM in their study. Tango's [7] study implemented several machine learning techniques for distraction detection. Results suggested that SVM had higher detection accuracy compared to two neural network methods, one fuzzy inference systems, and one traditional classifier, i.e., logistic regression.

Recently, a new machine learning tool, Extreme Learning Machine (ELM) [13,14], has gain much attention due to its simple structure, high generalization capability, and fast computation speed. ELM is a single-hidden layer feed forward neural network (SLFN), where the parameters of hidden mapping function are randomly generated and output weights between hidden neurons and output nodes are calculated analytically. The advantages of ELM are usually reflected in real applications as high accuracy, sensitivity and robustness. Previous work from our group has proven a higher accuracy gained by ELM compared to SVM in an on-road driving experiment, with 95% and 92% respectively [15].

However, all the distraction detection systems proposed so far are based on supervised learning, meaning that the training of such systems need to be 'supervised' by human experts by providing a target set for the training data containing distraction status. The supervised learning paradigm is only suitable for early stage research and may not be suitable for implementation in real driving cases, because of the huge cost and difficulty of creating target distraction status set, which requires additional subjective ratings by the driver [12], post-processing by the experimenters [16,7], or additional computation based on data from other sources [3]. On the other hand, data without labels of distraction status (unlabeled data) are in fact easy to collect without additional costs, e.g., collecting from driver's daily driving records.

Semi-supervised learning is therefore proposed to enable the detection systems to learn from these 'unlabeled' data efficiently, so as to reduce the cost of training and to adapt the detection system to individual drivers, bringing the system closer to real world applications. Semi-supervised learning methods are capable of learning the model with a limited number of labeled data and relatively large amount of unlabeled data. There are mainly two types of semi-supervised algorithms characterized by their underlying assumptions made on the data distribution, i.e., 1) cluster assumption, or equivalently low density assumption, and 2) manifold assumption [17]. Several semi-supervised learning algorithms have been proposed under each paradigm.

Recently Huang et al. [18] extend ELM [13] to semi-supervised learning under manifold regularization framework. The proposed Semi-Supervised Extreme Learning Machine (SS-ELM) not only performs well on data sets satisfying manifold assumption, but also inherits salient advantages of ELM, such as easy for implementation, fast training and predicting, and naturally capable of handling multi-class problems, making it more promising for real-time driver distraction detection system compared to other semi-supervised algorithms.

In this study, we aim to apply SS-ELM to driver distraction detection. We evaluate the detection performance of SS-ELM on a data set collected by eye tracking devices in on-road experiments and compare SS-ELM performance with that of purely supervised ELM. Testing accuracy and sensitivity are the evaluation measures used to assess the models. We expect the eye and head movement measures used in our study form manifold in high-dimensional space and satisfy the manifold assumption. SS-ELM method is thus able to effectively make use of unlabeled eye tracking data and becomes a better alternative to supervised learning methods, e.g., ELM, for distraction detection.

2   Extreme Learning Machine

Proposed by Huang [13,14], Extreme Learning Machine (ELM) is an effective and efficient learning scheme for "generalized" Single-hidden Layer Feedforward Neural networks (SLFNs) with randomly generated and fixed parameters associated with hidden neurons. Though simple in structure, such SLFNs retains the universal approximation capability [19], and therefore can be applied in many applications, such as regression and multi-class classification. ELM learns the model by solving a regularized least-square problem in closed form and thus avoids iterative tuning, which is considered more efficient than traditional learning algorithm for neural network, e.g., back-propagation.

ELM considers the driver distraction detection as a classification problem, where driver's state characterized by several input measures is classified into one of the states, i.e., normal driving or distracted driving. Given a set of N training data $\{X, Y\} = \{x_i, y_i\}_{i=1}^{N}$, where $x_i \in \mathbb{R}^d$ represents the input measures, such as eye and head movements, and $y_i \in \mathbb{R}^m$ is a binary vector with the entry, corresponding to the class that x_i belongs to, equals to one. Here d and m are the dimensions of input and output respectively.

The SLFNs with L hidden neurons and a sigmoid activation function is mathematically formulated in the following,

$$f_k = \sum_{i=1}^{L} h_i(x_k)\beta_i = h(x_k)\beta, k = 1, 2, \ldots, N, \tag{1}$$

where

$$h_i(x; a_i b_i) = \frac{1}{1 + \exp(-(a_i^T x + b_i))}. \tag{2}$$

Here a_i, b_i are the parameters of the sigmoid activation function, $h_i(x_k)$ is the output value of the i-th hidden neurons with respect to sample x_k, $h(x_k)$ is the vector form of output of all hidden neurons with respect to sample x_k, β are the output weights that connect the hidden layer with the output layer, f_k are the outputs corresponding to sample x_k. The predicted class corresponds to the output node with largest output value.

With the parameters between input nodes and hidden neurons fixed, the only free parameters that need to be determined are the output weights β. ELM solves the output weights by minimizing the sum of the squared losses of the prediction errors as well as the norm of the output weights, which can be formulated in the following

$$\min_{\beta \in \mathbb{R}^{L \times d}} \quad \frac{1}{2}\|\beta\|^2 + \frac{C}{2}\sum_{i=1}^{N}\|\xi_i\|^2 \tag{3}$$
$$\text{s.t.} \quad h(x_i)\beta = y_i^T - \xi_i^T, \quad i = 1, \ldots, N,$$

where the first term in the objective function is a regularization term to prevent the solution from overfitting problem, $\xi_i \in \mathbb{R}^L$ is the error vector caused by the i-th training sample, and C is a penalty coefficient on the training errors. This problem can be solved analytically.

3   Semi-Supervised Extreme Learning Machine

Semi-Supervised Extreme Learning Machine (SS-ELM) is a recently proposed semi-supervised learning algorithm based on ELM and manifold regularization framework [18]. Compared to its supervised origin, SS-ELM is proposed to improve the performance by incorporating both labeled and unlabeled data. Compared to other semi-supervised algorithms, SS-ELM inherits the salient advantages of ELM, such as high efficiency and naturally capable of handling multi-class problems.

Under manifold regularization framework, SS-ELM assumes that the high-dimensional input data from each class lie on a low-dimension data manifold, and the optimal separating hyperplane is 'smooth' with respect to the manifold. In other words, input data that are close along one manifold should have similar predicted class labels, which can be formulated as minimizing the following regularization term

$$L_m = \frac{1}{2} \sum_{i,j} w_{i,j} \| \boldsymbol{f}_i - \boldsymbol{f}_j \|^2, \tag{4}$$

where $\boldsymbol{f}_i$ and $\boldsymbol{f}_j$ are the predictions with respect to sample $\boldsymbol{x}_i$ and $\boldsymbol{x}_j$ and w_{ij} is the pair-wise similarity between two samples $\boldsymbol{x}_i$ and $\boldsymbol{x}_j$. This formulation penalizes large difference in the predicted class labels when two samples $\boldsymbol{x}_i$ and $\boldsymbol{x}_j$ are of high similarity. The above form can be further simplified into a matrix form

$$L_m = Tr(\boldsymbol{F}^T \boldsymbol{L} \boldsymbol{F}), \tag{5}$$

where $\boldsymbol{L} = \boldsymbol{D} - \boldsymbol{W}$ is known as the *graph Laplacian* and $\boldsymbol{D}$ is an diagonal matrix with its diagonal elements

$$d_{ii} = \sum_{j=1}^{l+u} \boldsymbol{w}_{i,j}.$$

It is also common practice to normalize $\boldsymbol{L}$ by $\boldsymbol{D}^{-\frac{1}{2}} \boldsymbol{L} \boldsymbol{D}^{-\frac{1}{2}}$ or replace $\boldsymbol{L}$ by $\boldsymbol{L}^p$, based on some prior knowledge. By incorporating this manifold regularization term into the formulation of supervised ELM, SS-ELM has the formulation as:

$$\min_{\boldsymbol{\beta} \in \mathbb{R}^{L \times d}} \quad \frac{1}{2} \| \boldsymbol{\beta} \|^2 + \frac{C_{y_i}}{2} \sum_{i=1}^{l} \| \boldsymbol{\xi} \|^2 + \frac{\lambda}{2} Tr(\boldsymbol{F}^T \boldsymbol{L} \boldsymbol{F})$$

$$\text{s.t.} \quad h(\boldsymbol{x}_i)\boldsymbol{\beta} = \boldsymbol{y}_i^T - \boldsymbol{\xi}^T, \quad i = 1, \ldots, l \tag{6}$$

$$\boldsymbol{f}_i = h(\boldsymbol{x}_i)\boldsymbol{\beta}, i = 1, \ldots, l+u,$$

where $\boldsymbol{L} \in \mathbb{R}^{(l+u) \times (l+u)}$ is the graph Laplacian, $\boldsymbol{F}$ is the output of ELM with each row corresponding to the output of one sample, λ is a tradeoff parameter, and l and u are number of labeled and unlabeled data respectively. To handle imbalanced data set, different penalty coefficients C_{y_i} are assigned to the prediction error of samples from different classes.

Similar to the supervised ELM, this problem has analytic solution in two forms. If the number of training samples are greater than the number of hidden neurons, the output weight $\boldsymbol{\beta}$ can be solved in the following form

$$\boldsymbol{\beta} = \left(\boldsymbol{I}_L + \boldsymbol{H}^T \boldsymbol{C} \boldsymbol{H} + \lambda \boldsymbol{H}^T \boldsymbol{L} \boldsymbol{H} \right)^{-1} \boldsymbol{H}^T \boldsymbol{C} \boldsymbol{Y}, \tag{7}$$

where $\boldsymbol{C} \in \mathbb{R}^{(l+u) \times (l+u)}$ is a diagonal matrix with

$$c_{ii} = \frac{C_0}{N_{y_k}}$$

. Here C_0 is the user specified penalty coefficient, $\boldsymbol{x}_i$ belongs to $\boldsymbol{y}_k$ which has N_{y_k} training samples. If the number of hidden neurons is greater than the number of training samples, the output weight should be computed in the alternative form

$$\boldsymbol{\beta} = \boldsymbol{H}^T \left(\boldsymbol{I}_{l+u} + \boldsymbol{C} \boldsymbol{H} \boldsymbol{H}^T + \lambda \boldsymbol{L} \boldsymbol{H} \boldsymbol{H}^T \right)^{-1} \boldsymbol{C} \boldsymbol{Y}. \tag{8}$$

In summary, both ELM and SS-ELM train the model in two steps: 1) compute the output matrix H of the hidden neurons with input weights and bias of the hidden neurons generated at random; and 2) solve the output weight β (using Equation (7) or Equation (8) for SS-ELM).

4   Model Construction

Data used in this study are from an on-road driving experiment with eleven experience drivers who participated and performed two types of driving tasks: normal driving and driving with secondary tasks, representing attentive and distracted driving respectively. The secondary tasks were designed to be visually intensive. A serial of images containing one to three bigger circles (as targets) and a group of smaller ones (as distractors) were displayed on a five-inch in-vehicle touch screen, mounted on the dashboard. The driver were asked to click on the bigger circles within 10 seconds after each image appeared and each secondary task lasted 60 seconds in duration. Each secondary task contains three levels of difficulty and each driver was required to repeat each level for three times. To avoid the learning effect, drivers were given time to practice the primary driving and the secondary tasks before the data collection commenced. Raw eye and head movement data was collected using $FaceLAB^{TM}$ eye tracker, which consists of two infrared cameras to record drivers' eye movements include drivers' head and gaze positions, gaze angles, blink and saccade, with a frequency of 60 Hz [20]. Pre-processing of data was conducted, including filtering and data reduction. Thirteen measures were used for model training including head position filtered x,y,z, head rotation angle filtered x,y,z, etc.

The processed data were then summarized using sliding windows, with the size of 10 seconds and 95% overlap, to create instances for later model construction. The features contained in each instance include the statistical eye/head movement measurements within the current 10 second window, such as the mean and standard deviation of head rotation angles. The summarized *instances* with feature information were then labeled based on the experimental condition, i.e. the instances were labeled as "normal" or "distracted", according to the experimental condition under which the instances belong to.

To construct and evaluate detection models, we 1) split the *instances* of each driver into three sets, i.e., labeled, unlabeled and test set; 2) constructed ELM models using only labeled set and SS-ELM models using both labeled and unlabeled sets; and 3) evaluated the performance of two types of models on the test sets.

More specifically, we split the *instances* of each driver into 4 folds, one fold was reserved for testing (named test set) and the rest three folds were used to train the model (training set). Each one of the 4 folds was used for testing once. This 4-fold cross validation process was repeated for 3 times, resulting 12 different splits in total. 60 data were randomly selected from the training set to form the labeled set, while the rest of each split formed unlabeled set. The regularization parameter C in ELM and C_0 in SS-ELM were both selected from $[10^6, 10^5, \ldots, 10^6]$ and optimized via 3 runs of a 4-fold cross-validation on the validation accuracy. For *instance* from all 11 drivers, we constructed 10 nearest-neighbor graph based on Euclidean distance with weights computed using Gaussian kernel (with width equal to 1). The resulting 1st degree *graph Laplacian* L without normalization was used for training the SS-ELM model.

We evaluate the performance of each model by comparing the predicted class with the ground truth, i.e., the experimental condition. Two evaluation measures were used to assess the model performance, i.e., testing accuracy and sensitivity.

5   Results and Discussion

There were 12 models of both ELM and SS-ELM constructed for each driver, based on different splits of data sets to rule out the possibility that one result may be an artifact created by a particular selection of data. The testing accuracy and sensitivity of the models were used as indicators for model performance. The performance of ELM and SS-ELM were first compared against the by-chance performance, i.e., 50% testing accuracy and zero sensitivity. The performances of 12 ELM and SS-ELM models averaged over all drivers were then statistically compared using paired t-test. The p-value of 0.05 was used as the criteria for statistical significance.

Table 1. Performance Comparison of SS-ELM

Performance Measures	ELM	SS-ELM	p-value
Testing Accuracy (%)	93.01 ± 4.43	95.49 ± 3.33	1.06e-13*
Sensitivity (%)	95.51 ± 3.98	97.61 ± 2.80	3.25e-12*

* denotes a statistical significance of $p < 0.05$.

From the results in Table 1, both ELM and SS-ELM achieved better-than-chance performance. ELM achieved an testing accuracy of 93.0% on average, while SS-ELM achieved an average testing accuracy of 95.5%. The sensitivity of ELM and SS-ELM were 95.5% and 97.6%. The sensitivity of the models are considered high, as there were only less than 5% and 3% of cases, where distracted state were not detected. This has also taken into account the conditions where even in a "performing secondary task" condition, a driver may not actually have engaged in the secondary tasks for some short duration, due to critical primary driving tasks (e.g. when being taken over by a lorry).

The comparison between ELM and SS-ELM by paired t-test within each driver suggests that SS-ELM significantly outperformed ELM for most of the time, both in terms of testing accuracy and sensitivity, i.e., for 7 out of 11 drivers (p < 0.05). Even in the cases where SS-ELM did not show a significant improvement, it achieved the same level of performance compared to supervised ELM. Cross-subject t-test comparison shows that SS-ELM significantly outperforms ELM overall, in terms of both testing accuracy (t = 8.18, p = 1.06e-13) and sensitivity (t = 7.55, p = 3.25e-12).

Assuming that the labeled training data are sampled temporally and independently and have no overlap among one another (to better represent the underlying distraction of the whole data set), 60 instances (created using 10 seconds windows) require 10 minutes data. Increasing the number of labeled training data may improve the performance, but is undesired because the collection of larger labeled data prolongs the training time and brings other costs in real world applications, such as labeling costs and drivers' emotional cost. In the contrast, by incorporating unlabeled eye and head movement data into model construction, the model developed using SS-ELM produces around 2% accuracy compared to the pure supervised ELM model. This is equivalent to reducing the error rate by 35.48%. The significant improvement of model accuracy by SS-ELM requires only additional unlabeled data, which can be easily retrieved from driving records.

6   Conclusion

In this paper, we proposed an SS-ELM based driver distraction detection method and evaluated its performance against its pure supervised counterpart - ELM. Both ELM and SS-ELM have several significant advantages over other algorithms, such as easy for implementation, fast training and predicting, and naturally capable of handling multi-class problems. Both algorithms were tested on data set collected from real on-road driving experiment and showed satisfactory detection accuracy, suggesting that these algorithms have the capability of handling noisy data in complicated real traffic situation and have the potential to be used in driver assistance system for safety enhancement. Compared with ELM, SS-ELM model achieved a significantly higher testing accuracy. Therefore we believe the underlying assumption of SS-ELM, i.e., manifold assumption, holds for this and other set of eye and head movement data for distraction detection problem. Our findings also show the potential of SS-ELM and other manifold based (semi-supervised and unsupervised) algorithms on improving the accuracy of driver distraction detection, and further alleviating the reliance on labeled data.

References

1. World Health Organization, "Road traffic injuries" (March 2013), http://www.who.int/mediacentre/factsheets/fs358/en/ (accessed September 2014)
2. Wilson, F.A., Stimpson, J.P.: Trends in fatalities from distracted driving in the United States, 1999 to 2008. American Journal of Public Health 100, 2213–2219 (2010)
3. Liang, Y., Reyes, M., Lee, J.: Real-time detection of driver cognitive distraction using support vector machines. IEEE Transactions on Intelligent Transportation Systems 8, 340–350 (2007)

4. Miyaji, M., Kawanaka, H., Oguri, K.: Effect of pattern recognition features on detection for driver's cognitive distraction. In: Proc. of Intelligent Transportation Systems (ITSC), pp. 605–610 (September 2010)

5. Hirayama, T., Mase, K., Takeda, K.: Detection of driver distraction based on temporal relationship between eye-gaze and peripheral vehicle behavior. In: Proc. of the 15th International IEEE Conference on Intelligent Transportation Systems (ITSC), pp. 870–875 (September 2012)

6. Jin, L., Niu, Q., Hou, H., Xian, H., Wang, Y., Shi, D.: Driver cognitive distraction detection using driving performance measures. Discrete Dynamics in Nature and Society 2012 (2012)

7. Tango, F., Botta, M.: Real-time detection system of driver distraction using machine learning. IEEE Transactions on Intelligent Transportation Systems 14, 894–905 (2013)

8. Sonnleitner, A., Treder, M.S., Simon, M., Willmann, S., Ewald, A., Buchner, A., Schrauf, M.: EEG alpha spindles and prolonged brake reaction times during auditory distraction in an on-road driving study. Accident Analysis & Prevention 62, 110–118 (2014)

9. Arun, S., Sundaraj, K., Murugappan, M.: Driver inattention detection methods: A review. In: Proc. of IEEE Conference on Sustainable Utilization and Development in Engineering and Technology (STUDENT), pp. 1–6 (October 2012)

10. Yang, Y., McDonald, M., Zheng, P.: Can drivers' eye movements be used to monitor their performance? a case study. IET Intelligent Transport Systems 6, 444–452 (2012)

11. Liang, Y., Lee, J.D., Reyes, M.L.: Nonintrusive detection of driver cognitive distraction in real time using Bayesian Networks. Transportation Research Record: Journal of the Transportation Research Board 2018(1), 1–8 (2007)

12. Wöllmer, M., Blaschke, C., Schindl, T., Schuller, B., Farber, B., Mayer, S., Trefflich, B.: Online driver distraction detection using Long Short-Term Memory. IEEE Transactions on Intelligent Transportation Systems 12(2), 574–582 (2011)

13. Huang, G.-B., Zhu, Q.-Y., Siew, C.-K.: Extreme learning machine: theory and applications. Neurocomputing 70(1-3), 489–501 (2006)

14. Huang, G.-B., Zhou, H., Ding, X., Zhang, R.: Extreme learning machine for regression and multiclass classification. IEEE Transactions on Systems, Man, and Cybernetics, Part B: Cybernetics 42, 513–529 (2012)

15. Yang, Y., Sun, H., Liu, T., Huang, G.-B., Sourina, O.: Drivers' workload detection in on-road driving environment using machine learning. In: Proc. of International Conference on Extreme Learning Machines (in press, 2014)

16. Oyini Mbouna, R., Kong, S., Chun, M.-G.: Visual analysis of eye state and head pose for driver alertness monitoring, pp. 1462–1469 (September 2013)

17. Chapelle, O., Schölkopf, B., Zien, A. (eds.): Semi-supervised learning, vol. 2. MIT Press, Cambridge (2006)

18. Huang, G., Song, S., Gupta, J., Wu, C.: Semi-supervised and unsupervised extreme learning machines. IEEE Transactions on Cybernetics PP, 1–1 (2014)

19. Huang, G.-B., Chen, L., Siew, C.-K.: Universal approximation using incremental constructive feedforward networks with random hidden nodes. IEEE Transactions on Neural Networks 17, 879–892 (2006)

20. Seeing Machines, Canberra, Australia. FaceLAB5 User Manual (2009)

Driver Workload Detection in On-Road Driving Environment Using Machine Learning

Yan Yang[1,*], Haoqi Sun[1,2,3,4], Tianchi Liu[3,**],
Guang-Bin Huang[3], and Olga Sourina[4]

[1] Energy Research Institute @ NTU (ERI@N),
Nanyang Technological University, Singapore
[2] Interdisciplinary Graduate School, Nanyang Technological University, Singapore
[3] School of Electrical and Electronic Engineering,
Nanyang Technological University, Singapore
[4] Fraunhofer IDM @ NTU, Nanyang Technological University, Singapore
{y.yang,tcliu,egbhuang,eosourina}@ntu.edu.sg,
hsun004@e.ntu.edu.sg

Abstract. Drivers' high workload caused by distractions has become one of the major concerns for road safety. This paper presents a data-driven method using machine learning algorithms to detect high workload caused by surrogate in-vehicle (IV) secondary tasks performed in an on-road experiment with real traffic. The data were collected using an instrumented vehicle while drivers performed two types of secondary tasks: visual-manual and auditory-vocal tasks. Two types of machine learning methods, support vector machine (SVM) and extreme learning machine (ELM), were applied to detect drivers' workload via drivers' visual behaviour (i.e. eye movements) data alone, as well as visual plus driving performance data. The results suggested that both methods can detect drivers' workload at high accuracy, with ELM outperformed SVM in most cases. We found that for visual intensive workload, using drivers' visual data alone achieveed an accuracy close to using the combination information from both visual and driving performance data. This study proves that machine learning methods can be used for real driving applications.

Keywords: Driving behaviour, Machine learning, Support Vector Machine (SVM), Extreme Learning Machine (ELM), Workload, Distraction, In-vehicle systems, Eye movements.

* To whom correspondence should be addressed. y.yang@ntu.edu.sg. The permanent address is with State Key Laboratory of Millimeter Waves, School of Information Science and Engineering, Southeast University, Nanjing 210096, China. Yan Yang's work was supported in part by the National Science Foundation of China under Grant No. 61101216, and in part by the 111 Project under Grant No. 111-2-05.
** Tianchi Liu's work was supported by a grant from Singapore Academic Research Fund (AcRF) Tier 1 under Project RG 80/12 (M4011092).

J. Cao et al. (eds.), *Proceedings of ELM-2014 Volume 2*,
Proceedings in Adaptation, Learning and Optimization 4, DOI: 10.1007/978-3-319-14066-7_37

1 Introduction

The prevalence of advanced in-vehicle information systems support drivers with more assistance but at the same time may also inevitably cause distraction, which has become one of the major concerns for road safety [4]. The increased workload induced by multi-tasking when using these systems is known to negatively affect drivers' performance, reaction time and critical event detection, which can lead to driver errors, and sometimes accidents. In order to mitigate the effects of distractions, monitoring drivers' workload in real-time via non-intrusive driver/vehicle information is required. When a driver is observed as being overloaded, some in car infotainment can be switched off, or further services can be provided accordingly.

Machine learning techniques are such tools that are able to capture, approximate and generalize the complex relationships in high dimensional feature distributions, which typically occur in driving performance evaluation. Therefore, they are proposed to mine the hidden patterns between high workload and drivers' performance, physiological changes for workload detection. In the past, machine learning has been successfully used to reveal the underlying patterns of brain states [15], drowsiness [12], decision making [19] and driver identification [26].

Some research efforts are found using machine learning to detect drivers' workload. For example, Liang and colleagues conducted a study in a driving simulator to detect mental workload when performing a short memory task: stock tracking secondary task, using SVM [11]. They achieved an accuracy of 95% when distinguishing between two driver statuses: normal/baseline driving and distracted driving. It needs to be noted that the impact of different types of secondary tasks, for example, visual-manual and auditory-vocal tasks, can have distinctively different effects on drivers' visual and driving behaviour [28]. Testing on mental tasks alone is not sufficient for future in-vehicle applications, and a comparison between both types of secondary tasks will provide further insight in the usefulness of these techniques. In addition, the recognition of a visual overload and mental distraction would evoke different mitigating approaches (e.g. declutter visual information or highlight extra visual aids). Also, it is necessary to test this concept in real road traffic environment, as data obtained in real traffic would contain much higher noise level from unstable surrounding traffic, compared to simulator settings. In this paper, two most common types of secondary tasks were performed in the experiment, and tested the capability of machine learning methods in distinguishing these different behaviour patterns. The hypotheses are: *i) Machine learning can recognize drivers' higher workload, compared to baseline. ii) The different patterns between mental and visual workload can be detected by machine learning techniques.*

This paper is structured as follows. Section 2 provides a review on both machine learning tools: ELM and SVM. Section 3 describes the research methods including database description, experimental setup and feature selection. Section 4 presents the results on parameter selection and an analysis on the accuracy. Discussion and conclusions are presented in Section 5.

2   Classification Methods

Two types of machine learning methods, Extreme Learning Machine (ELM)[9, 14, 3, 5, 22] and Support Vector Machine (SVM)[24, 10, 17, 21], are selected in the comparative study in this work. They are first reviewed and compared.

2.1   SVM

SVM is fundamentally a binary classification learning method, which generally achieves the separation between two classes by two steps: in the first step, the input data (i.e. training dataset $\{(\mathbf{x}_i, y_i)\}_{i=1}^N$ where $\mathbf{x}_i \in R^d$ and $y_i \in (-1, 1)$) are nonlinearly mapped into a higher dimensional feature space $\mathbf{Z}$ using $\Phi : \mathbf{x}_i \rightarrow \Phi(\mathbf{x}_i)$. In this way the possibility of linear separability between two different classes becomes higher. In the second step, within the higher dimensional feature space $\mathbf{Z}$, a linear decision hyperplane is constructed. This hyperplane is optimized to ensure that it maximizes the functional margins between the two different classes, i.e. maximizes $2/\|\mathbf{w}\|$ (better generalization ability) and minimizes the training errors: ξ (higher accuracy). Given $((\mathbf{x}_i, y_i)\}_{i=1}^N)$ being the training data, they are considered to be linearly separable, if there exists a vector $\mathbf{w}$ and bias b such that

$$\begin{aligned} \mathbf{w} \cdot \Phi(\mathbf{x}_i) + b \geq 1, \quad if \quad y_i = 1; \\ \mathbf{w} \cdot \Phi(\mathbf{x}_i) + b \leq -1, \quad if \quad y_i = -1, \end{aligned} \tag{1}$$

where $\mathbf{w}$ is the normal vector to the hyperplane (not necessarily normalized) and $\cdot$ denotes the dot product; and b is the bias.

The solution of the linear hyperplane is formulated into an optimization problem as following,

$$\begin{aligned} \min_{\mathbf{w},b} \quad & \frac{1}{2}\|\mathbf{w}\|^2 + C\sum_{i=1}^N \xi_i \\ \text{s.t.} \quad & y_i(\mathbf{w} \cdot \Phi(\mathbf{x}_i) + b) \geq 1 - \xi_i, \\ & \xi_i \geq 0, \quad i = 1, \ldots, N. \end{aligned} \tag{2}$$

The solution can be then found by iterations using Lagrange multipliers to obtain the weight vector $\mathbf{w}$ and the bias b of the optimal hyperplane. It should be noted that as SVM is originally proposed for binary classification, the multiclass classification problems are often solved by one-against-all (OAA) or one-against-one (OAO) methods [6].

2.2   ELM

As mentioned earlier, ELM is a Single Hidden Layer Feed-forward Neural Network (SLFN). Unlike conventional neural networks which use back-propagation to tune all parameters, ELM uses randomly generated weights $\mathbf{W}$ in input-to-hidden layer [7, 8]; and analytically calculates the hidden layer output matrix as well as the output, without iteratively gradient descent. In detail, the random weights map input data $\mathbf{X}_{M \times N}$ to ELM feature space (i.e. hidden layer input matrix). When the number of hidden nodes is large enough, after this random mapping, the relevant structure of the input data is still preserved, according to Johnson-Lindenstrauss Lemma. After the random mapping, nonlinear activation functions $g(\mathbf{x})$, such as sigmoid or radius basis function are applied to generate a higher dimension space (i.e. the hidden layer output matrix $\mathbf{H}_{M \times \tilde{N}}$) where the possibility of linear separability becomes high, according to Cover's theorem. The process can be described by following function for the calculation of the hidden layer output matrix $\mathbf{H}_{M \times \tilde{N}}$:

$$\mathbf{H} = g(\mathbf{XW} + \mathbf{1b}^{\top}), \tag{3}$$

where $\mathbf{X}_{M \times N}$ is the input, M is the sample number and N is the number of features; $\tilde{N}$ is the number of hidden layer neurons; $\mathbf{W}_{N \times \tilde{N}}$ is the randomly generated input weight matrix and g is the nonlinear activation functions such as sigmoid or radius basis. For samples of $(\mathbf{x}_i, t_i)$, ELM is to minimize the training errors as well as the norm of the output weights, i.e.

$$\min_{\beta} \quad \frac{1}{2}\|\beta\|^2 + \frac{C}{2}\sum_{i=1}^{N}\|\mathbf{H}\beta - \mathbf{T}\|^2. \tag{4}$$

The hidden layer output weight matrix is then calculated by the following function:

$$\beta^* = (\mathbf{H}^{\top}\mathbf{H} + \frac{1}{C}\mathbf{I})^{-1}\mathbf{H}^{\top}\mathbf{T}, \tag{5}$$

where β^* is the optimal norm least-square solution for output weight and $\mathbf{T} = [t_1, \ldots, t_n]^{\top}$ is the target.

2.3   Comparison between SVM and ELM

Based on the review and comparison between ELM and SVM, some advantages of ELM become apparent. They are summarized as following:

First, the hidden parameters in the input-to-hidden layer of the algorithm are randomly generalized. Compared to the mechanism applied in SVM, where all parameters are tuned throughout the network, this randomization avoids the cost of tuning the parameters, which results in significantly shorter training time compared to other machine learning methods like back-propagation and SVM.

Second, the hidden layer outputs in ELM are analytically calculated. Conversely in other learning methods, this step is generally achieved by many iterative learning steps. This analytical approach of ELM can reach the global minimum, compared to SVM which may reach local minimum and/or large errors due to early stop in the iteration process when searching for optimized solution. ELM also finds the unique solution with small norm of output weights, therefore guarantees a good generalization capability of the model learnt. According to Bartlett's theory, for neural networks reaching smaller training error, the smaller the norm of output weights, the better the generalization capability [1]. ELM ensures the norm least-square solution being unique, which again avoids the problem of local minimum. The analytical approach does not require to set learning rate. As a small rate can be very slow in computation, and a large rate can cause the learning process to be unstable and diverge. Direct calculation of β avoids over-training problem which often happens in other learning algorithms. If a model is over-trained, it will lead to bad generalization performance. Other algorithms would often require extra validation and to set suitable stopping methods.

Third, ELM can classify multiple classes directly, compared to SVM which traditional divides multiple classes into two (one-against-others) in each step, and then achieve the classification by multiple steps.

3 Methods

The database used in this paper, feature selection and experiment setup are described in this section.

3.1 Database Description

An on-road experiment was conducted using an Instrumented Vehicle (IV). Drivers' eye movements and vehicle dynamic data were recorded when they were asked to perform two sets of surrogate in-vehicle secondary tasks, auditory-vocal and visual-manual, induced through auditory stimulus and an on-board touch screen respectively. Each secondary task lasted about 60 seconds. During normal driving, i.e. driving without performing secondary tasks, baseline data were also collected. For more detailed description on experimental setup, the secondary tasks and experiment procedure, see [27].

The vehicle dynamic data were collected from the Controller Area Network (CAN Bus), including speed, steering wheel movements and readings from other IV sensors [13], e.g. vehicle lane position. The vehicle dynamics data were logged at various sample rates and later re-sampled to 10 Hz. $Facelab^{TM}$ 5.0 eye tracking system was used to record drivers' visual behaviour. The system measurements included drivers' head and gaze positions, gaze angles, blink and saccade, with a frequency of 60Hz [20].

Vehicle and eye movement's data were first preprocessed to remove noise. This includeed smoothing and excluded abnormal values due to technical errors for the speed, steering wheel movements, and lane positions based on human limitations of movement. Visual behaviour data were screened first, and where the data reported as highest quality lower than 80% of the entire drive, the data from this test run (1 participant) was excluded from further analysis. The final database used in this study consisted of a sample of 9 subjects.

3.2 Feature Selection and the Experiment Setup

ELM with sigmoid activation function and SVM with sigmoid kernel were applied for driver status recognition. Vehicle dynamics and drivers' eye movement data were used in this study to train the machine learning models for the detection of drivers' high workload. Vehicle dynamics have historically been used as an indicator of performance impairment [2], and eye movements were also suggested to be related to drivers' workload related information processing, event detection, and visual load, as a measure of drivers' physiological information [18]. Both vehicle and eye signals were firstly segmented into fixed a time window with a certain overlap. Twelve combinations of window size and overlap were used to draw instances: 4 different window sizes (1, 2, 5 and 10 seconds) with three overlaps (20%, 50% and 80%) respectively. The features were then extracted from each segment. Three combinations of features were fed into classifiers and tested: vehicle only, eye only, and eye plus vehicle. The features used for the classification included steering wheel angles, speed, left/right lane offsets for vehicle dynamics and 3-D angles of head rotation, blink frequency, percentage of eye closure, pitch/yaw angles for left and right eye gaze for eye movements. Another objective of the experiment was to understand the capability of machine learning tools to detect different types of workload: mental, visual and to differentiate the two. Therefore, three types of classification were set, including: two types of two-class classification (i.e. high mental workload vs. baseline driving and high visual workload vs. baseline driving) and a three-class classification, i.e. mental vs. visual workload. vs. baseline driving.

Finally, for each subject, 108 separate models were built for the driver state classification: 12 (window size & overlap) × 3 (feature sets) × 3 (types of classification) = 108 models were built for each participant for both ELM and SVM respectively. For each model, all features were first normalized, and the instances were drawn based on the experiment setting. The instances were then divided into three parts, each consisting baseline, high mental and visual workload, later used as training, cross validation and testing samples, with the proportions of each state set to be equal within each part. Nested cross validation was conducted within the 3 partitions for the parameter selection [23], to avoid over-training and to optimize the model performance. Following the nested cross validation approach, the whole learning, validating and testing process requires two loops on all three parts, with outer loop for testing data selection and inner loop for training and validation data selection. In detail, for each model, 2/3 of the total instances were used for the learning process, within which half instances were used for training the model, and the other half for cross validation. The parameters which gained the highest average accuracy in cross validation were used as final model parameters for the later testing process. After cross validation, parameters assuring the best balance of learning and accuracy were obtained for both SVM and ELM, and the model performance was then tested on the rest 1/3 instances.

4   Results

The accuracy of detection was used as the measurement of classifiers' performance. All analysis was conducted in R [16] using ANOVA, with repeated

ANOVA by ranks employed in cases where the normality assumption was violated. The evaluation of the machine learning methods was first compared against by-chance performance, i.e. 50% accuracy for two classes and 33% for three classes - see Figure 1. And then the accuracy achieved from ELM and SVM were compared across all combination of settings. Later the effects of number of classes, features combinations were assessed individually for model construction using accuracy as reference.

4.1 The Model Performance of ELM and SVM

The performances of both classifiers with all feature and instance combinations were higher than by chance. The performance across all 12 different combinations of window sizes and overlaps was first compared between ELM and SVM. The statistical results suggested that ELM significantly outperformed SVM in each feature set (eye, vehicle, eye plus vehicle) and types of classification (2-class mental, 2-class visual, and 3 classes), by 4.1% at an average of all classification conditions (87.0% for ELM vs. 82.9% for SVM), $F(1, 8) = 34.29$, p <0.01. When we looked at the best performance across all features, window size and overlap combination, ELM achieved 94.4% accuracy for visual workload, 81.5% for mental workload and 88.1% for 3 classes. SVM generally fell slightly lower accuracy, but still achieved 74.7% for mental workload and 82.5% for three classes, with

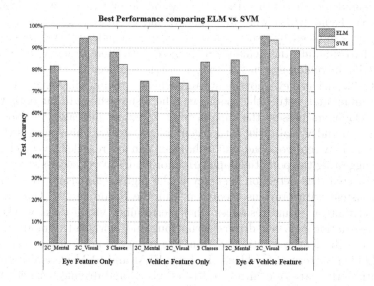

Fig. 1. Comparison of the best performance between ELM and SVM

2C_Mental: 2−classes classification between high mental workload and baseline driving;
2C_Visual: 2−classes classification between high visual workload and baseline driving;
3_Classes: 3−classes classification between high mental, visual workload and baseline.

a slightly higher accuracy of 95.0% for visual workload. Figure 1 shows the best performance of ELM and SVM averaged across all subjects.

4.2   Features for Model Construction

The selection of feature significantly impacted the model accuracy ($F(2, 16) =$ 18.28, p< 0.01). In detail, the feature set of eye plus vehicle achieved highest accuracy of 76.4% for mental workload, 91.0% for visual workload and 82.3% for three classes. Using eye movement information alone also achieved nearly as good results with 74.5% for mental workload, 90.0% for visual workload and 81.0% for three classes. The use of vehicle features alone showed the lowest performance for all three types of classification, with 68.9% for mental workload, 72.4% for visual workload and 73.6% for three classes, and therefore is not recommended for further application. In addition, although the results compared between "eye data only" and "eye plus vehicle" complied with the conventional belief of "the more the features, the higher the performance of the models", the advantage of adding vehicle features on to eye features appeared to be limited.

5   Discussion and Conclusion

This study proposed to use machine learning as real-time tools for drivers' high workload detection. In this study, the capabilities of two machine learning methods (SVM and ELM) were investigated and evaluated. The workload in this study was induced by two types of surrogate secondary tasks, which were designed to simulate the demands associated with mental and visual load while operating in-vehicle systems. Two types of data collected in the field study with real traffic, drivers' eye movements and vehicle dynamics information, were used for training and testing the classifiers. The results showed that both ELM and SVM classifiers possess the capability of detecting increased workload caused by mental and visual tasks with acceptable accuracy. This suggests the hidden patterns between workload and behaviour can be revealed by machine learning, suggesting the potential application for future in-vehicle assistance systems. As expected, the detection accuracy is the highest for visual workload, which is consistent with previous theory on Multiple Resource [25]. Due to the high competition for visual resources when performing visual tasks whilst driving, the association between drivers' behaviour and workload is more direct, and therefore the features extracted from the observation of drivers can represent the hidden pattern of workload the best. ELM and SVM were also able to differentiate the pattern changes in drivers' visual and driving behaviour between all three types workload: low in normal driving, visual and mental workload.

In conclusion, this study proposed a data-driven method to detect drives' workload using less features and as little calculation cost as possible. The performance of ELM showed a high accuracy and good potential in real-time applications, due to the low computational cost and fast calculation capacity.

References

1. Bartlett, P.L.: The sample complexity of pattern classification with neural networks: the size of the weights is more important than the size of the network. IEEE Transactions on Information Theory 44(2), 525–536 (1998)
2. Brookhuis, K.A., Waard, D.D., Fairclough, S.: Criteria for driver impairment. Ergonomics 46(5), 433–445 (2003)
3. Chacko, B.P., Krishnan, V.V., Raju, G., Anto, P.B.: Handwritten character recognition using wavelet energy and extreme learning machine. International Journal of Machine Learning and Cybernetics 3(2), 149–161 (2012)
4. Gelau, C., Stevens, A., Cotter, S.: Impact of ivis on driver workload and distraction: Review of assessment methods and recent findings. Deliverable D 2 (2004)
5. Gomathi, M., Thangaraj, P.: A computer aided diagnosis system for detection of lung cancer nodules using extreme learning machine. Intl. J. Engg. Sci. & Technol. 2(10), 5770–5779 (2010)
6. Hsu, C.W., Lin, C.J.: A comparison of methods for multiclass support vector machines. IEEE Transactions on Neural Networks 13(2), 415–425 (2002)
7. Huang, G.B., Chen, L., Siew, C.K.: Universal approximation using incremental constructive feedforward networks with random hidden nodes. IEEE Transactions on Neural Networks 17(4), 879–892 (2006)
8. Huang, G.B., Zhou, H., Ding, X., Zhang, R.: Extreme learning machine for regression and multiclass classification. IEEE Transactions on Systems, Man, and Cybernetics, Part B: Cybernetics 42(2), 513–529 (2012)
9. Huang, G.B., Zhu, Q.Y., Siew, C.K.: Extreme learning machine: theory and applications. Neurocomputing 70(1), 489–501 (2006)
10. John, J., Pramod, K., Balakrishnan, K.: Unconstrained handwritten malayalam character recognition using wavelet transform and support vector machine classifier. Procedia Engineering 30, 598–605 (2012)
11. Liang, Y., Reyes, M.L., Lee, J.D.: Real-time detection of driver cognitive distraction using support vector machines. IEEE Transactions on Intelligent Transportation Systems 8(2), 340–350 (2007)
12. McDonald, A.D., Schwarz, C., Lee, J.D., Brown, T.L.: Real-time detection of drowsiness related lane departures using steering wheel angle. In: Proceedings of the Human Factors and Ergonomics Society Annual Meeting, vol. 56, pp. 2201–2205. Sage Publications (2012)
13. McDonald, M., Brackstone, M.: The role of the instrumented vehicle in the collection of data on driver behaviour (1997)
14. Mohammed, A.A., Minhas, R., Jonathan Wu, Q., Sid-Ahmed, M.A.: Human face recognition based on multidimensional pca and extreme learning machine. Pattern Recognition 44(10), 2588–2597 (2011)
15. Müller, K.R., Tangermann, M., Dornhege, G., Krauledat, M., Curio, G., Blankertz, B.: Machine learning for real-time single-trial eeg-analysis: from brain–computer interfacing to mental state monitoring. Journal of Neuroscience Methods 167(1), 82–90 (2008)
16. R Core Team: R: A language and environment for statistical computing (2012)
17. Ramírez, J., Górriz, J., Salas-Gonzalez, D., Romero, A., López, M., Álvarez, I., Gómez-Río, M.: Computer-aided diagnosis of Alzheimer's type dementia combining support vector machines and discriminant set of features. Information Sciences 237, 59–72 (2013)

18. Recarte, M.A., Nunes, L.M.: Effects of verbal and spatial-imagery tasks on eye fixations while driving. Journal of Experimental Psychology: Applied 6(1), 31 (2000)
19. Rosenfeld, A., Zuckerman, I., Azaria, A., Kraus, S.: Combining psychological models with machine learning to better predict people's decisions. Synthese 189(1), 81–93 (2012)
20. Seeing Machines: FaceLAB5 User Manual. Canberra, Australia (2009)
21. Subasi, A., Ismail Gursoy, M.: Eeg signal classification using pca, ica, lda and support vector machines. Expert Systems with Applications 37(12), 8659–8666 (2010)
22. Sun, H., Sourina, O., Yang, Y., Huang, G.-B., Denk, C., Klanner, F.: Machine learning reveals different brain activities in visual pathway during TOVA test. In: Mao, K., Cambria, E., Cao, J., Man, Z., Toh, K.-A. (eds.) Proceedings of ELM-2014 Volume 1. PALO, vol. 3, pp. 245–262. Springer, Heidelberg (2014)
23. Varma, S., Simon, R.: Bias in error estimation when using cross-validation for model selection. BMC Bioinformatics 7(1), 91 (2006)
24. Wang, C., Lan, L., Zhang, Y., Gu, M.: Face recognition based on principle component analysis and support vector machine. In: 2011 3rd International Workshop on Intelligent Systems and Applications (ISA), pp. 1–4. IEEE (2011)
25. Wickens, C.D.: Engineering psychology and human performance. HarperCollins Publishers (1992)
26. Wu, J.D., Ye, S.H.: Driver identification using finger-vein patterns with radon transform and neural network. Expert Systems with Applications 36(3), 5793–5799 (2009)
27. Yang, Y.: The effects of increased workload on driving performance and visual behaviour. Ph.D. thesis, University of Southampton (2011)
28. Yang, Y., Reimer, B., Mehler, B., Wong, A., McDonald, M.: Exploring differences in the impact of auditory and visual demands on driver behavior. In: Proceedings of the 4th International Conference on Automotive User Interfaces and Interactive Vehicular Applications, pp. 173–177. ACM (2012)

Author Index